D0849349

Handbook Of Electronics Packaging Design and Engineering

To my Father,
with many thanks.

Handbook Of Electronics Packaging Design and Engineering

BERNARD S. MATISOFF, P.E., CMfgE

VNR VAN NOSTRAND REINHOLD COMPANY
NEW YORK CINCINNATI TORONTO LONDON MELBOURNE

Library of Congress Catalog Card Number: 81-2362
ISBN: 0-442-20171-0

Manufactured in the United States of America

Published by Van Nostrand Reinhold Company Inc.
135 West 50th Street, New York, N.Y. 10020

Van Nostrand Reinhold Publishing
1410 Birchmount Road
Scarborough, Ontario M1P 2E7, Canada

Van Nostrand Reinhold Australia Pty. Ltd.
17 Queen Street
Mitcham, Victoria 3132, Australia

Van Nostrand Reinhold Company Limited
Molly Millars Lane
Wokingham, Berkshire, England

15 14 13 12 11 10 9 8 7 6 5 4 3 2 1

Library of Congress Cataloging in Publication Data

Matisoff, Bernard S.
Handbook of electronics packaging design and engineering.
Includes index.
1. Electronic apparatus and appliances–Design and construction–Handbooks, manuals, etc. I. Title.
TK7870.M386 621.381'046 81-2362
ISBN 0-442-20171-0 AACR2

PREFACE

The *Handbook of Electronics Packaging Design and Engineering* has been written as a reference source for use in the packaging design of electronics equipment. It is designed to provide a single convenient source for the solution of recurring design problems.

The primary consideration of any design is that the end product meet or exceed the applicable product specifications. The judicious use of uniform design practices will realize the following economies and equipment improvements:

- *Economics of design.* Uniform design practices will result in less engineering and design times and lower costs. They will also reduce the number of changes that may be required due to poor reliability, maintainability, or producibility.
- *Improved design.* Better designs with increased reliability, maintainability, and producibility will result from the use of uniform design practices.
- *Production economies.* Uniform designs employing standard available tools, materials, and parts will result in the cost control of manufacturing.

The *Handbook* is intended primarily for the serious student of electronics packaging and for those engineers and designers actively engaged in this vital and interesting profession. It attempts to present electronics packaging as it is today. It can be used as a training text for instructional purposes and as a reference source for the practicing designer and engineer.

During my years as an engineer involved in all the phases of electronics packaging–both design and manufacturing–I have attempted to review carefully all books and articles published on this subject. One characteristic seems common to most of them: an overemphasis of the theoretical. Obviously, comprehensive coverage of theory is needed. Moreover, there is also a need for a book that separates practical and applicable information from the vast field of theory and presents it in a readily accessible format. The theoretical presentations herein are considered sufficient (1) for the routine solutions of the daily problems encountered and (2) for establishing the size and complexity of those problems so that the engineer or designer will know when the assistance of a specialized consultant is required.

Today there are numerous techniques for packaging electronics systems, and due to the increasingly stringent customer requirements, their number continues to grow. Engineers can no longer simply document the components—they must select the most efficient packaging techniques as well.

The physical attributes of electronics equipment are fast becoming recognized as significant in the performance and the reliability of contemporary systems. Practical applications of materials, processes, and components in the evaluation of electronics packaging can be made to yield a new order of performance and reliability at reasonable costs. Modern electronics equipment must be designed to function reliably in the specific environments defined by the intended applications. Extreme environments, characterized by mechanical shock and vibration, high temperatures, or radio-frequency interference, for example, will present many problems. To achieve the most efficient design, the engineer must review all of the requirements and generate a design plan that will provide the foundation for the product design.

A systematic approach to any electronics packaging task must be taken. The engineer must consider all of the design parameters as well as their effect on each other. The basic design parameters to be considered in the early planning stages are:

- Physical
- Environmental
- Repairability and maintainability
- Manufacturability
- Appearance
- Economics

Reliable performance depends on the ability of a structure to secure all parts and assemblies together in a good working relationship. The optimal system is obtained through judicious integration of the circuit design, structural techniques, component applications, heat transfer, and the vibrational dynamics within realistic operational requirements.

The designer must be thoroughly familiar with the many processes and techniques available in order to fabricate parts at the lowest cost, as well as with the advantages and limitations of each of the processes. The parts and materials selected by the designer—and the method of assembly—must be capable of withstanding and functioning in the specified environments.

All things are designed and constructed to extend human capability in some way. Therefore, the engineer must remember that humanity is the central reason for all designs. The design object is to provide people with the accessories they need to achieve a specified goal.

BERNARD S. MATISOFF, P.E., CMFGE

CONTENTS

1

ELECTRONICS PACKAGING DESIGN AND ENGINEERING

The electronics packaging design and engineering fields perform tasks of ever-increasing importance in the electronics industry. They can be either the strength or the weakness of electronics systems engineering. Much depends on how its purpose and functions are interpreted within the engineering organization.

The range of technical activity encompassed by electronics packaging design and engineering includes structures, mechanisms, materials, finishes, appearance, utility, serviceability, and environmental reliability. It also provides the solution to the three-dimensional jigsaw puzzle of creating a whole from parts while maintaining electrical integrity.

In consumer products, the packaging designer or engineer determines the competitive price position of the finished product. The configuration of parts and the mechanical design of the equipment usually determine whether fabrication and assembly operations will be economical or costly.

The military services are interested in design from the standpoint of reducing manufacturing times. In the design of military equipment, cost must necessarily defer to optimum performance and maximum reliability.

ESSENTIAL QUALIFICATIONS

Electronics packaging design and engineering contributes toward achieving the desired end results by providing these essential qualifications:

1. *Interpretation of specifications*. Commercial and military requirements, other than those pertaining to electrical circuits, are best interpreted by an electronics packaging engineer. The highly competitive market demands that commercial equipment be attractive, functional, reliable, reasonable in cost, and able to survive shock and vibration loads experienced during handling and shipment. Military equipment specifications impose yet more severe requirements, includ-

ing extreme conditions of shock and vibration, temperature, and reliability. The correct translation of these requirements into design solutions requires the knowledge of an electronics packaging engineer.

2. *Conception of design approach*. As a result of the engineer's training in the areas of human engineering and aesthetic design, three-dimensional perceptive ability, and intimate knowledge of constructional techniques, the electronics packaging engineer is particularly suited to conceive a design approach that will meet the requirements imposed on the equipment.

3. *Value and cost consciousness*. A review made at the conceptual phase based on the existence of several approaches will determine the ultimate cost of the product. The electronics packaging engineer will employ standardization in all permissible areas in order to achieve minimum cost. The electronics packaging engineer's contributions are a major factor in the final cost of the product.

4. *Specialized knowledge in noncircuit areas*. In the design of electronic equipment, the engineer must resolve many problems that require the knowledge of an engineer well trained and experienced in the fields of mechanical engineer-

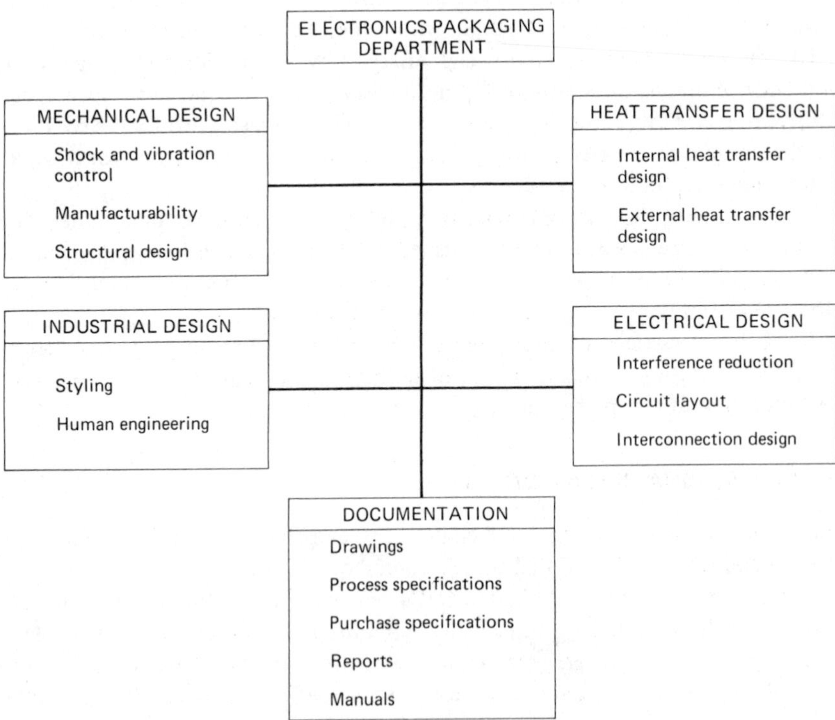

Figure 1-1. Functions performed by the electronics packaging department.

ing, heat transfer analysis, industrial design, and production techniques. The electronics packaging designer and engineer provides this knowledge.

5. *Ability to carry the design through production.* It is necessary that the electronics packaging engineer's design activity plans for the specific methods of fabrication and assembly of an equipment, which would prove most suitable and economical. The electronics packaging engineer is therefore best qualified to understand and help resolve manufacturing problems as they arise on the production floor.

FUNCTIONS OF ELECTRONICS PACKAGING ENGINEERING DEPARTMENT

Mechanical Design

A factor often overlooked in design is that equipment, if it is to be of complete service to the user, must arrive at its destination in the same condition in which it left the factory. The electronics packaging engineer must foresee and evaluate the various conditions that can arise during transportation and storage.

Design for Production

It is the electronics packaging engineer's responsibility to create a design that will meet the customer's requirements at the least cost. To realize this goal, the engineer must consider a multitude of factors that contribute to both the quality and the cost of the end product. The engineer must select the raw materials from which the parts of the assembly will be fabricated. These materials differ in their physical properties, of course, as well as in their behavior under varying environmental conditions. The engineer must therefore select the most suitable materials for the particular applications on the basis of weight, strength, electrical properties, thermal characteristics, and the cost of the raw stock and fabrication. Electronics packaging engineers provide the theoretical and practical knowledge of shop practices and limitations, which enable them to create a design that can be produced economically.

Heat Transfer Design

The thermal design is a significant factor in the reliability of an electronics design. Applied to the solution of thermal problems, the heat transfer techniques may be divided into two areas. The first, internal heat transfer, deals with those factors that must be considered in the removal of heat from individual power-dissipating devices (the transfer of heat to the external shell of the equipment). The second area, external heat transfer, takes into consideration the transfer of heat from the external shell to an ultimate heat sink.

Electrical Design

The reliable electrical performance of any equipment depends on the success of the circuit design effort and the electronic packaging engineer's ability to translate the proven circuit into practical equipment. Electronics packaging engineers, although not expected to design circuits, must be able to converse intelligently with their electrical counterparts on circuit performance and problems. They must also be familiar with the techniques of construction that are in the best interest of circuit performance.

Industrial Design

The electronics packaging engineer is responsible for the styling and human engineering aspects of the equipment. The styling of a successful product is seldom the result of pure artistic talent. Before the first sketch is made, the electronics packaging engineer must be aware of the objectives of the marketing and manufacturing departments. The marketing department lists two main objectives: (1) the product must appeal to the prospective buyer, and (2) it must look better than competitive products. The manufacturing department requires: (1) ease of assembly for labor savings, (2) methods using existing equipment and manufacturing capabilities, (3) a minimum variety of parts for material handling and inventory economics, and (4) interchangeable parts for the same reason. Working within the scope of these aims, the engineer must give the product the most pleasing appearance possible.

When applying the principles of human engineering, the electronics packaging engineer will find it convenient to consider human operators of equipment as components. Both human beings and components have capabilities and limits, and both are subject to malfunctions under adverse operating conditions for which they were not designed. To insure proper system functioning, the engineer must consider the conditions under which the components operate most efficiently and incorporate these conditions into the systems whenever feasible. The electronics packaging design engineer takes into account all of the aspects related to the overall functioning of the man-machine complex.

Documentation

The electronics packaging engineer's responsibilities do not end when the equipment is shipped to the customer. In addition to all of the previously mentioned activities, the engineer must also contribute to the proper documenting of all drawings, reports, operator's manuals, and specifications required to manufacture, maintain, and operate the equipment.

ELECTRONICS PACKAGING ENGINEER

The qualifications of good electronics packaging engineers are many. They may be electronics engineers who like to build things, mechanical engineers who like electronics, or industrial engineers who seriously use the engineering approach. They should be thoroughly familiar with the principles of physics, mechanics, electronics, human engineering, and aesthetics. They should have an excellent three-dimensional perception and possess an active imagination. A knowledge of the practical limitations of drafting and shop practices is also required.

The need to develop the technical background of individual engineers and to extend it into as many different fields of engineering as possible is evident. Only in this way can engineers contribute to the productivity and efficiency of their groups. Certainly, a major factor responsible for the increase of their knowledge is the amount of effort they put forth toward achieving the required technical background. Much can be accomplished through individual endeavor.

2

PROJECT PLANNING

To make the most effective use of time, the electronics packaging engineer should begin each task with a formal plan. The engineer should set milestones to be met at specific dates and then make every effort to meet them. Unforeseen problems sometimes cause setbacks in the original plan; at such times, the engineer should simply replot the original plan, noting the cause for delay, and then proceed. The history recorded during the task will be useful in estimating future tasks.

The electronics packaging engineer making the plan should always be truthful, basing estimates on real capabilities. The engineer must never proceed with a plan generated solely from the dates supplied on a sales order if it is known that the plan cannot be accomplished due to a lack of personnel, for example, or dependence on external organizations. (For instance, parts required from a vendor cannot be delivered in time to meet a planned schedule.) Management and the customer should be informed at the earliest possible time if it is not possible to meet the original dates. Alternate plans can then be made to reschedule the customer requirements, to increase personnel, or to use external support.

The engineer should begin each assignment by reviewing the task to be performed. The requirements of the customer and the capabilities of the organization must be analyzed, and the environmental conditions, both for operating and nonoperating modes, must be reviewed. The engineer should then analyze the installation of the equipment and its proximity to other equipment and consider the effects of radio or magnetic frequency interference and methods of control. Next, the engineer should make notes on special design considerations, such as maintenance requirements. Finally, after the task is thoroughly understood, the engineer should generate a system packaging design study.

SYSTEM DESIGN STUDY

A system packaging design study is conducted to define the requirements of the equipment to be designed (Figure 2-1). Out of this study will come the mechan-

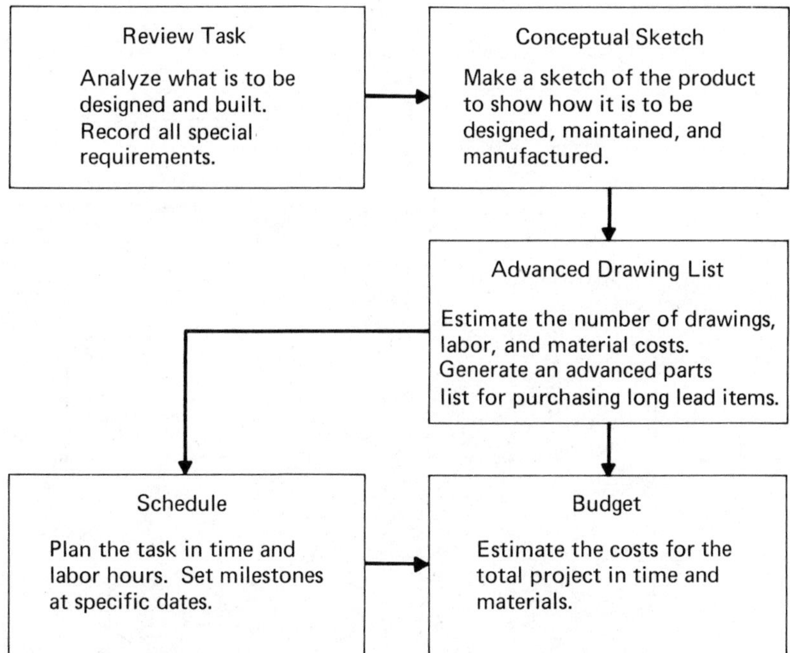

Figure 2-1. Project planning and control.

ical product description, which is used to establish the cost of the program and to formulate a design plan. The design plan clearly defines the product size, weight, maintenance philosophy, producibility, functional requirements, and the operating environments. (A system may be required to function in different environments at the same time, such as in high temperatures, chemical fumes from nearby equipment, dust, etc.) The design plan is generated from the customer's specifications or, in the case of an internal program, from the marketing department. When the product to be designed is to be used in military applications, it will require close review of the military specifications. For "best commercial" products, the military specifications should be used as a guide only. Other specifications available for commercial products that should be considered are the Underwriters Laboratories (UL) standards and the Occupational Safety and Health Administration (OSHA) requirements.

After the product requirements have been reviewed, a simple freehand sketch is useful in establishing an advanced drawing list and in developing the conceptual design (Figure 2-2). The sketch does not need to be fully developed, as the design will be refined until all the requirements are met. The sketch is a record of the thoughts that arise during the development phase. The design details are defined during the layout phase.

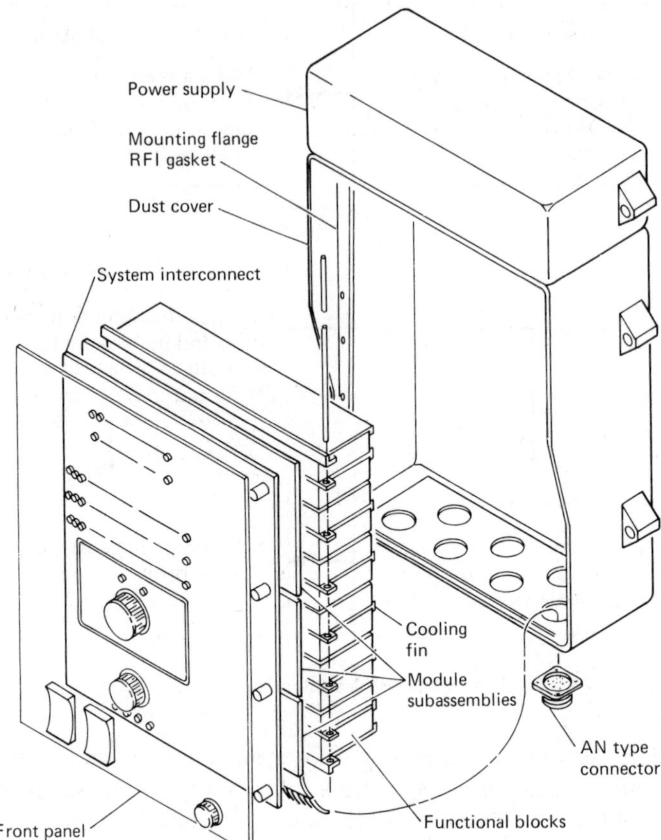

Figure 2-2. Typical digital system.

ADVANCED DRAWING LIST

With the sketch as a guide, the engineer generates an *advanced drawing list* (ADL). The engineer considers how many drawings will be required to document the system being designed. The ADL is used to estimate the number of drawings, such as the assemblies, subassemblies, and detail parts that will be required to complete the design effort (Figure 2-3). The ADL graphically defines the product by using a system structure. The part number indenture levels indicate the using assemblies. For example, indenture level 1 uses items from indenture level 2, and level 2 uses items from level 3 to build up the various assemblies. Indentured parts lists are thoroughly covered in any standard drafting manual. The ADL is also used to provide the input for all other planning.

Advanced Drawing List Micro Logic Resolver

Line	Rev.	Part No. Indenture Level								Title	Material	Qty	Hours	Material Cost	Notes
		1	2	3	4	5	6	7	8						
		1	0	0	1	0	0			Assembly-micro logic resolver		1	90	—	in house
			1	0	0	1	0	1		Dust cover assembly		1	36	—	in house
				1	0	0	1	0	2	Dust cover		1	36	225^{00}	
				1	0	0	1	0	3			2	30	50^{00}	
			1	0	0	1	0	4		Power supply assembly		1	90	—	in house
				1	0	0	1	0	5	Cover		1	36	190^{00}	
				1	0	0	1	0	6	Base		1	30	75^{00}	
				1	0	0	1	0	7	Printed circuit assembly		1	383	275^{00}	
			1	0	0	1	0	8		Front panel assembly		1	90	—	in house
				1	0	0	1	0	9	Front panel		1	36	120^{00}	
			1	0	1	0	0	1		Interconnect assembly		1	36	—	in house
				1	0	1	0	0	2	Printed circuit board		1	200	275^{00}	
				1	0	1	0	2	0	Cable		1	30	125^{00}	
			1	1	0	0	1			Module assembly A		12	36	—	in house
				1	1	0	0	0	2	Printed circuit assembly		12	90	275^{00}	
			1	0	2	0	1	1		Fin, thermal		14	30	135^{00}	
		1	1	1	1	0	1			Schematic		Ref.	45	—	

Figure 2-3. Advanced drawing list.

ENGINEERING JOB ESTIMATE NOTES

DEPT. ENGINEERING

PROJECT ENGINEER BERNIE MATISOFF

DATE E.J.E. ISSUED	E.J.E. NO.	REV.

APPROVAL:	G. LEADER	DATE	DIR.	DATE

J.O.		QUANTITY	MODEL 1001	CUSTOMER U.S.N.
R.F.Q.		DESCRIPTION MINIATURIZED/RUGGEDIZED MICRO LOGIC RESOLVER FOR SHIPBOARD APPLICATIONS		
S.O.				
REF.				

E.J.E. ITEM REF.	—NOTES—
1	UNIT MUST HAVE RFI/EMI SEALS ON ENCLOSURE AND ALL FRONT PANEL DEVICES (SWITCHES, INDICATORS, & METERS)
2	FINISH MUST PROVIDE PROTECTION FROM SALTWATER EXPOSURE
3	UNIT MUST HAVE MOISTURE SEALS
4	MUST MATE WITH INTERFACE CONNECTORS SPECIFIED IN CUSTOMER SPEC.
	UNLESS OTHERWISE SPECIFICALLY INDICATED THIS QUOTE IS BASED UPON:
	1. INSPECTION & ACCEPTANCE AT CUSTOMERS PLANT.
	2. NO ENVIRONMENTAL QUALIFICATION.
	3. NO CERTIFICATION TO MIL SPECS OF MATERIALS, PARTS AND PROCESSES.

INCLUDE DETAIL LIST OF PARTS REQUIRED FROM PRODUCTION IN THESE NOTES. SHEET ______ OF ______ SHEETS.

Figure 2-4. Engineering job estimate notes.

It gives the estimated labor hours for each design document required, as well as the estimated costs of any materials needed to build an engineering or prototype unit. During the conceptual phase, a record (a sketch and the ADL) should be kept of any requirement that is not standard design practice (Figure 2-4).

STANDARD DRAFTING HOURS

Standard drafting hours, which are a means of estimating drafting costs, may vary from one organization to another, based on the experience and performance of each individual design and drafting group. Table 2-1 can be used as a reference. Using the information from the job labor and material plan (Figure 2-5), it can be updated to conform to the standard hours of any organization.

JOB LABOR AND MATERIAL PLAN

The job labor and material plan (JLM) is the schedule planned from the information given in the ADL. It establishes the time required to complete each of the various tasks. The JLM is coordinated with the customer's requirements and establishes the number of personnel required to meet the delivery date. In addition, the JLM provides a means of tracking and recording the actual hours and costs expended on a given project, which will be used to establish actual standard hours for future planning. The electronics packaging engineer enters information on the JLM in the areas in which he or she has authority. Estimates for the electrical engineer and the electronics technician should be provided by the electrical project engineer, unless the packaging engineer is responsible for the building of the engineering unit. In such cases, the packaging engineer should review the estimates with those more experienced in the areas of electrical design and assembly.

THE PLANNING PHASE

The last step in the planning phase is the budget. The cost estimate form (Figure 2-6) is used to compute and record the cost of labor in the various disciplines, and of materials. The cost estimate, which provides for all of the design and manufacturing costs, allows the sales department to establish a price for the customer. In most cases, the electronics packaging engineer will not have to complete items 5 through 12 of the cost estimate form.

DOCUMENTATION TRACKING AND PARTS CONTROL

Two additional forms may be used for detail tracking. The documentation tracking list is a means of controlling individual drawings as they are released

Table 2-1. Drafting and Check Standard Hours.

MECHANICAL

Drawing Size	Cable Assembly and Check		Mechanical Assembly and Check		Mechanical Detail and Check		Notes
A	4.5	1	4.5	1	4.5	1	
B	12.0	3	9.0	2	7.5	2	
C	15.0	4	18.0	5	15.0	4	
D	30.0	8	36.0	9	30.0	8	
E	48.0	12	90.0	23	48.0	12	

ELECTRICAL

Drawing Size	Schematic and Check		Logic and Check[a]		Wire List and Check[a]		Notes
A	14.0	4	9.0	2	1	.5	
B	18.0	5	12.0	3	–	–	
C	36.0	9	24.0	6	–	–	
D	45.0	11	30.0	8	–	–	
E	54.0	14	36.0	9	–	–	

PRINTED CIRCUIT

	Layout[b]		Tape-Up[b]		Silk Screen		Solder Mask		Fabrication		Assembly	
Size	**Drawing**	**Check**	**Drawing**	**Check**	**Drawing**	**Check**	**Drawing**	**Check**	**Drawing**	**Check**	**Drawing**	**Check**
Two-Sided												
10 × 12	158	63	119	38	12	3	2	1	8	2	24	6
12 × 20	185	74	139	42	24	6	2	1	8	2	32	8
16 × 20	205	82	154	46	30	8	2	1	8	2	40	10
Multisided												
10 × 12	232	93	174	52	12	3	2	1	8	2	24	6
12 × 20	330	132	248	74	24	6	2	1	8	2	32	8
16 × 20	395	158	296	89	30	8	2	1	8	2	40	10

[a] No. sheets × Hours/Sheets = Total time.
[b] Layout and tape-up check times include two people.

Customer[1] Quantity[2] Model[3] Date[4] Engineer[5]

Job No.[6] Quotation No.[7] Task[8] Revision Date[9]

Original Estimate[11]	Week Ending[10]																		
	Mechanical Engineer	Plan																	
		Act.[12]																	
	Electrical Engineer	Plan																	
		Act.																	
	Des. Hours	Plan																	
		Act.																	
	Drafting Hours	Plan																	
		Act.																	
	Mechanical Technician	Plan																	
		Act.																	
	Electrical Technician	Plan																	
		Act.																	
	Material $[13]	Plan																	
		Act.																	
	Total $[14]	Plan																	
		Act.																	

[1] Name of customer.
[2] Number of units to be built.
[3] Model number or identification number.
[4] Start date.
[5] Responsible engineer (originator).
[6] Charge number for accounting.
[7] Quotation or sales order number.
[8] Task description.
[9] Last date of revision.
[10] Week ending date.
[11] Estimated hours per week.
[12] Actual hours per week.
[13] Material costs per week.
[14] Weekly total expenditures.

Figure 2-5a. Job labor and material plan, vs. actual. Sample format.

Customer[1] Debtronics Quantity[2] 4 Model[3] MLR Date[4] 5-1-79 Engineer[5] B. Matisoff

Job No.[6] 2756 Quotation No.[7] 27001 Task[8] Engineering Revision Date[9]

Original Estimate[11]	Week Ending[10]		3-3	3-10	3-17	3-24	3-31	4-7	4-14	4-21	4-28	5-5	5-12	5-19	5-26	6-2	6-9	6-16
331	Mechanical Engineer	Plan	30	30	30	20	20	20	20	20	20	20	20	20	20	16	16	9
		Act.[12]																
	Electrical Engineer	Plan																
		Act.																
331	Des. Hours	Plan	—	—	40	40	40	40	40	20	20	20	20	20	11	10	10	
		Act.																
993	Drafting Hours	Plan	—	—	—	80	80	80	80	80	80	80	80	80	80	80	80	33
		Act.																
216	Mechanical Technician	Plan								36	36	36	36	36	36			
		Act.																
402	Electrical Technician	Plan											67	67	67	67	67	67
		Act.																
1745	Material $[13]	Plan				Dust cover flange		275.00		Power supply		540.00	F.P. Inter.	520.00	Mod. Fin	410.00		
		Act.																
	Total $[14]	Plan																
		Act.																

[1] Name of customer.
[2] Number of units to be built.
[3] Model number or identification number.
[4] Start date.
[5] Responsible engineer (originator).
[6] Charge number for accounting.
[7] Quotation or sales order number.
[8] Task description.
[9] Last date of revision.
[10] Week ending date.
[11] Estimated hours per week.
[12] Actual hours per week.
[13] Material costs per week.
[14] Weekly total expenditures.

Figure 2-5b. Job labor and material plan, vs. actual. Plan shows estimated hours and expenditures.

COST ESTIMATE

PREPARED BY	DATE	PAGE
LOU BRESKIN	2-24-79	1 OF 1

PROJECT/PROPOSAL NAME & NO.	CUSTOMER NAME & NO.		DATE RECEIVED
MICRO-LOGIC RESOLVER			2-2-79
ITEM NAME OR NUMBER	QUANTITY 2	CONTRACT TYPE	DELIVERY DATE 6-24-79

1. MATERIALS & SUBCONTRACTING	MATERIALS	MAT'L. HANDLING	MAT'L. HANDLING	OTHER DIR. CHGS.	SUBCONTRACTS	TOTAL
	1745.00	@ 18 %	314.00			

DIRECT LABOR CATEGORIES	MAN HRS.	RATE	ENGINEERING $	PRODUCTION $	$	TOTAL
ELECT. ENG'R		17.00				
MECH. ENG'R	331	17.00	5627.00			
DESIGNER	331	11.00	3641.00			
DRAFTING	993	8.00	7944.00			
MECH. TECH	216	6.50	1404.00			
ELECT. TECH	402	6.50	2613.00			
PRINTED CIRCUITS/CABLE					950.00	
MECH. PARTS					795.00	
2. SUB-TOTAL DIRECT LABOR						
3. DIRECT LABOR CONTINGENCY						
4. TOTAL DIRECT LABOR						
5. OVERHEAD RATES			@ %	@ %	@ %	
6. OVERHEAD DOLLARS						

7. FACTORY COST (lines 1 + 4 + 6) ____
8. G & A ____ % OF FACTORY COST (line 7) ____
9. TOTAL COST (lines 7 & 8) ____
10. FEE ____ % OF TOTAL COST (line 9) ____
11. SUB-TOTAL (lines 9 & 10) ____
12. ____

TOTAL PRICE (lines 11 & 12) ____

☐ PROPOSAL ☐ PROJECT

APPROVALS	DATE
DEPARTMENT MANAGER	DATE
ACTIVITY DIRECTOR	DATE
VICE PRESIDENT ENGINEERING	DATE
ACCOUNTING	DATE
MARKETING	DATE
VICE PRESIDENT FINANCE	DATE
PRESIDENT	DATE

Figure 2-6. Cost estimate form.

Table 2-2. Checklist for Labor Cost Estimates.

Engineering
System design
Circuit design
Packaging
Layouts
Wiring (general design)
Cooling design
Framework design
Procurement
Proposal writing
Component test equipment design
Package test equipment design
System tests
Power supply design
Supervision and planning
Travel
Instruction books
Periodic reports
Production
Manufacturing of packages
Test of packages
Test of components
Wiring
Layout
Procurement
Shop work
Assembly
Packing and shipping
Breadboarding
System testing
Drafting
System block diagrams
Logical block diagrams
Schematics
Artwork, assemblies, and details for packages
Instruction book diagrams
Mechanical drawings–framework, chassis, etc.

through the drafting system (Figure 2-7). The list may be used for details as they are released separately or for the release of complete assembly packages. The documentation tracking list may also be used as an indentured parts list showing the component breakdown of each assembly (Figure 2-7). The second form, parts control, is a means of controlling the purchase of the materials and

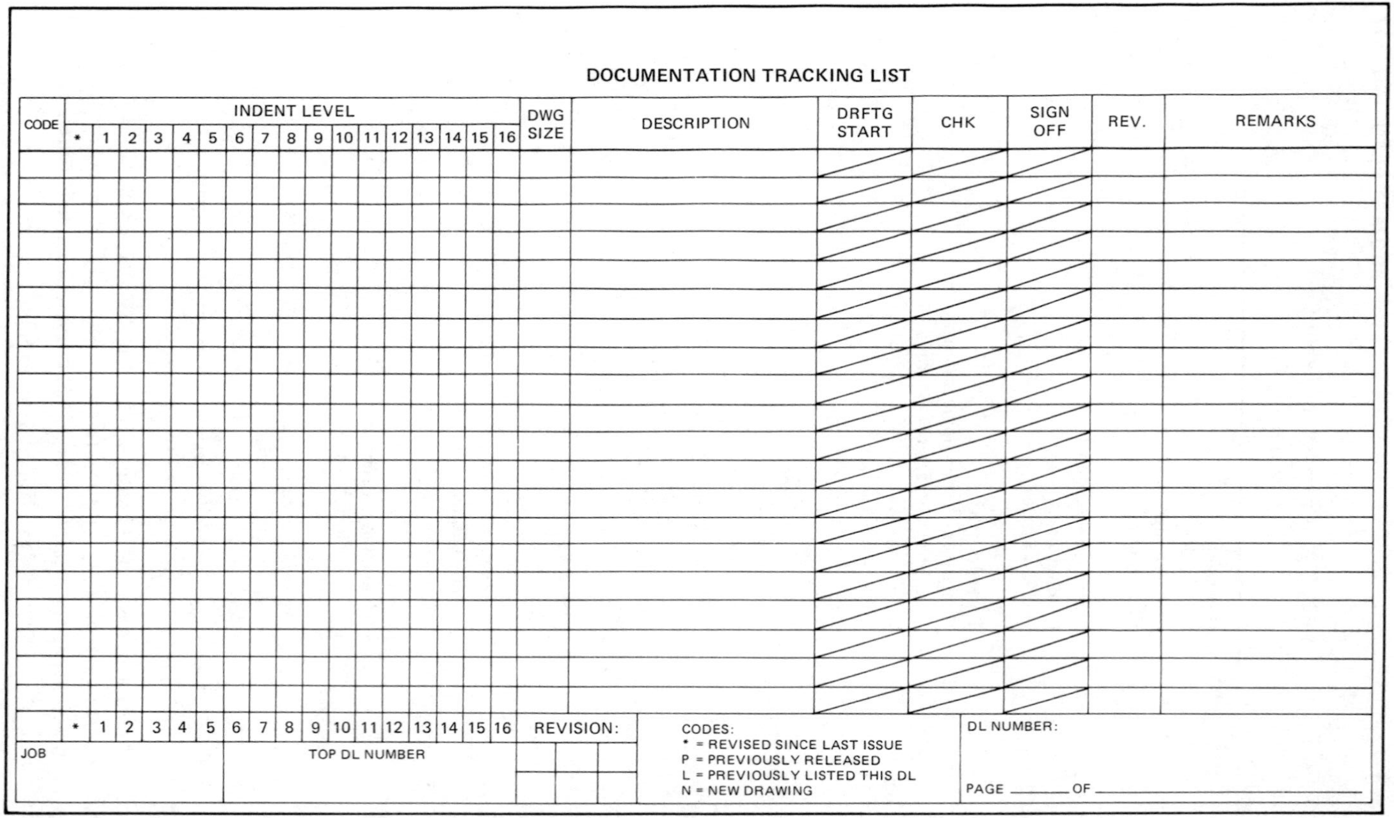

DOCUMENTATION TRACKING LIST

CODE	INDENT LEVEL																	DWG SIZE	DESCRIPTION	DRFTG START	CHK	SIGN OFF	REV.	REMARKS
	*	1	2	3	4	5	6	7	8	9	10	11	12	13	14	15	16							
	*	1	2	3	4	5	6	7	8	9	10	11	12	13	14	15	16	REVISION:						

JOB

TOP DL NUMBER

CODES:
* = REVISED SINCE LAST ISSUE
P = PREVIOUSLY RELEASED
L = PREVIOUSLY LISTED THIS DL
N = NEW DRAWING

DL NUMBER:

PAGE ______ OF ______

Figure 2-7. Documentation tracking list.

JOB NO. PAGE

ITEM	QTY	DESCRIPTION	PART NO.	DATE ORDERED	GEN. PURCHASE REQ. NO.	PURCHASE ORDER NO.	VENDOR	DATE REQ.	DATE PROMISED	DATE REC. PARTIAL	DATE REC. TOTAL

Figure 2-8. Parts control.

parts for the engineering or prototype build of the equipment being designed (Figure 2-8). It is important to coordinate the build of the engineering model with the design and drafting effort. The dates of availability of the parts and materials are critical in scheduling the proper personnel for the project. The packaging engineer must work closely with the purchasing department to insure that scheduled personnel are not idle due to lack of materials. Every packaging engineer should always remember that planning pays off in dollars (Figure 2-8).

DESIGN STUDY

The following is a typical example of a design study.

Mechanical Design: Micro Logic Resolver

Project No. ________________ **Title** ________________ **Date** __________

Engineer ________________ **Customer** ________________

Description. A micro logic resolver to be designed using TO-5 micro logic devices. The unit must be capable of functioning reliably when installed on shipboard platforms such as missile launchers and gun mounts. Military specifications (MIL-E-16400 and MIL-E-5400) shall be used as design requirements in addition to the specific requirements of the Naval Bureau of Weapons. The unit shall be modular in construction with provisions for maximum accessibility during the development phase for maintenance and testing. The changeover from the development phase to the production unit must be accomplished with a minimum of redesign.

Major Design Parameters

1. The system shall incorporate a modular technique to achieve simplicity and maximum producibility and maintainability.
2. All electrical circuits shall be broken down to optimize the number of repeatable circuits and simplify the system. The modules (functional blocks) shall be as large as practical to minimize the number of system interconnections. Most of the interconnections should be accomplished within the functional blocks.
3. Conductive cooling techniques must provide adequate cooling under extreme high temperature conditions.
4. The system should be divided into major functions to facilitate manufacturing and maintenance.

Product Description. The micro logic resolver shall consist of groups of functional blocks that are to be interconnected into submodule assemblies, which in turn will be assembled into the logic module (Figure 2-2). The basic functional block (Figure 2-9) consists of a welded matrix containing the TO-5 microelectronic devices. The TO-5 devices are installed on dimensionally variable increments of three devices in width only. In addition, variations of the functional blocks using cordwood welded modules with conventional components may be employed. The peripheral circuitry which will use only conventional components will be packaged in standard cordwood modules and assembled onto a welded matrix similar to the TO-5 devices.

The first assembly of the functional block into the submodules will be done by conventional point-to-point wiring on a breadboard built by the engineering department and will be used to verify the logic design. The external connector assembly is to be a flat cable cast directly into the submodule matrix and terminating in a round cable with hermetically sealed AN connectors on a connector panel. The submodules, when completely assembled onto the front panel, will be installed into an aluminum cover, and the front panel will be bolted into position.

Environmental protection is provided by applying a hard anodized finish prior to painting. All of the exposed parts are hermetically sealed, and connector covers are provided for protection when the system is not in service.

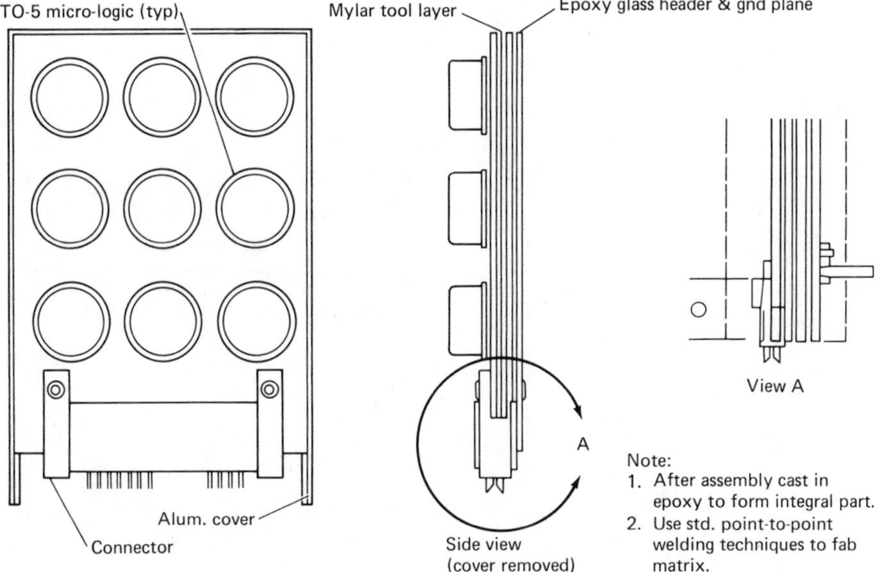

Figure 2-9. Plug-in functional block.

Radio-frequency interference (RFI) control is accomplished by designing in a minimum number of mechanical joints. The front panel and the switch indicator assemblies are to be sealed with RF gasket materials (Figure 2-2). With the entire assembly mounted onto the front panel and then mounted into a deep drawn aluminum cover, the only mechanical seams are those around the front panel and the connectors in the base of the cover. In addition to the RF gaskets, all of the connectors will be RFI-filtered.

Heat transfer will occur through fins inserted between the rows of functional blocks and will provide a conductive thermal path directly to the external surfaces. Encapsulation of the functional blocks in a silicon potting compound will be employed to eliminate thermal "hot spots" and will considerably improve the heat transfer. In addition, the use of nickel leads throughout the matrix will improve the removal of heat and provide additional dissipation surfaces.

The level of field service will be at the functional block level. In the production units, these blocks will be removed by unwrapping the wire-wrap connections and rewrapping a new block in position. The defective functional blocks will be returned to the factory for repair.

3

HUMAN FACTORS ENGINEERING

The human engineering design guidelines presented in this chapter are directed toward hardware problems involving man-machine interfaces. When the engineer or designer cannot find an adequate solution herein, a human factors specialist should be consulted.

The protection of the operating and maintenance personnel must be considered in a design. There are numerous methods of incorporating safeguards into a design; many of these are implicit and are routine practice. However, certain procedures are of such importance that they warrant special attention here.

The basic function of human factors engineering is to establish the design criteria, to achieve success through the integration of the human into the system, and to create a design that embraces the maximum in effectiveness, simplicity, efficiency, reliability, and safety in operation and maintenance.

Each design should insure operability and maintainability by at least 90% of the user population. In measuring a given population, most of the persons measured will fall somewhere in the middle of the distribution while a small number will appear at each end of the distribution. Since it is not practical from a cost standpoint to design for the entire population, it is common practice to design for all those people between the 5th and 95th percentiles. The anthropometric data presented in this chapter are nude-body measurements expressed in inches and represent United States Army personnel (1966), United States Navy aviators (1964), and United States Air Force flight personnel (1967). When using the information presented in this chapter, the following must be considered:

- Nature, frequency, and difficulty of the related tasks
- Position of the body while performing the task
- Mobility and flexibility requirements imposed by the task
- Need to compensate for obstacles, projections, etc.

ANTHROPOMETRIC APPLICATIONS

Accesses, safety clearances, and passageways that must accommodate the passage of the body or parts of the body should be based on the values of the 95th percentile (i.e., larger body dimensions). Reaching distances, displays, test points, and controls that are limited by body extensions (length of reach) should be based on the 5th percentile values.

The life support and biomedical factors that will affect the performance of the operating and maintenance personnel should be given consideration in every design.

- Maximum temperature within the range of human endurance
- Safe range of acoustic noise, vibration, and electrical shock
- Protection from radiological, toxicological, electromagnetic, and visual hazards
- Adequate space for the operator and the equipment, and free space for the required movements of operation and maintenance
- Adequate physical, visual, and auditory communication between the operators and the equipment, and between other operators
- Provisions against injury
- Adequate illumination
- Provisions for minimizing psychological stress and fatigue
- Adequate emergency systems for complete shutdown of the system

Figures 3-1 through 3-10 and Tables 3-1 and 3-2 provide the recommended dimensions to be used in designing equipment for 90% of the user population.

ILLUMINATION

Adequate illumination should be provided wherever operator and maintenance personnel are working. Glare and reflections should be reduced. Table 3-3 provides recommended lighting requirements for specific tasks.

CONTROL-DISPLAY RELATIONSHIP

Controls are the means with which the operator communicates with a system; for example, the means of turning the system on or off or of putting various types of information into the system for instructions to perform a given task. Displays are the means with which the system communicates with the operator; for example, the dials, gauges, and lamps that inform the operator of a machine's functioning and the results of the operations performed by the system.

The relationship of a control to its associated display should be immediately apparent and unambiguous to the operator. The relationship should be func-

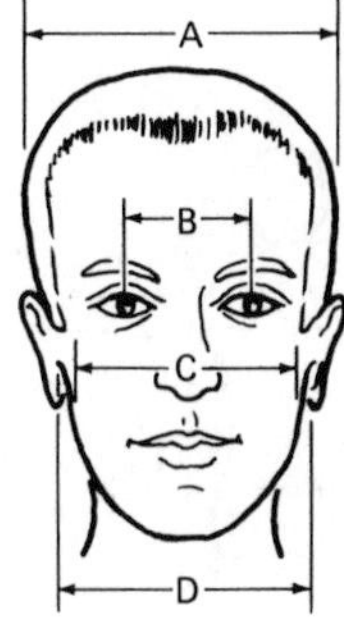

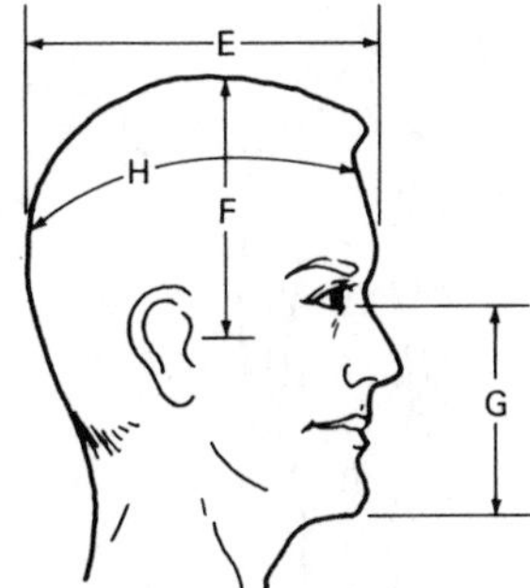

Sym	Dimension	Fifth Percentile			Ninety-Fifth Percentile		
		USA	USN	USAF	USA	USN	USAF
A	Head breadth	5.65	5.69	5.80	6.47	5.47	6.50
B	Interpupillary breadth	2.15	2.39	2.24	2.67	2.75	2.71
C	Face breadth	5.15	4.87	5.27	5.88	5.79	5.94
D	Bitragion breadth	4.95	5.17	5.26	5.69	5.85	5.98
E	Head length	7.19	7.38	7.39	8.14	8.24	8.27
F	Head heigth	4.69	4.74	4.89	5.72	5.57	5.69
G	Face length	4.31	4.23	4.35	5.17	5.48	5.13
H	Head circumference	21.07	21.74	21.74	23.16	23.57	23.59

Figure 3-1. Head and face dimensions.

tionally effective—that is, it should require a minimum of mental involvement by the operator. Therefore, the control should be located directly under or to the right of the associated display so that the operator's hand, when setting the control, will not obscure the operator's vision. Good control-display relationships are accomplished by the use of such design considerations as proximity, similarity of groupings, coding, framing, and labeling (Figure 3-11).

The sequence of operation is from left to right or from top to bottom. The most frequently used grouping of a control-display should be located so that they are readily accessible to the operator. Groups having different functions should be set apart by coding or framing. The display indicator should direct and guide the appropriate control response clearly, and the response of a display indicator to control movements should be consistent, predictable, and compatible with the operator's expectations. For example, when a dial is turned, the indicator responds with the movement. The time lag between control activation and display response should be kept to a minimum.

In normal applications, the display should cease to move after control movement has stopped. Whenever possible, controls and displays should be designed so that the movement of the display and that of its associated control occur in

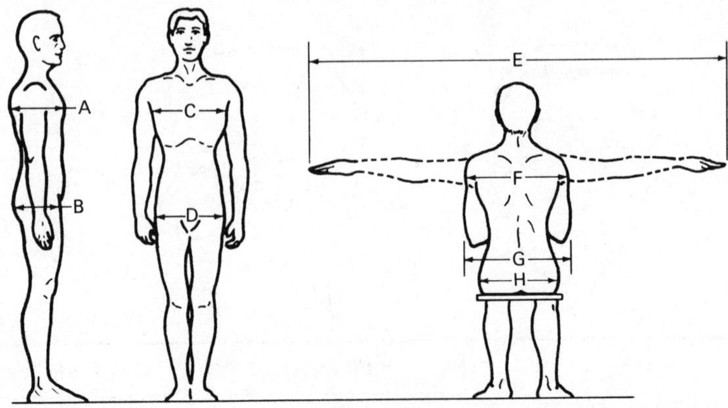

Sym	Dimension	Fifth Percentile			Ninety-Fifth Percentile		
		USA	USN	USAF	USA	USN	USAF
A	Chest depth	8.0	8.2	8.4	10.5	10.6	10.9
B	Buttock depth	–	8.3	8.2	–	10.7	10.8
C	Chest breadth	10.8	11.6	11.6	13.5	14.3	14.4
D	Hip breadth, standing	11.9	12.6	12.7	14.4	14.9	15.2
E	Span	–	–	65.9	–	–	75.6
F	Shoulder breadth	16.3	17.3	17.4	19.6	20.3	20.7
G	Forearm-forearm breadth[a]	15.7	–	19.0	21.1	–	23.9
H	Hip breadth, sitting	12.1	13.1	13.5	15.1	15.9	16.4

[a] Army and Air Force dimensions not compatible.

Figure 3-2. Breadth and depth dimensions.

the same direction. When a rotary control and a linear display are in the same plane, that part of the control adjacent to the display should move in the same direction as the moving part of the display. When levers are used, both the lever and the display pointer should travel through parallel arcs.

When there is a direct linkage between a control and a display, a rotary control should be employed if the indicator moves through an arc exceeding 180°. If the arc is less than 180°, a linear control may be used provided that the path of control movement parallels the average path of the indicator movement, and the indicator and control move in the same relative direction.

Control-display ratios for continuous-adjustment controls should be selected in a manner that will minimize the total time required to perform the desired control movement. When a wide range of display movement is required, a small movement of the control should yield a large movement of the display element. Conversely, when a fine adjustment is required, a large movement of the control should yield a small movement of the display element.

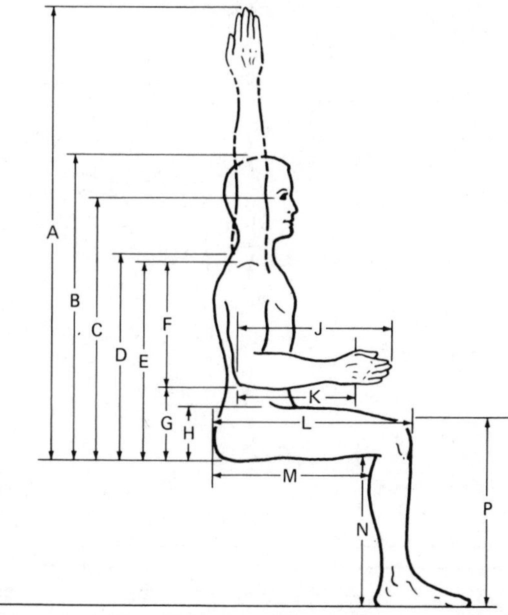

Sym	Dimension	Fifth Percentile			Ninety-Fifth Percentile		
		USA	USN	USAF	USA	USN	USAF
A	Vertical reach	50.7	–	–	53.2	–	–
B	Sitting height	33.3	34.2	34.7	38.1	33.4	33.8
C	Eye height sitting	28.6	29.7	30.0	33.3	33.6	33.9
D	Mid-shoulder height	22.5	–	23.7	26.6	–	27.3
E	Shoulder height	–	22.0	22.2	–	25.5	25.9
F	Shoulder-elbow length	13.3	13.4	13.1	15.7	15.6	15.3
G	Elbow rest height	–	7.6	8.2	–	10.9	11.6
H	Thigh clearance height	–	–	5.6	–	–	7.4
J	Elbow-fingertip length	17.4	17.9	–	20.4	20.4	–
K	Elbow-grip length	–	–	12.8	–	–	14.9
L	Buttock-knee length	21.6	22.5	22.1	25.3	25.8	25.6
M	Buttock-popliteal length	17.6	18.2	18.2	20.9	21.4	21.5
N	Popliteal height	15.6	15.9	15.8	18.8	18.8	18.7
P	Knee height	19.6	20.3	20.4	23.1	23.5	23.6

Figure 3-3. Seated body dimensions.

DISPLAYS

The information displayed to the operator should be limited to those items that are necessary to enable him or her to perform specific actions or to make specific decisions. This information should only be as precise as required. It should be

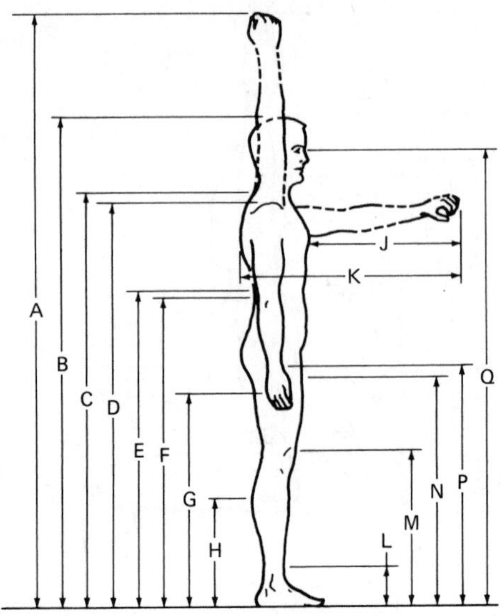

Sym	Dimension	Fifth Percentile USA	Fifth Percentile USN	Fifth Percentile USAF	Ninety-Fifth Percentile USA	Ninety-Fifth Percentile USN	Ninety-Fifth Percentile USAF
A	Overhead reach height	–	–	76.8	–	–	87.6
B	Stature	64.5	66.2	65.9	73.1	73.9	73.9
C	Cervical height	54.8	56.0	56.1	63.0	63.3	63.7
D	Shoulder height	52.6	–	53.4	60.7	–	60.9
E	Elbow height	–	–	41.3	–	–	47.3
F	Wallet height	38.4	39.4	38.9	45.3	45.3	45.0
G	Knuckle height	–	–	27.7	–	–	32.4
H	Calf height	12.2	–	12.6	15.7	–	15.5
J	Depth of reach						
	One arm	–	–	20.2	–	–	26.8
	Both arms	–	–	19.2	–	–	24.5
K	Functional reach	29.5	29.3	29.1	35.8	34.0	34.3
L	Ankle height	–	–	4.7	–	–	6.2
M	Kneecap height (top)	18.8	19.3	19.1	23.0	22.8	22.4
N	Crotch height	29.6	30.6	30.8	35.7	36.0	36.2
P	Wrist height	–	–	31.6	–	–	36.7
Q	Eye height	–	–	61.2	–	–	69.1

Figure 3-4. Standing body dimensions.

presented to the operator in directly usable form which will not require any mental translation.

The maximum viewing distance to displays located close to the associated controls is limited by the operator's ability to reach the control and should not

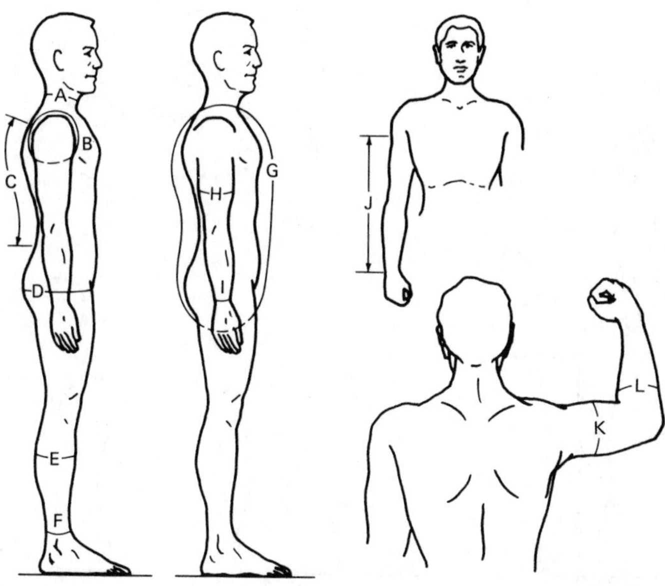

	Dimensions	Fifth Percentile USA	Fifth Percentile USN	Fifth Percentile USAF	Ninety-Fifth Percentile USA	Ninety-Fifth Percentile USN	Ninety-Fifth Percentile USAF
A	Neck circumference	13.5	14.0	13.9	16.1	16.4	16.4
B	Arm scye circumference	15.6	16.1	17.2	19.8	19.5	20.9
C	Waist back length	15.6	–	17.0	20.0	–	20.1
D	Hip circumference	33.5	35.2	35.3	41.6	41.8	42.5
E	Calf circumference	12.8	13.4	13.2	16.2	16.3	16.1
F	Ankle circumference	8.1	8.1	8.0	9.9	9.7	9.7
G	Vertical trunk circumference	59.3	61.6	61.7	70.3	70.2	71.0
H	Biceps circumference, relaxed	10.0	–	10.6	13.5	–	13.7
I	Wrist circumference	6.2	6.2	6.4	7.3	7.2	7.6
J	Sleeve inseam length	17.4	16.8	17.5	20.9	19.9	20.8
K	Biceps circumference, flexed	11.0	11.4	11.5	14.6	14.5	14.4
L	Forearm circumference, flexed	10.3	10.7	10.7	13.0	13.0	12.8

Figure 3-4. *(Continued)*

exceed 28 in. Otherwise, there is no maximum limitation except that imposed by the available space, provided that the display can be read adequately. The minimum viewing distance should never be less than 13 in. and, preferably, not less than 20 in.

Visual displays should provide the operator with a clear indication of equip-

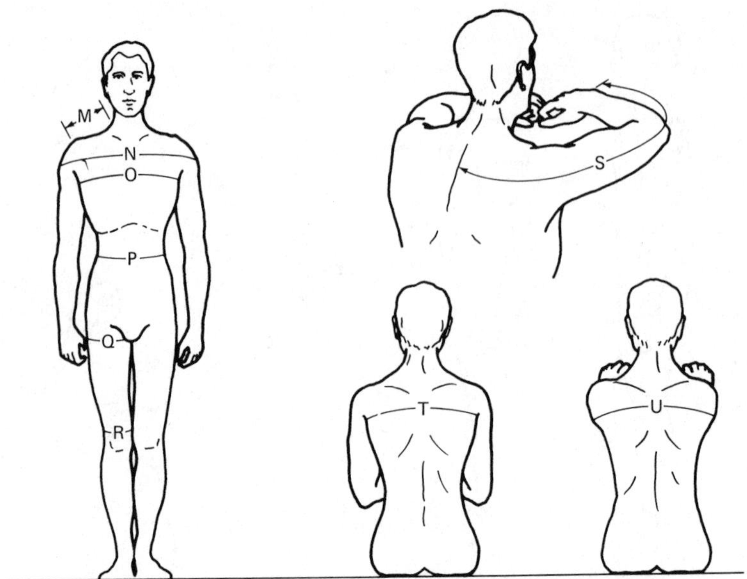

	Dimensions	Fifth Percentile			Ninety-Fifth Percentile		
		USA	USN	USAF	USA	USN	USAF
M	Shoulder length	5.0	–	5.8	7.6	–	7.4
N	Shoulder circumference	40.7	42.3	42.7	48.9	49.4	50.3
O	Chest circumference	33.1	35.2	34.9	41.7	42.7	43.1
P	Waist circumference	27.4	29.3	29.8	37.8	37.8	39.4
Q	Upper thigh circumference	18.9	20.0	20.3	25.1	25.2	26.0
R	Lower thigh circumference	13.6	14.1	15.1	18.6	18.0	19.6
S	Sleeve length	31.3	32.8	33.5	36.4	37.4	38.1
T	Interscye breadth	13.4	14.2	12.8	17.4	18.2	17.7
U	Interscye maximum[a]	18.3	19.1	22.3	23.1	23.6	26.2

[a]USAF dimensions not comparable with USA and USN dimensions.

Figure 3-5. Circumference and surface dimensions, in inches.

ment or system conditions. Therefore, the display must be designed to accommodate the particular conditions under which it will be used, the method of use, and the purpose it is intended to serve.

Displays should be arranged in relation to one another according to their sequence of use or the functional relationships of the components that they represent. Those displays most frequently used should be grouped together and located in the optimum visual zone. Displays should be designed and located so that they can be read to the required degree of accuracy by personnel in the

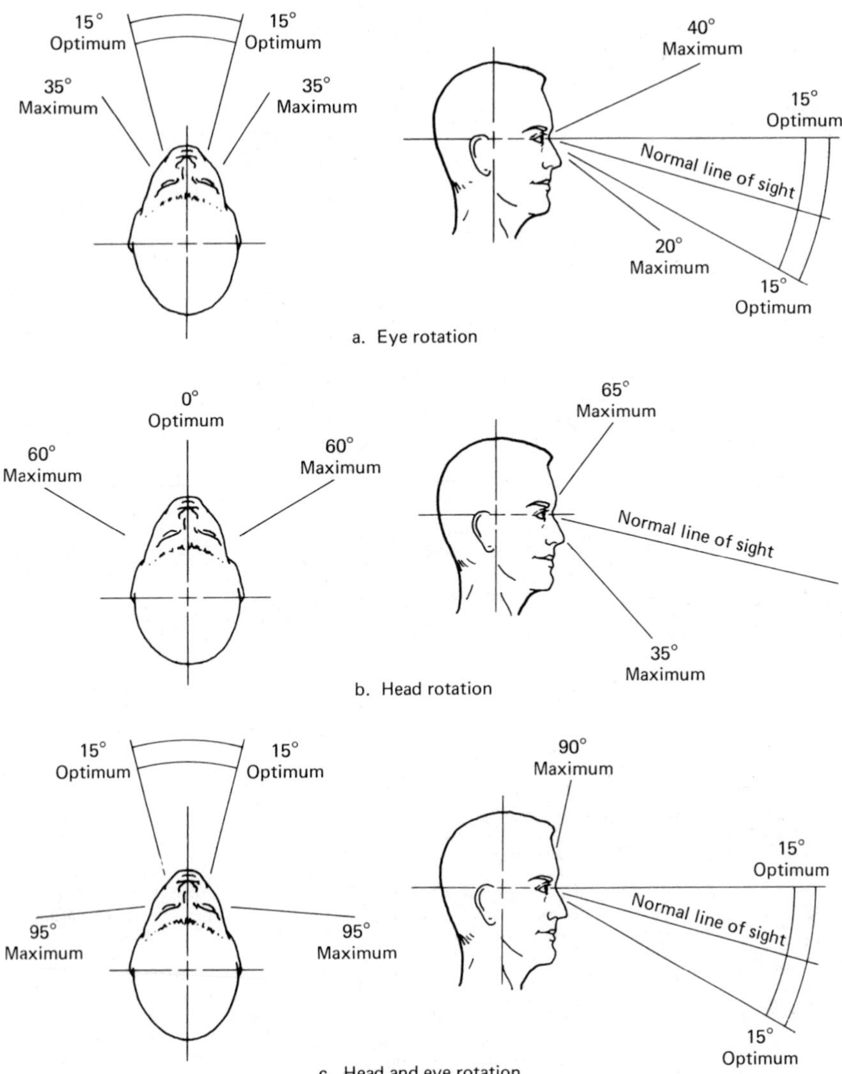

Figure 3-6. Horizontal and vertical visual fields.

normal operating or servicing position. Display faces should be oriented perpendicular to the normal line of sight of the operator. To minimize parallax, display orientation that is less than 45° from the line of sight should be avoided.

Displays should be arranged or shielded to minimize the reflectance of ambient illumination from the glass or plastic face. Reflectance of instruments or consoles in windshields or other enclosures should be avoided.

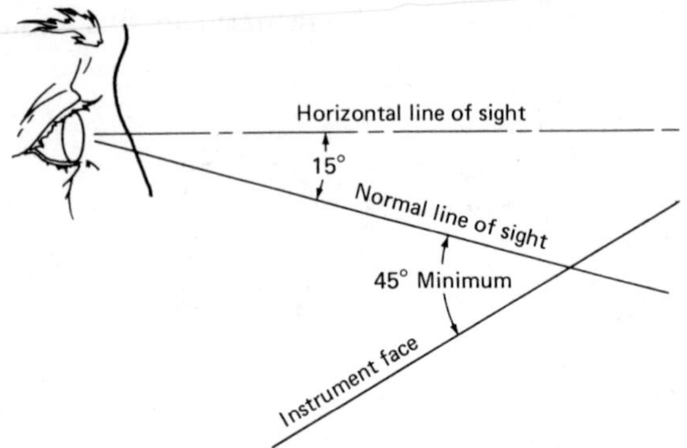

Figure 3-7. Lines of sight.

Opening Configuration	Dimensions (Inches) A	Dimensions (Inches) B	Maintenance Task	Opening Configuration	Dimensions (Inches) A	Dimensions (Inches) B	Maintenance Task
	4.2	4.7	Grasping small objects (up to 2-1/16 inches in diameter)		5.2	4.5	Using pliers or similar hand tools
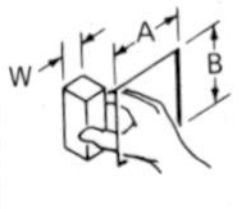	W +1.75	5.0	Grasping large objects (2-1/16 inches in width) with one hand		5.3	6.1	Using T-handle wrench with freedom to turn hand through 180 degrees
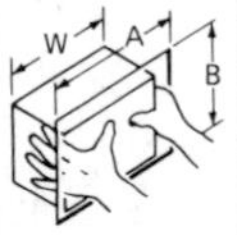	W +3.0	5.0	Grasping large objects with two hands, with hands extended through opening up to fingers		10.6	8.0	Using open-end wrench with freedom to turn wrench through 60 degrees
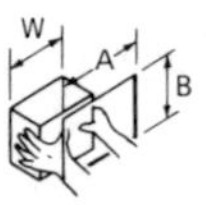	W +6.0	5.0	Grasping large objects with two hands, with hands extended through opening up to wrists		4.8	6.1	Using Allen wrench with freedom to turn wrench through 60 degrees
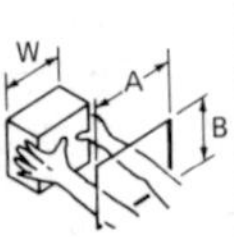	W +6.0	5.0	Grasping large objects with two hands, with arms extended through opening up to elbows		3.5	3.5	Using test probe or similar device
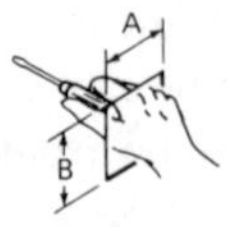	4.2	4.6	Using common screwdriver with freedom to turn hand through 180 degrees				

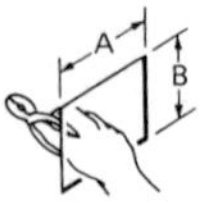

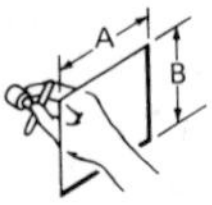

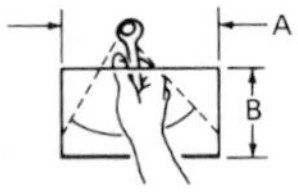

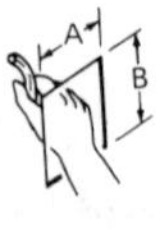

Figure 3-8. Recommended access dimensions.

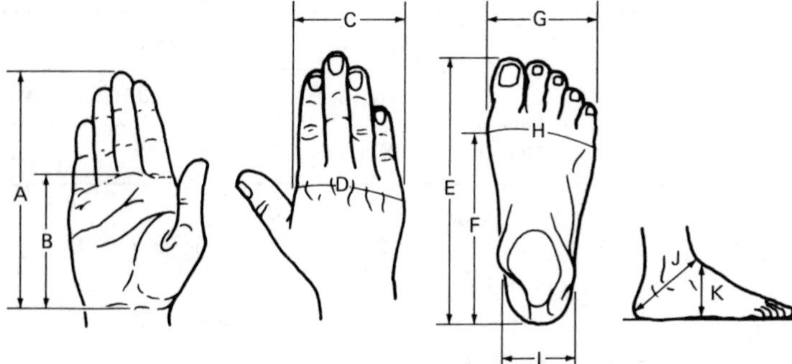

	Dimensions	Fifth Percentile			Ninety-Fifth Percentile		
		USA	USN	USAF	USA	USN	USAF
	Hand Dimensions						
A	Hand length	6.90	6.98	7.00	8.13	8.09	8.07
B	Palm length	3.77	–	3.92	4.59	–	4.63
C	Hand breadth	3.20	3.25	3.24	3.83	3.80	3.78
D	Hand circumference	7.81	7.76	7.89	9.28	9.06	9.11
	Foot Dimensions						
E	Foot length	9.71	9.73	9.89	11.41	11.29	11.44
F	Instep length	7.06	–	7.18	8.41	–	8.42
G	Ball-of-foot breadth	3.53	3.58	3.54	4.24	4.58	4.18
H	Ball-of-foot circumference	8.86	–	9.02	10.78	–	10.62
I	Heel breadth	2.42	–	2.40	3.02	–	2.87
J	Heel-ankle circumference	12.38	–	12.47	14.53	–	14.30
K	Instep circumference	9.43	–	9.38	11.56	–	10.95

Figure 3-9. Hand and foot dimensions, in inches.

When the maximum degree of dark adaptation of the eyes of the operator is required, a low-level red light should be used for panel illumination. Low-intensity white light should be used when dark adaptation is not critical.

Color Coding of Indicator Lamps

Simple visual indicators that illuminate to convey qualitative information to the operator are frequently color-coded for ready recognition. According to the commonly used color code:

- Red should be used to alert the operator that the system or some part of it is inoperative (e.g., "no-go," "failure," and "malfunction").

Configuration	Type of Handle	Dimensions (in.) Bare Hand			Gloved Hand			Mittened Hand		
		X	Y	Z	X	Y	Z	X	Y	Z
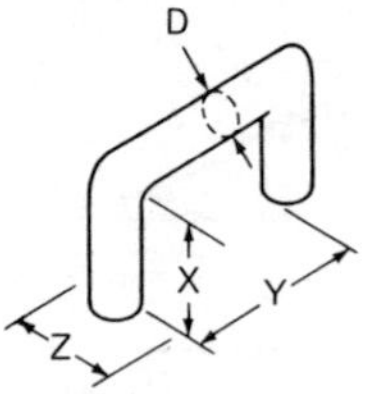	Two-finger bar	1.25	2.5	3.0	1.5	3.0	3.0	–	–	–
	One-hand bar	2.0	4.5	3.0	3.5	5.25	4.0	3.5	5.25	6.0
	Two-hand bar	2.0	8.5	3.0	3.5	10.5	4.0	3.5	11.0	6.0
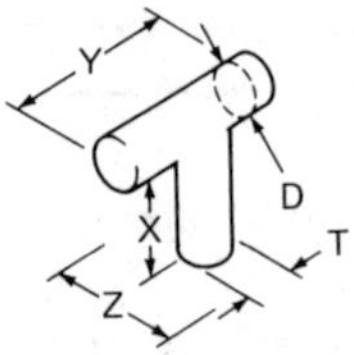	T-bar	1.5	4.0	3.0	2.0	4.5	4.0	–	–	–
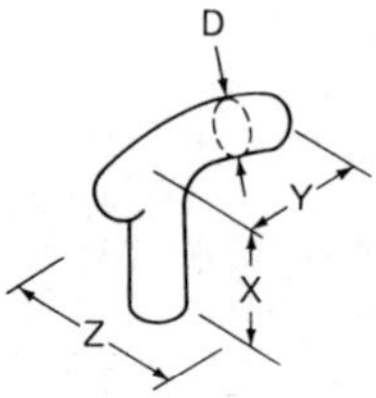	J-bar	2.0	4.0	3.0	2.0	4.5	4.0	3.0	5.0	6.0
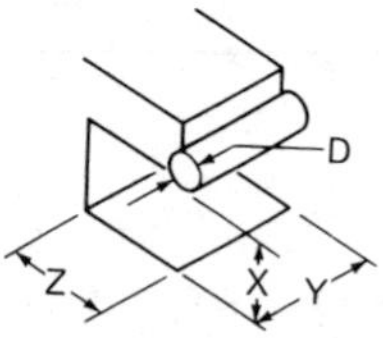	Two-finger recess	1.25	2.5	2.0	1.5	3.0	2.0	–	–	–
	One-hand recess	2.0	4.25	3.5	3.5	5.25	4.0	3.5	5.25	5.0
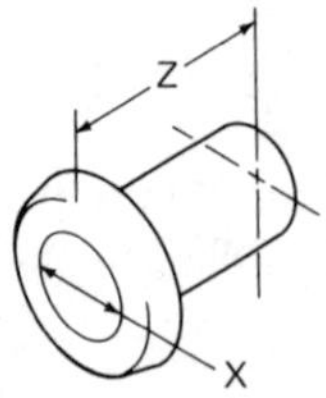	Fingertip recess	0.75	–	0.5	1.0	–	0.75	–	–	–
	One-finger recess	1.25	–	2.0	1.5	–	2.0	–	–	–

Figure 3-10. Recommended handle dimensions.

Table 3-1. Age and Weight.

	Fifth Percentile			Ninety-Fifth Percentile		
Age/Weight	USA	USN	USAF	USA	USN	USAF
Age (yr)	18.6	–	22.4	31.5	–	42.4
Weight (lbs)	126.3	140.3	140.0	201.9	203.6	211.0

- Flashing red indicators should be used to denote emergency conditions only. The flash rate should be three to five flashes per second, with "on" and "off" times approximately equal. Such indicators should be designed so that, if the flasher mechanism fails, the indicator will remain illuminated.
- Amber indicators should be used to advise the operator of the existence of a marginal condition (e.g., "caution," "recheck," and "delay").
- Green indicators should be used to convey to the operator that the monitored equipment is operating satisfactorily or is within tolerance limits (e.g., "go," "ready").
- White indicators should be used to denote system conditions that have no "right" or "wrong" implications, such as alternative functions or transitory conditions.

Transilluminated Indicators

Transilluminated indicators should be used to display qualitative information to the operator, such as information that requires the operator's immediate reaction and information to which the operator's attention must be called. The brightness of transilluminated displays should be at least 10% greater than the ambient or background brightness. Three general types of transilluminated displays are

Table 3-2. Weight Limits for One-Man Lift.

Height of Lift Above Ground (ft)	Maximum Weight of Item (lb)
5	35
4	50
3	65
2	80
1	85

Table 3-3. Specific Task Illumination Requirements.

Work Area or Type of Task	Illumination Level (Foot-Candles)[a, b] Recommended	Minimum
Assembly, general		
coarse	50	30
medium	75	50
fine	100	75
precise	300	200
Benchwork		
rough	50	30
medium	75	50
fine	150	100
extra fine	300	200
Business-machine operation (calculator, digital, input, etc.)	100	50
Console surface	50	30
Corridors	20	10
Circuit diagram	100	50
Dials	50	30
Electrical equipment testing	50	30
Emergency lighting	–	3
Gauges	50	30
Hallways	20	10
Inspection tasks, general		
rough	50	30
medium	100	50
fine	200	100
extra fine	300	200
Machine operation, automatic	50	30
Meters	50	30
Missiles		
repair and servicing	100	50
storage areas	20	10
general inspection	50	30
Office work, general	70	50
Ordinary seeing tasks	50	30
Panels		
front	50	30
rear	30	10
Passageways	20	10
Readings		
large print	30	10
newsprint	50	30
handwritten reports, in pencil	70	50
small type	70	50
prolonged reading	70	50

Table 3-3. *(Continued)*

Work Area or Type of Task	Illumination Level (Foot-Candles)[a, b] Recommended	Minimum
Recording	70	50
Repair work		
general	50	30
instrument	200	100
Scales	50	30
Screw fastening	50	30
Service areas, general	20	10
Stairways	20	10
Storage		
inactive or dead	5	3
general warehouse	10	5
live, rough or bulk	10	5
live, medium	30	20
live, fine	50	30
Switchboards	50	30
Tanks, containers	20	10
Testing		
rough	50	30
fine	100	50
extra fine	200	100
Transcribing and tabulation	100	50

[a] As measured at the task object or 30 in. above the floor.
[b] As a guide in determining illumination requirements, the use of a steel scale with $\frac{1}{64}$-in. divisions requires 180 foot-candles of light for optimum visibility.

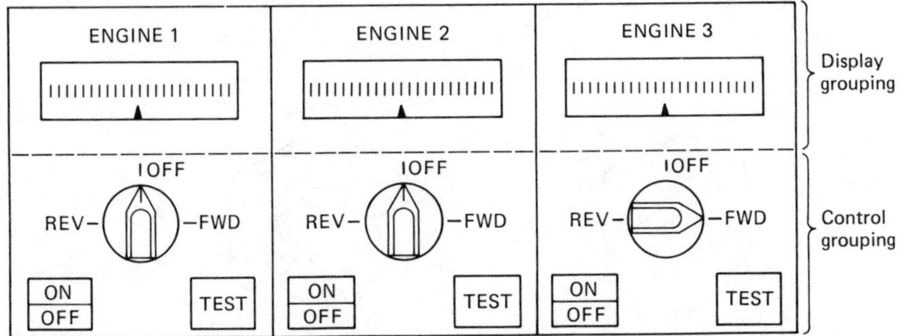

Figure 3-11. Example of functional grouping.

available:

1. Single- and multiple-legend indicators that present such information as meaningful words, numbers, symbols, or abbreviations
2. Simple indicator lamps
3. Transilluminated panel assemblies that present qualitative status or system-readiness information

Unless design considerations dictate otherwise, legend indicators should be used rather than simple indicator lamps. Except for warning and caution indicators, which must be highly visible at all times, the lettering on single-legend indicators should be visible whether or not the indicator is energized. Multiple-legend indicators should be designed to conform to the following:

1. When a rear legend is illuminated, it should not be obscured by the front legends.
2. Rear-legend plates should be placed in a manner that minimizes parallax.
3. Rear legends should be equal in apparent brightness and contrast to front legends.

Simple Indicator Lamps

Simple indicator lamps should be used when design considerations preclude the use of legend indicators. There should be sufficient space between simple round indicator fixtures to permit unambiguous labeling, indicator interpretation, and convenient bulb removal and replacement. Single-status indicators (Figure 3-12)

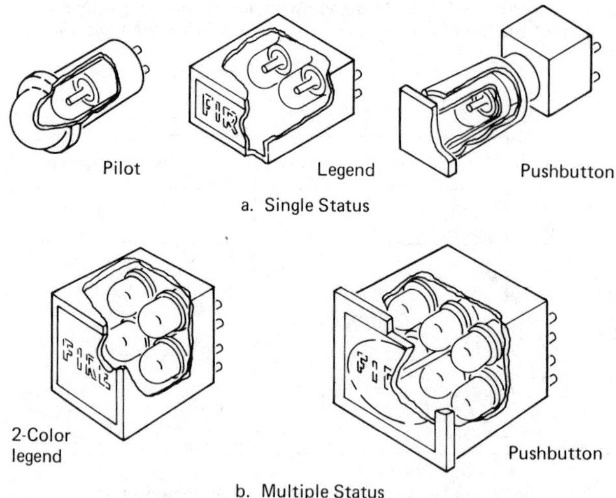

Figure 3-12. Typical status indicators. *a*, single status; *b*, multiple status.

provide discrete, and positive information by using a single-color illumination of display areas. A single-status display of this type may employ a solitary miniature lamp (e.g., a pilot light) or several lamps. Banks of small units of this type may be used as binary or digital readouts. Indicators that illuminate to identify a blown fuse are also classified as single-status displays.

Multiple-Status Indicators

Multiple-status indicators (Figure 3-12b) provide as many as three discrete and positive indications by means of color change within a single display field. For instance, a green indicator signals that a system is operating properly. The same indicator will turn yellow when a malfunction occurs or red when an emergency system "shutdown" is required. Multiple-status displays may employ various combinations of lamps and color filters; the number and position of such elements are determined by both equipment engineering and human factors criteria.

Illuminating Indicators

Illuminating indicators are used to present both qualitative and discrete information means of controlled light sources. The illumination of one or more lamps behind a translucent display surface, or a change in color, is used to indicate status, equipment, or action required.

Segmented and Dot Matrices

Segmented matrices are random-access displays in which alphanumeric characters are formed by the independent illumination of individual segments incorporated into one plane of a single matrix. The segments may be illuminated by light-emitting diodes (LED) or incandescent, neon, or electroluminescent lamps; the characters appear as dotted or broken-line figures against a contrasting dark background. A logic circuit can be used to decode binary information and energize appropriate display segments. Character legibility is improved by increasing the number of available segments. The dot matrix, which is also used for alphanumeric data presentations, is particularly adaptable to LED sources. Figure 3-13 shows typical segmented and dot matrices.

Back-Lighted Belt Displays

A back-lighted belt display employs a motor-driven continuous belt on which character spaces (usually up to 64) are provided. These devices are capable of binary coding, data storage, and electrical and visual readout. One character at

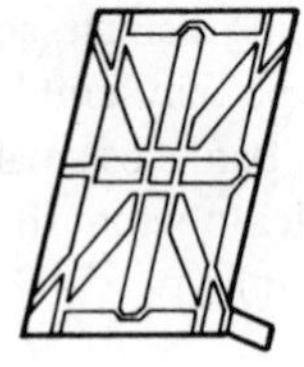

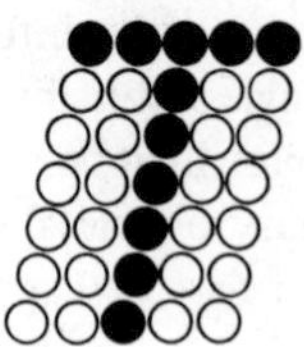

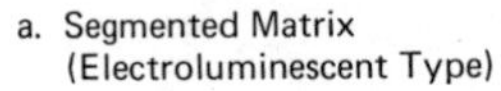

a. Segmented Matrix
(Electroluminescent Type)

b. Dot Matrix
(LED and Incandescent Type)

Figure 3-13. Segmented matrices.

a time is positioned in the viewing window, and multiple units are therefore required to form a message or multidigit readout. The tape background is opaque, and the transparent characters on the tape are back-lighted (Figure 3-14).

Projection Displays

Projection displays employ the principles of light projection, in combination with artwork transparencies or sliding templates, to present words, numerals, symbols, etc., on a common display screen. Such devices may consist of a number of miniature optical systems packaged in a single unit, or of a single optical system using sliding templates for light interference. The image brightness of such displays ranges from moderate to low. Furthermore, optical centering of lamp filaments may be required when units are installed or repaired. Figure 3-15 shows a typical projection display.

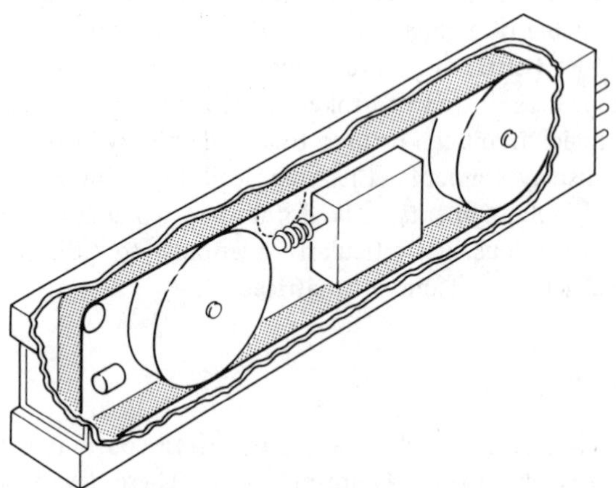

Figure 3-14. Back-lighted belt display.

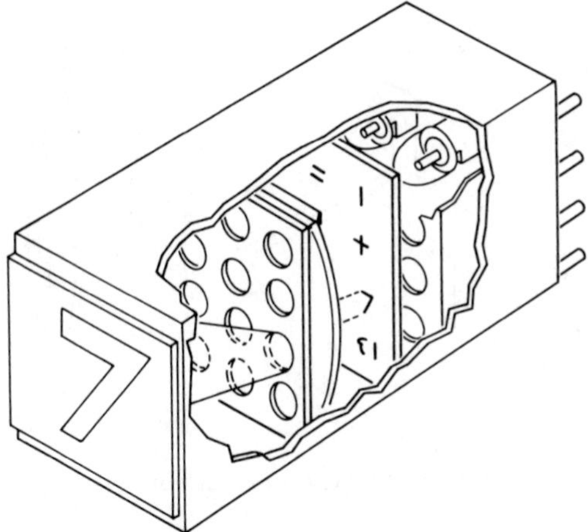

Figure 3-15. Projection display.

Scalar Displays

Scalar displays employ graduated scales in conjunction with an indexing element, such as a pointer, hairline, column of mercury, or bubble, to display quantitative (and, in some instances, qualitative) information. In such displays, the indication of change may be implemented by the motion of the scale, the indexing element, or both. Devices in this category include electrical indicating instruments, mechanical gauges, integral dials, or straight scales, which may be part of a larger display system (e.g., a recorder). Scalar displays may also use moving indexing elements and may be either circular or straight. The exposed length of the pointer and its movement facilitate rapid and qualitative reading. The use of vertical or horizontal straight scales makes it possible to fit more scales on crowded panels than would be feasible if circular or curved scales were employed. Figures 3-16 and 3-17 illustrate some typical scalar displays. Table 3-4 provides a relative evaluation of basic types of indicators when employed for various applications.

Zone Marking and Coding

On many indicators, zone markings are used to indicate various operating conditions such as desirable operating range, danger limits, and caution areas. Two commonly used methods for coding zone markings are color coding and pattern coding (Figure 3-18).

The brightness of transilluminated displays should be compatible with the expected ambient illumination and should be at least 19 percent greater than the

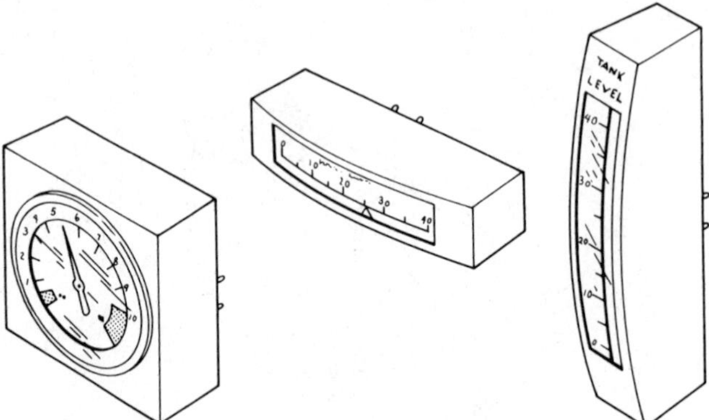

Figure 3-16. Types of scalar displays.

surrounding brightness. The indicator brightness should not, however, be greater than 300 percent of the surrounding brightness reflections. Reflections of any surrounding light should be prevented from making the indicators appear to be illuminated when they are not. When indicators are to be used under varied ambient lighting, a variable control should be used. Controls should be spaced as required to avoid interference and minimize the possibility of inadvertent actuation of controls that could affect safety or performance. The proper spacing of controls within a given panel area should be determined by careful analysis of the following factors:

1. Requirements for simultaneous or sequential use
2. Control size and range of movement
3. Necessity for blind reaching
4. Effects of inadvertent actuation

4	3	7	0	0

a. Direct-reading counter

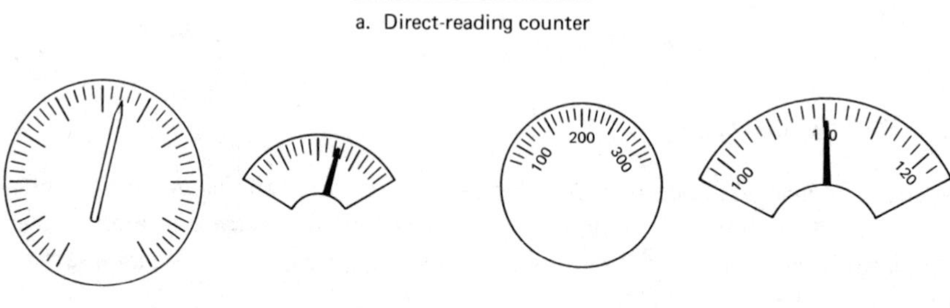

b. Moving pointer, fixed scale

c. Moving scale, fixed index

Figure 3-17. Basic types of scale indicators.

Table 3-4. Evaluation of Indicator Types.

Function	Moving Pointer	Moving Scale	Counter
Quantitative indication	Fair	Fair	Good. Requires minimum reading time with minimum reading error.
Qualitative and check indication	Good. Location of pointer and change in position easily detected.	Poor. Difficult to judge direction and magnitude of pointer derivation.	Poor. Position changes difficult to detect.
Setting	Good. Direct relation between pointer motion and that of setting knob. Pointer position change aids monitoring.	Fair. Somewhat ambiguous relation between pointer motion and that of setting knob.	Good. Most accurate for monitoring numerical settings, but less direct relation between counter and setting knob motion.
Tracking	Good. Pointer position readily monitored and controlled. Provides simple relationship to manual control motion and some information concerning rate.	Fair. Not readily monitored and relationship to manual control motion is somewhat ambiguous.	Poor. Not readily monitored and relationship to manual control motion is ambiguous.
General	Good. However, it requires greatest exposed and illuminated area on panel. Scale length is limited.	Fair. Offers saving in panel space because only small section of scale need be exposed and illuminated. Long scale is possible.	Fair. Most economical in panel and illuminated area. Scale length limited only by number of counter drums. Difficult to illuminate properly.

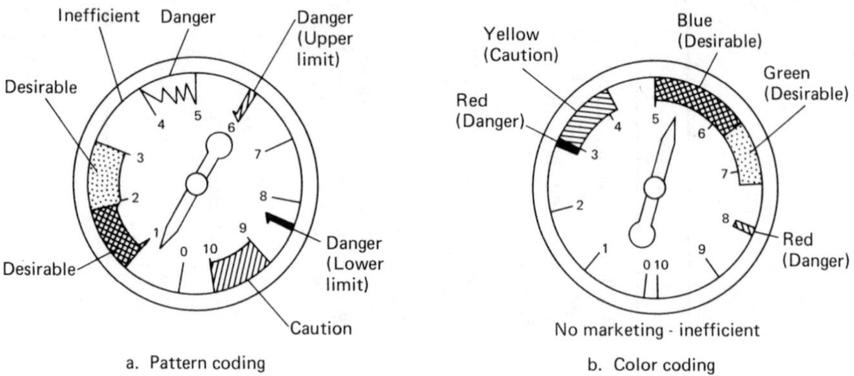

Figure 3-18. Coding of instrument range markers.

Rotary Switches

Discrete-action rotary controls (i.e., rotary selector switches and thumbwheel controls) are particularly useful in applications where many parameters must be controlled individually. Rotary selector switches are available with as few as three to as many as 24 control positions. They are operated by applying force to the switch knob until the switch snaps into the subsequent position. Activation is indicated by an audible click and a tactile detent action. Thumbwheel selector switches are also applicable where panel space is limited (Figure 3-21).

Thumbwheel controls are useful where the particular function requires a digital control input device and a readout of these manual inputs for verification. The use of thumbwheel controls for any other purpose is discouraged. Thumbwheels may be either discrete or continuous, depending on the application. Detent indexing units should provide 10 positions (0 through 9) in digital output.

Continuous adjustment rotary knobs should be selected when it is necessary to change, align, or adust a continuous variable in a precise manner over a relatively short range with minimal force requirements. A movable knob with a fixed scale

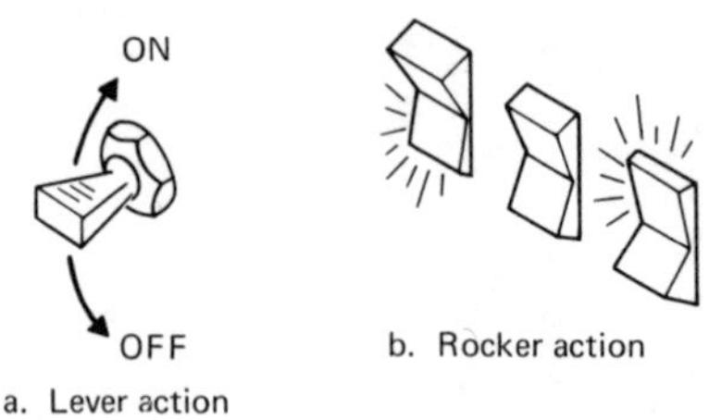

Figure 3-19. Typical toggle-switches.

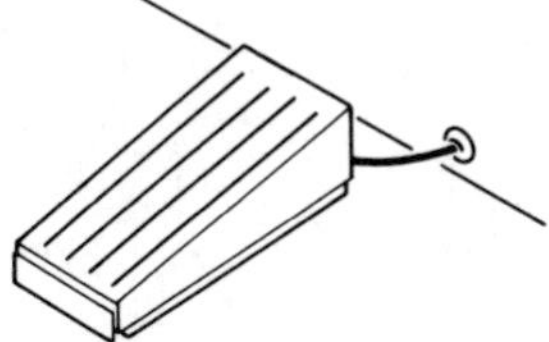

Figure 3-20. Foot-operated push-button switch.

is preferred for most applications. If positions of nonmultirevolution controls must be distinguished, a pointer or marker should be available on the knob. They have an unlimited span of control movement and can be used for either coarse or fine positioning over a wide range of adjustments. Continuous position controls can be either single or ganged (Figure 3-22).

Controls and displays should be clearly, simply, and directly labeled with the basic information needed for proper identification, utilization, actuation, or manipulation of the element. The type of labeling to be used should be selected on the basis of:

1. Accuracy of identification required
2. Time available for recognition
3. Distance at which the label will be read
4. Illumination level and color characteristics
5. Criticality of the function being labeled

Labels and information thereon should be oriented horizontally to facilitate rapid and accurate reading from left to right. Vertical orientation of labels should

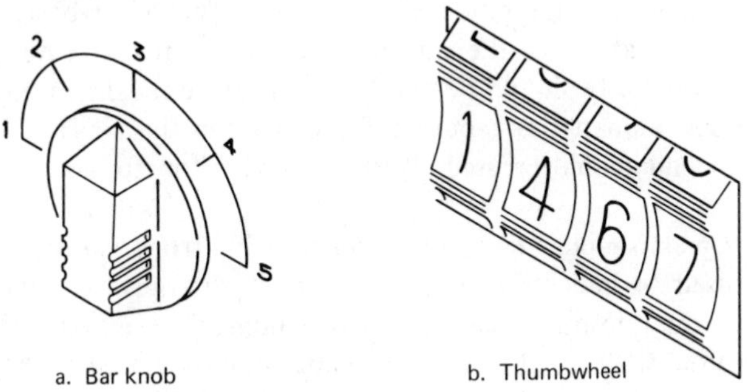

a. Bar knob b. Thumbwheel

Figure 3-21. Rotary selector-switch types.

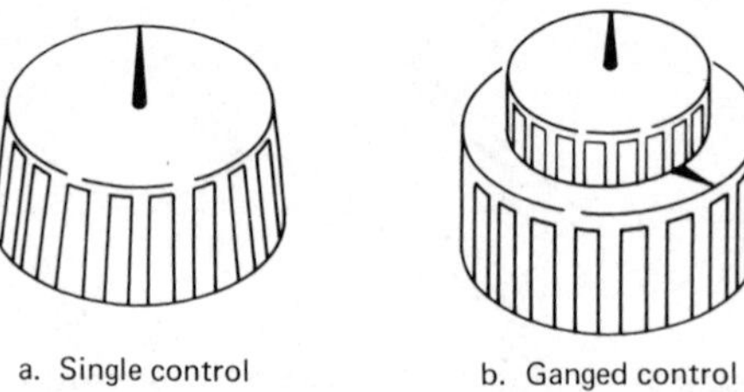

Figure 3-22. Continuous position controls.

be used only when labels are not critical to equipment performance or personnel safety and where space is severely limited. Labels should be placed on or very near the items that they identify to preclude confusion with other items or other labels. Whenever possible, the label should be located above the control to which it refers.

The height of the letters and numerals to be used in a label should be determined by the required reading distance and illumination. Table 3-12 presents a range of values for the height of characters for use with a 28-in. viewing distance for low and high ambient brightness conditions. Table 3-13 lists the character height for various viewing distances where the brightness is normally above 1 foot-lambert.

Auditory Display Devices

Audio signals should be provided to warn personnel of impending danger and to alert the operator to a critical change in the system status or that a critical action must be performed. The audio alarm system should require a positive action to reset it. The major advantage of the audio alarm is its high attention value–it can attract the attention of the operator regardless of the direction in which he or she is looking. The amplitude of the audio alarm should be at least equal to and preferably greater than the ambient noise level. If this is not possible, the signal pitch or quality should be adjusted to provide maximum discrimination. When tone discrimination is required, frequencies in the range of 1250 to 2500 cycles per second should be used. Warbling tones allow for good signal discrimination.

Caution signals should be readily distinguishable from warning signals. They should be used to indicate conditions of which the operator should be aware, even though they do not necessarily require immediate action. When auditory devices are used with visual devices, they are supportive, that is, they are used to alert the operator to the appropriate visual display (Table 3-5).

The sound pressure level should be at least 20 db above the ambient noise

Table 3-5. Functional Evaluation of Audio Signals.

Function	Tones (Periodic)	Complex Sounds (Nonperiodic)	Speech
Quantitative Indication	Poor (maximum of 5 to 6 tones absolutely recognizable).	Poor (interpolation between signals is inaccurate).	Good (minimum time and error in obtaining exact value in terms compatible with response).
Qualitative Indication	Poor to fair (difficult to judge approximate value and direction of deviation from null setting unless presented in close temporal sequence).	Poor (difficult to judge approximate deviation from desired value).	Good information concerning displacement, direction, and rate presented in form compatible with required response.)
Status Indication	Good (permits start-and-stop timing; continuous information where rate of change of input is low).	Good (especially suitable for irregularly occurring signals such as alarms).	Poor (inefficient, more easily masked; presents repeatability problem).
Tracking	Fair (null position easily monitored; signal-response compatibility poses problem).	Poor (required qualitative indications difficult to provide).	Good (meaning intrinsic in signal).
General	Good for automatic communication of limited information. Meaning must be learned. Easily generated.	Some sounds available with common meaning (e.g., firebell). Easily generated.	Most effective for rapid (but not automatic) communication of complex multidimensional information. Meaning intrinsic in signal and context when standardized. Minimum of new learning required.

level, but should not exceed the maximum exposure level of 120 db. As sound approaches 120 db, it becomes uncomfortable to hear. At 130 db, sound becomes annoying, and at 140 db the sensation becomes painful.

When the ambient noise level exceeds 100 db, the operator should hear audio signals on headphones. Such audio signals should never be so startling that they confuse the operator.

CONTROLS

Selection criteria for controls should include functional suitability, consistency of movement, ease of identification, operator efficiency, and control characteristics appropriate for the task (Table 3-6).

The general design criteria relating to controls can be divided into two major categories:

1. Physical characteristics, which include shape, color, and size
2. Operating characteristics, such as control displacement and mechanical resistance

Sufficient space should be provided between controls to permit efficient selection and operation; the possible necessity for blind reaching should be considered.

Controls should be selected so that the direction of movement of the control is consistent with the related movement of an associated display, equipment, or component. Movement of the control forward, clockwise, to the right, or upward should be used to energize the equipment, cause the quantity to increase, or cause the equipment or component to move forward, clockwise, to the right, or up. Conversely, movement of the control backward, counterclockwise, to the left, or downward should deenergize the equipment or component, cause the quantity to decrease, or cause the equipment or component to move backward, counterclockwise, to the left, or down.

All controls that have sequential relationships, are concerned with a particular function of operation, or are operated simultaneously should be grouped together, along with the associated displays. The most critical and frequently used controls should be grouped together and located in the optimum reach and visibility zones. Controls should be arranged so that the sequence of operation is not conducive to inadvertent or accidental actuation of any control.

Controls may be classified in categories relative to the type of movement involved in their actuation. There are four basic categories of controls from which selection can be made, depending on the specific application:

1. Discrete-adjustment linear controls such as push-button switches, toggle switches, and detent levers

Table 3-6. Control Selection Guide.

Characteristic	Type of Control										
	Discrete Adjustment					Continuous Adjustment					
	Rotary Selector Switch	Thumb-Wheel	Hand Push-Button	Foot Push-Button	Toggle Switch	Knob	Thumb-Wheel	Hand-Wheel	Crank	Pedal	Lever
Large forces can be developed	–	–	–	–	–	No	No	Yes	Yes	Yes	Yes
Time required to set control	Short to medium	–	Very short	Short	Very short	–	–	–	–	–	–
Recommended number of positions	3 to 24	3 to 24	2	2	2 or 3	–	–	–	–	–	–
Space requirement for location and operation	Medium	Small	Small	Large	Small	Small to medium	Small	Large	Medium to large	Large	Medium to large
Desirable limits of control movement	270°	–	$\frac{1}{8}$ to 1-$\frac{1}{2}$ in.	$\frac{1}{2}$ to 4 in.	120°	Unlimited	180°	±60°	Unlimited	Small[a]	±90°
Effectiveness of coding	Good	Poor	Fair to good	Poor	Fair	Good	Poor	Fair	Fair	Poor	Good
Effectiveness of visual identification of control position	Fair to good	Good	Poor[b]	Poor	Fair to good	Fair to good[c]	Poor	Poor to fair	Poor[d]	Poor	Fair to good
Effectiveness of nonvisual identification of control position	Fair to good	Poor	Fair	Fair	Good	Poor to good	Poor	Poor to fair	Poor[d]	Poor to fair	Poor to fair

Table 3-6. *(Continued)*

Characteristic	Type of Control										
	Discrete Adjustment					Continuous Adjustment					
	Rotary Selector Switch	Thumb-Wheel	Hand Push-Button	Foot Push-Button	Toggle Switch	Knob	Thumb-Wheel	Hand-Wheel	Crank	Pedal	Lever
Effectiveness of check-reading to determine control position when part of group of like controls	Good	Good	Poor[b]	Poor	Good	Good[c]	Poor	Poor	Poor[d]	Poor	Good
Effectiveness of simultaneous operation with like controls in an array	Poor	Good	Good	Poor	Good	Poor	Good	Poor	Poor	Poor	Good
Effectiveness as part of a combined control	Fair	Fair	Good	Poor	Good	Good[e]	Good	Good	Poor	Poor	Good

[a] Except for rotary pedals, which have unlimited range.
[b] Except when control is backlighted and illuminates upon activation.
[c] Applicable only when control makes less than one revolution. Round knobs must also have pointer attached.
[d] Assumes that control makes more than one revolution.
[e] Effective primarily when mounted on single axis concentric with other knobs.

2. Discrete-adjustment rotary controls such as bar and pointer knobs and snap-action thumbwheels
3. Continuous-adjustment rotary controls such as knobs, thumbwheels, handwheels, and cranks
4. Continuous-adjustment linear controls such as joysticks, levers, and pedals

See Tables 3-7 and 3-8.

Linear Controls

Since discrete-action linear controls may be set at any one of a limited number of exact positions. They are useful in applications that involve turning equipment on and off, selecting modes of operation, and choosing appropriate meter scales. The two types of discrete-action linear controls commonly used are push-button switches and toggle switches.

Push-Button Switches

Push-button switches are two-position controls that may be used individually or for special applications, such as keyboards and matrices, in combination. Two types of hand-operated push-button switches are recommended: latching (push-in and lock/push-off) and momentary contact (push-on/release off). Push-button switches should be used when a control (or any array of controls) is needed for momentary contact or for additional visual indication, particularly when the button is used frequently. Push-button switches may be square or round, transilluminated or nonilluminated. The push-button surface should normally be concave to fit the finger. When that is impractical, the surface should provide a high degree of frictional resistance to prevent slipping. Control activation should be indicated by a "snap" sensation, a discernible click, or an indicator lamp.

Toggle Switches

Toggle switches may be either two- or three-position controls and may be momentary (i.e., spring-loaded) or locking in their action. These switches normally employ either a lever of rocker type of action. Lever-action toggle switches may be positive-action or spring-loaded controls. Rocker-action switches employ two control surfaces that join at an obtuse angle; the depressed surface indicates the present position of the control. Switch position is changed by applying force to the opposite surface until that face is depressed and locked into position (Figure 3-19).

Toggle switches should be used for functions that require two discrete positions and where space limitations are severe. Toggle switches with three or more

Table 3-7. Control Size Criteria.

Control Type and Application	Dimension	Control Size (in.) Minimum	Maximum
Pushbutton, fingertip operated	Diameter	$\frac{1}{2}$	a
Pushbutton, thumb or palm operated	Diameter	$\frac{3}{4}$	a
Pushbutton, foot operated	Diameter	$\frac{1}{3}$	a
Toggle switch	Tip diameter	$\frac{1}{8}$	1
	Lever arm length	$\frac{1}{2}$	2
Rotary selector switch	Length	1	a
	Width	a	1
	Depth	$\frac{5}{8}$	a
Continuous adjustment knob, finger or thumb operated	Depth	$\frac{1}{2}$	1
	Diameter	$\frac{3}{8}$	4
Continuous adjustment knob, hand or palm operated	Depth	$\frac{3}{4}$	a
	Diameter	1-$\frac{1}{2}$	3
Crank, for rate application	Radius	$\frac{1}{2}$	4-$\frac{1}{2}$
Crank, for force application	Radius	$\frac{1}{2}$	20
Handwheel	Diameter	7	21[b]
	Rim thickness	$\frac{3}{4}$	2
Thumbwheel	Diameter	1-$\frac{1}{2}$	a
	Width	a	a
	Protrusion	$\frac{1}{8}$	a
Lever handle, finger operated	Diameter	$\frac{1}{2}$	3
Lever handle, hand operated	Diameter	1-$\frac{1}{2}$	3
Crank handle	Grasp area	3	a
Pedal	Length	3-$\frac{1}{2}$	c
	Width	1	c
Valve handle	Diameter	3 inches per inch of valve size	

[a]No limit set by operator performance.
[b]Dependent on space available.
[c]Two-hand grasp.

positions should be used only when the use of rotary switches is not feasible. When the prevention of accidental actuation is of primary concern, a channel guard, lift-to-lock, or other equivalent means should be provided. In general, toggle switches should be oriented vertically, with the down position "off."

Table 3-8. Control Displacement Criteria.

Control	Condition	Displacement	
		Minimum	Maximum
Pushbutton	Thumb or fintertip operation	$\frac{1}{8}$ in.	1 in.
	Foot operation (normal)	$\frac{1}{2}$ in.	
	Heavy boot	1 in.	
	Ankle flexion only		2-$\frac{1}{2}$ in.
	Leg movement		4 in.
Toggle switch	Between adjacent positions	30°	
	Total displacement		120°
Rotary selector switch	Between adjacent detents (visual	15°	
	Between adjacent detents (nonvisual)	30°	
	For facilitating performance		40°
	Special engineering required		90°
Continuous adjustment knob	Determined by desired control/display ratio		
Crank	Determined by desired control display ratio		

Foot-Operated Push-Button Switches

Like the hand-operated variety, foot-operated push-button switches are two-position controls. The momentary-contact foot-operated push-button switch is normally recommended. Actuation of the control should be indicated to the operator by an associated display. Foot-operated switches should be employed only in instances where the operator is expected to have both hands occupied at the time of push-button actuation, or where task sharing among limbs is required. Because foot-operated switches are extremely susceptible to accidental or inadvertent actuation, their use should be limited to noncritical applications. Such switches should be designed to be operated with the ball of the foot, not the heel. In general, foot-operated controls should be located in the position where they can best be reached and actuated with optimum speed, accuracy, and comfort (Figure 3-20).

Where more than two or three controls are involved, the application of coding methods should be considered to ensure that the controls may be:

1. readily identified and distinguished from one another
2. selected as rapidly as possible without error
3. manipulated quickly and accurately

Coding should be used as a means of standardizing control applications.

Control coding can be accomplished by systematic application of one or more variations: shape, size, mode of operation, location, and labeling. The selection of coding method shall be determined, for a given application, by the relative advantages and disadvantages of each type of coding (Table 3-11 and Figure 3-23).

Prevention of Inadvertent Actuation

Controls shall be designed and located so that they are not susceptible to accidental movement. Particular attention should be given to critical controls, the inadvertent operation of which might result in damage to equipment, injury to personnel, or degradation of system functions. In situations where controls must be protected from accidental activation, one or more of the following methods (see Figure 3-24) should be used:

1. Recess, shield, or otherwise surround the control with physical barriers.
2. Cover or guard the control.

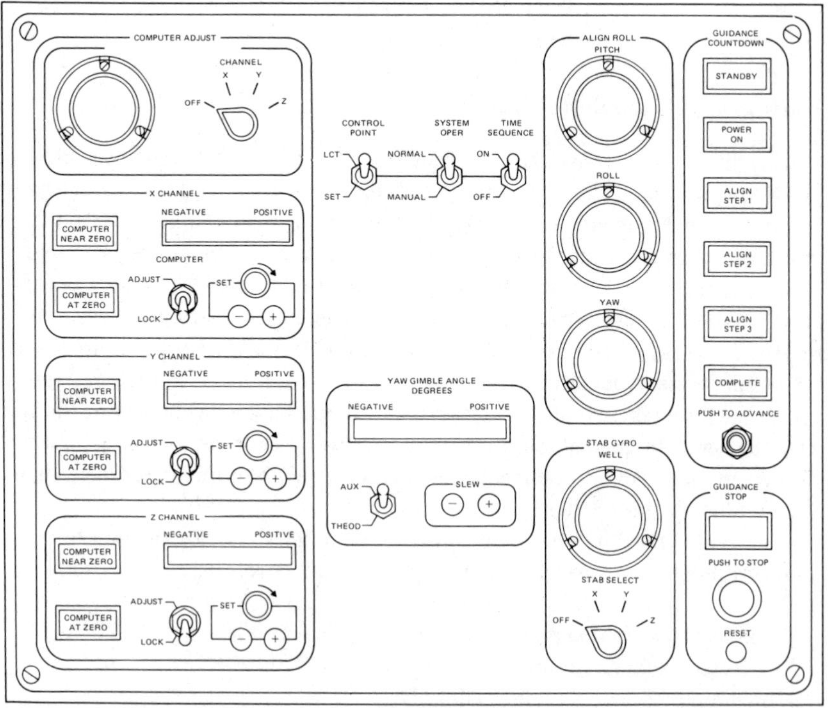

Figure 3-23. Example of standard labeling practices.

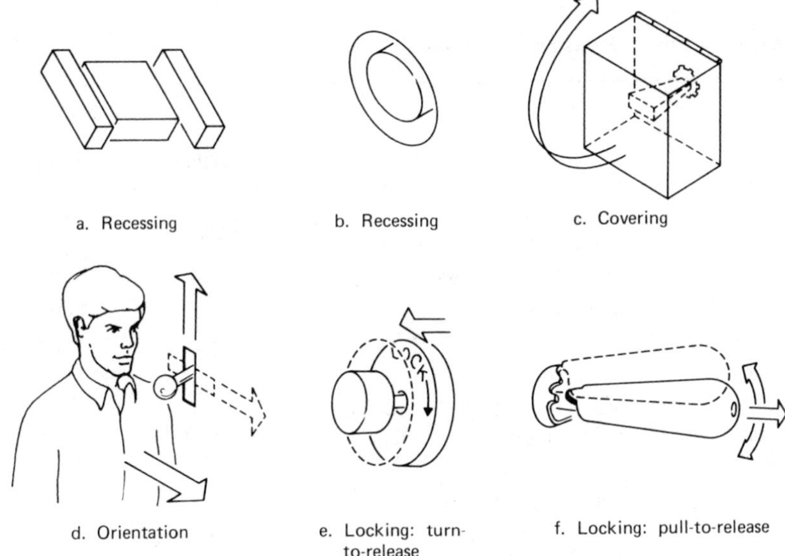

Figure 3-24. Techniques for preventing inadvertent actuation.

3. Provide the control with interlocks so that extra movement or prior operation of a related or locking control is required.
4. Provide the control with mechanical resistance so that a definite or sustained effort is required for operation.
5. Isolate the control to reduce the probability of inadvertent actuation.

DESIGNING FOR MAINTAINABILITY

Maintainability is that characteristic of a design which allows all maintenance operations to be performed economically, in a reasonable period of time, with a maximum of ease and accuracy. The system must be kept in, or returned to, operating condition by average technical skills. It is important for the designer to keep in mind that the minor decisions such as chassis layout, modularization, and the location of test points can seriously affect the reliability and maintenance capability of the system. The following considerations should always be kept in mind during the design phase:

- Whenever frequent adjustments are required, knobs rather than screwdriver adjustments should be selected.
- A reference scale or some other means of feedback should always be provided.
- When blind screwdriver adjustments are unavoidable, mechanical guides

Table 3-9. Control Resistance Criteria.

Control	Condition	Resistance Minimum	Resistance Maximum
Pushbutton	Fingertip operation	10 oz	40 oz
	Foot operation, normally off control	4 lb	20 lb[a]
	Foot operation, rested on control	10 lb	20 lb[a]
Toggle switch	Finger operation	10 oz	40 oz
Rotary selector switch	Torque	1 in.-lb	6 in.-lb
Continous adjustment knob	Torque, fingertip <1-in. diameter	b	$4\frac{1}{2}$ in.-oz
	Torque, fingertip >1-in. diameter	b	6 in.-oz
Crank	Rapid, steady turning		
	<3-in. radius	2 lb	5 lb
	5-to-8-in. radius	5 lb	10 lb
	Precise settings	$2\frac{1}{2}$ lb	8 lb
Handwheel[c]	Precision operation (<3-in. radius)	b	b
	Precision operation (5-to-8-in. radius)	$2\frac{1}{2}$ lb	8 lb
	Resistance at rim (one-hand)	5 lb	30 lb
	Resistance at rim (two-hand)	5 lb	50 lb
Thumbwheel	Torque	1 in.-lb	3 in.-lb
Lever handle	Finger grasp	12 oz	40 oz
	Hand grasp (one-hand)	2 lb	–
	Hand grasp (two-hand)	4 lb	–
	Fore-aft motion (along median plane):		
	One-hand, 10 in. forward SRP	–	30 lb
	One-hand, 16 to 24 in. forward SRP	–	50 lb
	Two-hand, 10 to 19 in. forward SRP	–	60–100 lb
	Lateral motion:		
	One-hand, 10 to 19 in. forward SRP	–	20 lb
	Two-hand, 10 to 19 in. forward SRP	–	30 lb
Pedal	Foot operation, normally off control	4 lb	–
	Foot operation, rested on control	10 lb	10 lb
	Ankle flexion only	–	10 lb
	Leg movement	–	180 lb

[a] Ankle flexion.
[b] Not established.
[c] For valves: 60 in.-lb torque per inch of valve size (20 in.-lb torque per inch of handle diameter).
SRP = Seat reference point.

Table 3-10. Control Separation Criteria.

Control	Condition	Separation (Inches)*	
		Minimum	Preferred
Pushbutton	One finger, random order	½	2
	One finger, sequential order	¼	1
	Different fingers, random or sequential order	½	½
Toggle switch	One finger, random order	¾	2
	One finger, sequential order	½	1
	Different fingers, random or sequential order	⅝	¾
Rotary selector switch	One hand, random order	1	2
	Two hands, simultaneous	3	5
Continuous adjustment knob	One hand, random order	1	2
	Two hands, simultaneous	3	5
Crank	One hand, random order	2	4
	Two hands, simultaneous	3	5
Handwheel	Two hands, simultaneous	3	5
Thumbwheel	One finger	¼	–
Lever handle	One hand, random order	2	4
	Two hands, simultaneous	3	5
	Maximum simultaneous operation, one hand span	–	6
Pedal	One foot, random order	4	6
	One foot, sequential order	2	4

*Edge-to-edge measurement.

should be provided to prevent the screwdriver from falling into the machine.

- Sensitive adjustments should always be protected from accidental disturbances during routine maintenance.
- Only interconnecting wiring and structural members should be permanently attached to the chassis or frame.
- Components should be mounted so that no other equipment must be removed for access.
- Components frequently removed for checking should be mounted on slides, hinges, or rollout racks.

Table 3-11. Summary of Control Coding Methods.

Coding Method	Description	Principal Advantages	Disadvantages and Limitations	Considerations in Use
Shape Coding	Shape coding permits both tactile and visual identification of controls. Primarily useful where controls cannot be seen, it is also employed as a supplement to other coding methods.	1. Controls can be identified without vision. 2. Double coding since both tactile and visual discrimination are possible. 3. A relatively large number of discernible shapes is available.	1. No standardized meaning is attached to shapes. The code must be learned. 2. Tactile identifcation is less rapid than visual. 3. Tactile identification requires that the hand be bare or only lightly gloved.	1. Tactile discrimination is slower than visual identification. 2. Speed of identification decreases as number and range of shapes increases. 3. Where gloves must be worn, useful only for visual discrimination. 4. Certain shapes may conflict with mode of operation requirements.
Color Coding	Allows for discrimination of controls by their color. Only eight colors can be discriminated absolutely (i.e., without reference to other colors for comparison). These are red, orange, yellow, green, blue, and violet, plus black and white.	1. Allows for excellent and rapid control discrimination. 2. Is compatible with and can be used to supplement other coding methods.	1. Adequate white ambient illumination required. 2. Controls must be located within visual area of operator. 3. Coding may fail if personnel are color deficient. 4. Coding associations must be learned.	1. White ambient illumination required for color illumination. 2. Some operators may be deficient in color vision. 3. No single standard system of color meaning. Code must be learned. 4. Important and frequently used controls should be coded with highly visible colors.
Size Coding	Size coding permits both tactile and visual identification. Primarily useful where controls cannot be seen and size differences become unmistakable.	1. Controls can be discriminated without visibility. 2. Double coding since both tactile and visual discrimination are possible.	1. The number of size categories is limited to three. 2. Separately placed control knobs consume panel space. 3. The associations of size and function must be learned. 4. Size may conflict with control movement requirements.	1. The number of coding categories is limited. No more than three sizes should be used. 2. Considerable panel space is required. 3. Large knobs should control major functions or have large gain valves; smaller knobs, minor functions or lesser gains. 4. Concentric mounting is feasible where knobs control the same function at different sensitivities.

Mode of Operation Coding	Uses the mechanics and dynamics of control operation as the basis for identification. Discrimination is purely tactile and kinematic. The only coding method which cannot allow visual identification.	1. Controls can be discriminated without visibility. 2. Compatible with other coding methods. 3. Compatibility between control operation and function is encouraged.	1. Operation or attempted operation of control is required for identification. 2. Cannot be used where faulty selection and/or operation will compromise the mission or equipment. 3. Gloves may impede identification.	1. Mode of operation, per se, is rarely a sufficient coding method. 2. Useful as backup coding to confirm initial control selection. 3. An attempt to operate a control must be made for the coding to be effective. 4. Generally associated with shape and size coding.
Location Coding	Permits identification of controls by position. This basic and primary coding method allows for identifying groups of controls and control sequences as well as individual controls.	1. Panel layout is readily learned. 2. Facilitates identification under low illumination conditions. 3. Grouping can follow function and sequence.	1. Location coding, per se, is rarely a sufficient coding method. 2. Difficult to standardize because of panel variations. 3. Space requirements may limit use or impede control operation.	1. Controls should be grouped by similar function or purpose. 2. Controls should be grouped according to sequence of operation. 3. Frequently used controls should be grouped together. 4. Frequently used and important controls should be located in optimal, center panel area.
Labeling	Conveys meaning by use of language. Makes possible the identification of any control uniquely. Facilitates initial learning of control function as well as subsequent control identification.	1. Labels allow for meaningful and unique control description. 2. Learning of coding system is not required. 3. Control identification is rapid.	1. Requires visual conditions suitable for reading labels. 2. Panel space within the visual area of the operator is required. 3. Limited to operators fluent in the same language.	1. Panel illumination must be sufficient to read labels. 2. Labels must be adjacent to and, where possible, above controls. 3. Labels should be brief and avoid all but very common abbreviations. 4. Labels should depict the controlled function, not the instrument name. 5. Lettering should be horizontal.

Table 3-12. Label Size vs. Illumination.

	Character Height (in.)	
Type of Markings	Low Intensity (<1 fl)	High Intensity (>1 fl)
Critical, with position variable (e.g., numerals on counters, moving scales, etc.)	0.20 to 0.30	0.12 to 0.20
Critical with position fixed (e.g., numerals on fixed scales, controls, etc.)	0.15 to 0.30	0.10 to 0.20
Noncritical	0.05 to 0.20	0.05 to 0.20

- Replaceable components should be mounted as plug-in units rather than solder-connected.
- Guide pins should be provided for components requiring alignment during installation.
- Components should be installed with enough space provided for test probes, soldering irons, and other tools required for testing and maintenance operations.

Those units most critical to system operation and those that require rapid maintenance should be the most accessible.

Cables and Connectors

The selection of the various cables and connectors to be used in a design should be based on the following considerations:

- Quick-disconnect connectors or plugs requiring a single turn should be selected whenever possible.

Table 3-13. Label Size vs. Viewing Distance.

Viewing Distance	Character Height (in.)
20 in. or less	0.09
21 to 36 in.	0.17
3 to 6 ft	0.34
6 to 11 ft	0.68
12 to 20 ft	1.13

- Keyed connectors or connectors of different sizes should be used to eliminate the possibility of inadvertently making incorrect connections.
- Connectors should be located sufficiently apart to permit a firm gripping for removal.
- Painted strips, arrows, or marks may be used to indicate the correct connector orientation.
- Cables should fan out at junctions to facilitate checking.
- Cables should be routed in a manner to preclude use as a handle or pulling device.

HUMAN ENGINEERING CHECKLIST

The following checklist should be used, as applicable, by the human factors engineer in determining the acceptability of engineering drawings.

- Does equipment represent the simplest design consistent with functional requirements?
- Has feedback on control response been provided?
- Are functionally related controls and displays located in close proximity to each other?
- Is the control-display ratio commensurate with the operator task?
- Is control movement consistent, predictable, and compatible with operator expectations?
- Are controls labeled clearly with respect to function, system status, and operating condition?
- Is displayed information presented to the operator in a directly usable form?
- Are display faces oriented properly to the operator's line of sight?
- Are displays arranged, mounted, or shielded in a manner to minimize reflectance of ambient light?
- Is the display viewing distance correct?
- Is visual coding clear? Are codes standardized within equipment groupings?
- Is the type of display compatible with the type of information to be presented?
- Is legend-lamp lettering legible and visible to the operator?
- Are scale-indicator selections compatible with the corresponding functions?
- Have auditory warnings been included, where necessary, to alert personnel of impending danger or critical change in system status?
- Do control selections satisfy the criteria of functional suitability, consistency of movement, ease of identification, etc.?

- Is direction of control movement consistent with related movement of the associated display, equipment component, or vehicle?
- Are frequently used and critical controls grouped together in the optimum reach and visual zones?
- Are controls arranged in a logical, sequential order?
- Has control coding been selected properly with respect to shape, size, texture, etc.?
- Are critical controls design and located to prevent inadvertent actuation?
- Is space between controls sufficient to avoid interferences?
- Are controls and displays clearly and simply labeled with basic information required for identification, use, and manipulation of the element?
- Are labels oriented horizontally?
- Are labels positioned to avoid confusion with other labels or items?
- Have 95th percentile anthropometric values been used for gross-dimension items (accesses, hatches)?
- Have 5th percentile anthropometric values been used for limiting dimensions (e.g., reach distance, lifting capacity)?
- Is the operator's position (seated, standing) at the work station appropriate to the functions and tasks to be performed?
- Does the console configuration provide adequate display and control surfaces?
- Do adjustable items (e.g., seats) accommodate the 5th through the 95th percentile range of personnel?
- Has operator seating been provided where required?
- Where ladders or stairs are used, do they conform to all human engineering and safety criteria?
- Have all environmental factors (e.g., light, heat, vibration, airflow, radiation) been considered?
- Do maintenance adjustment controls provide appropriate feedback?
- Are sensitive adjustment controls protected against inadvertent activation?
- Are frequently removed components mounted on slides, hinges, or rollout racks for accessibility?
- Is sufficient space provided for maintenance personnel to use required tools (e.g., test probes, soldering iron)?
- Are instructions, cautions, warnings, and schematics displayed in easily viewed locations?
- Have fasteners of the proper type, materials, etc., been selected?
- Are captive fasteners used whenever feasible?
- Have the number and diversity of fasteners been minimized?
- Are handles provided on all removable and carried units?
- Are handles located in relation to center of gravity of the item?
- Have quick-disconnect connectors been used where required?

- Are connectors and receptacles designed to preclude incorrect mating (e.g., keying, alignment pins)?
- Are accesses provided where frequent maintenance operations are likely?
- Are access openings large enough to admit the necessary components and the hands, arms, etc., of maintenance personnel?
- Are there enough test points and test jacks? Are they accessible?
- Are test points correctly labeled by symbol or name?

4

FABRICATION PROCESSES

The designer must have a good knowledge of the manufacturing processes used in the production of electronics packaging, the compatibility of the materials used with the processes, and the equipment available for each operation. All materials and processes used have certain restrictions; the designer must be aware of these to avoid the possibility of designing hardware that cannot be manufactured by current standard processes or equipments. A thorough understanding of the fabrication processes enables the designer to select the most economical process and thus is also a means of controlling the costs. This chapter provides some of the basic manufacturing procedures available and the restrictions that must be considered in designing modern electronics equipment that will be manufacturable and salable at competitive prices.

SHEET METAL

Sheet-metal parts can be made from various metals; normally the metals used are aluminum alloys, magnesium, titanium, copper, beryllium, and low-carbon, corrosion-resistant, and alloy steels. The most common metals used in electronics packaging are aluminum alloys due to their compatibility with modern manufacturing methods, availability, and cost advantages.

Shearing is the most economical method of cutting sheet metal and should be used whenever possible. The thickness limitations listed in Table 4-1 generally provide an acceptable surface finish, with only a deburring operation required. Heavier gauges will require that the additional material be left all around for trimming after the shearing operation.

Corner cutoffs and bend reliefs must be designed in such a manner that the metal will not crack during the forming operation. The relief cutout must always extend behind the intersection of the flange lines to avoid cracking. A straight cut in the flat pattern is most ıfective (Figure 4-3). Where a straight cut is not

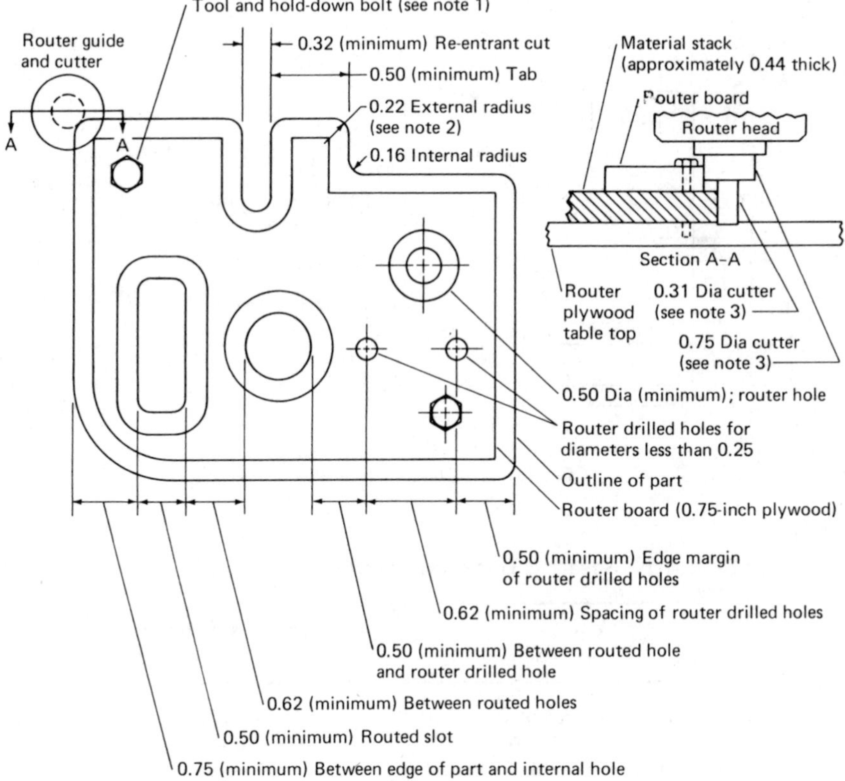

Figure 4-1. Routing limitations.

practical, a notch should be used (see Figure 4-4). (See Table 4-2 for standard dimensions.)

Bend reliefs are not required of parts that are draw-formed. On parts where it is necessary to use a notch on two-way bends, the size of the notch may be calculated by:

$$ND = T - R_b - 0.239R_n - 0.022 \quad (4.1)$$

BRAKE FORMING

Brake forming is the simplest method available for the forming operation where straight sections such as angles, channels, or hats are required. No appreciable

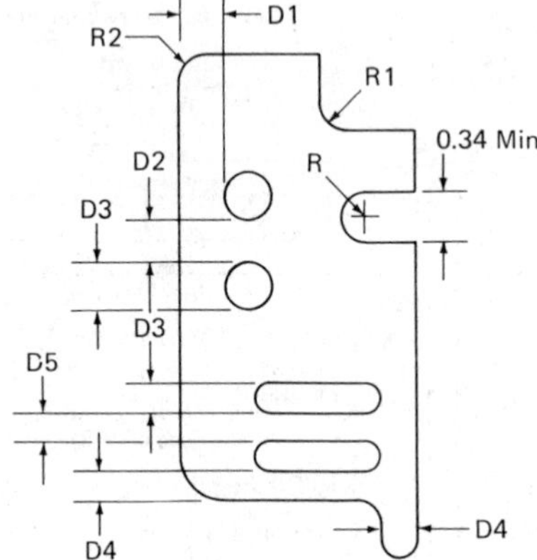

	Minimum Dimensions		
R1	Greater of 2T or 0.06		
R2	2T		
D1 or D2	Thickness	Nonferrous	Stl & Ti
	Thru 0.063	0.12	0.12
	0.064-0.125	Greater of 0.12 or 1.5T	2T
D3	1T (see diameter-thickness ratio)		
D4 or D5	Thickness		
	Thru 0.032	0.06	
	0.033-0.125	2T	

Figure 4-2. Blanking limitations.

Table 4-1. Shearing Thickness Limitations.[a]

Material	Thickness Limitation (in.)
Aluminum alloy (except 7075-T6)	0.250
Aluminum alloy 7075-T6	0.100
Corrosion-resistant molybdenum steel	0.125
Stainless steel (soft)	0.160
Stainless steel (hard)	0.125
Magnesium	0.050
Titanium (commercially pure)	0.125[b]

[a]Blanking, piercing, perforating, notching, trimming, and shaving are essentially shearing operations. Blanking and piercing are related operations that are performed using a punch or die. Blanking cuts the material to the desired contour, whereas piercing removes material from the parts being formed. Blanking is fast and accurate for producing irregular shapes but is relatively expensive and should be used only when the quantities are high enough to warrant the cost of tooling. Piercing holes of a diameter less than the thickness of the metal will result in excessive punch breakage. Routing is used to profile flat sheet-metal parts in nonferrous alloys. Routing is performed by following a template. Although routing is less expensive than blanking, it has certain limitations. These include the inability of routing square corners inside of the part and when attempting square corners outside a guide and cutter of the same diameter will be required. See Figures 4-1 and 4-2.

[b]Reduce on gauge for alloyed materials.

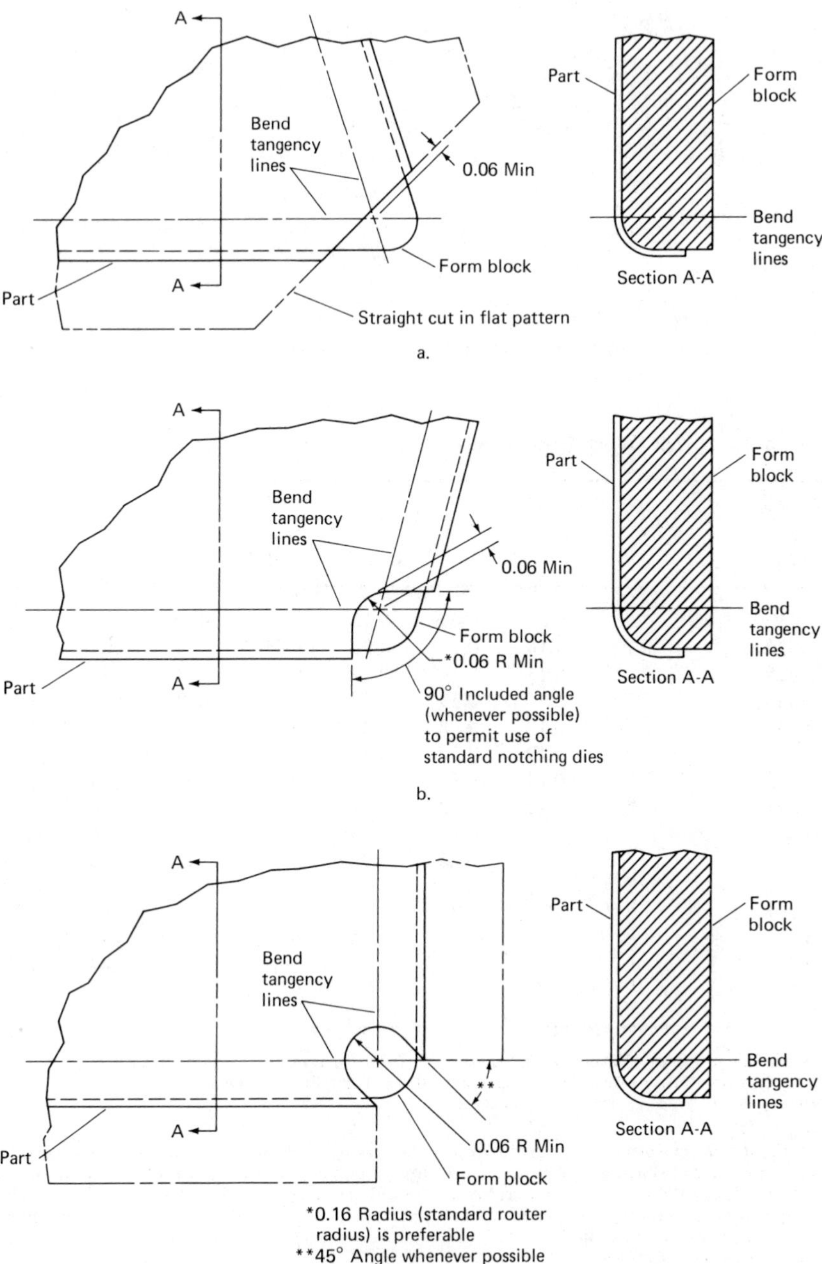

Figure 4-3. Corner cutoffs for bend relief.

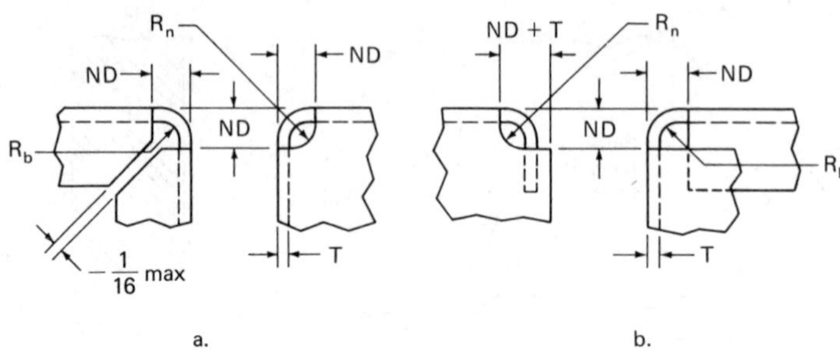

Note: $ND = T + R_b + 0.293R_n + 0.022$

Figure 4-4. Notch relief for formed corners.

Table 4-2. Notch Depth Dimensions.[a]

Material Thickness (T)	Bend Radius (R_b)							
.06	0.03	0.06	0.09	0.12	0.16	0.19	0.22	0.25
0.020	0.12	0.16	0.19	0.22	0.25	0.28	0.31	0.34
0.025	0.12	0.16	0.19	0.22	0.25	0.28	0.31	0.34
0.032	0.12	0.16	0.19	0.22	0.25	0.28	0.31	0.34
0.040	0.16	0.19	0.22	0.25	0.28	0.31	0.34	0.38
0.050	a	0.19	0.22	0.25	0.28	0.31	0.34	0.38
0.063	b	0.19	0.22	0.25	0.28	0.31	0.34	0.38
0.071	b	0.22	0.25	0.28	0.31	0.34	0.38	0.41
0.080	b	b	0.25	0.28	0.31	0.34	0.38	0.41
0.090	b	b	0.25	0.28	0.31	0.34	0.38	0.41
0.100	b	b	0.28	0.31	0.34	0.38	0.41	0.44
0.125	b	b	b	0.31	0.34	0.38	0.41	0.44
0.160	b	b	b	0.31	0.34	0.38	0.41	0.44
0.190	b	b	b	b	0.38	0.41	0.44	0.50
0.250	b	b	b	b	b	b	0.50	0.50

[a]Values given are for ND when R_n is equal to 0.16 for combinations of thickness and bend radii shown. (Other values may be calculated from the Equation 4.1).
[b]Not recommended.

Notes

1. For titanium forming, (a) avoid abrupt changes in contour across such sections as hats, angles, and zees; (b) avoid closely spaced or double joggles; (c) for hydropress forming, keep the percentage of shrink or stretch as low as possible, and avoid shrink flanges; and (d) avoid bending to the minimum allowable radius, using as large a bend radius as possible or feasible.
2. All dimensions are in inches.
3. T = material thickness.
4. Sheet-metal thicknesses are in accordance with MIL-STD 204A except for 0.190 in., which is included because material is still supplied in this gauge.
5. All gauges less than 0.016-in. shall have the same bend radius as 0.016-in.
6. Decimal equivalents of fractions are in accordance with MIL-STD-8.

alteration in thickness, length, or width of the stock occurs when the proper bend radius is used (Tables 4-3 through 4-6).

GRINDING

In all grinding operations, small amounts of material are removed by a mechanical action of abrasive materials. Grinding produces very close tolerances and fine surface finishes ($\overset{63}{\surd}$ to $\overset{8}{\surd}$). Because of the slow rate of material removal inherent in grinding operations, the parts are usually "rough-machined" to remove the bulk of the material by such techniques as turning or milling. For parts that are to be hardened, the rough-machining is done prior to the heat treatment while the material is soft. The heat-treated part is then finished by the grinding operation.

MACHINING

In a limited sense, the term *machining* relates to cutting methods such as turning or milling. However, as used today, the term includes abrasive methods such as grinding and displacement methods such as knurling. Machined parts normally represent a higher cost per part than do parts produced by other methods. It is important for the designer to consider the economy of production; this involves visualizing the number of parts to be made and the complexity of the machine tool operations required. Complicated machine configurations should be avoided whenever possible due to the inherent increase of the manufacturing costs. The optimum design of machined parts requires careful consideration of all of the factors involved. These factors must be evaluated and incorporated into details so that a minimum of machining operations is required to produce the desired configuration for the proper fit and function.

The following materials are listed in order of their ease of machinability.

1. alloy steels
2. chromium plate
3. copper alloys
4. aluminum alloys
5. magnesium alloys

CUTTING

For cutting operations such as turning or boring, the following general rules hold true:

1. For ferrous alloys, machinability is inversely proportional to the tensile strength. (The higher the tensile strength, the more difficult it is to machine.)

Table 4-3. Bend Radii for Aluminum Alloys.

Material Thickness (T)	Alloy Designation										
	1100-O	2024-O	2024-T3 2024-T4	5052-O	5052-H32	5052-H34	6061-O	6061-T4	6061-T6	7075-O	7075-T6
0.016	0	0	0.05	0	0	0	0	0.02	0.02	0.03	0.08
0.018	0	0	0.05	0	0	0	0	0.02	0.03	0.03	0.10
0.020	0	0	0.05	0	0	0	0.02	0.02	0.03	0.03	0.12
0.022	0	0	0.05	0	0	0	0.02	0.02	0.03	0.03	0.14
0.025	0	0	0.05	0	0	0	0.03	0.03	0.04	0.03	0.16
0.028	0	0	0.09	0	0	0	0.03	0.03	0.05	0.03	0.18
0.032	0	0.03	0.09	0	0.03	0.03	0.03	0.03	0.05	0.03	0.20
0.036	0	0.04	0.09	0.03	0.06	0.03	0.04	0.04	0.07	0.04	0.22
0.040	0	0.04	0.16	0.03	0.04	0.03	0.06	0.05	0.08	0.04	0.25
0.045	0	0.05	0.18	0.03	0.05	0.06	0.05	0.06	0.09	0.05	0.28
0.050	0	0.05	0.20	0.03	0.05	0.06	0.05	0.06	0.09	0.05	0.31
0.056	0	0.06	0.23	0.03	0.06	0.08	0.06	0.08	0.11	0.06	0.36
0.063	0	0.06	0.25	0.06	0.06	0.09	0.06	0.09	0.12	0.06	0.40
0.071	0	0.07	0.28	0.06	0.07	0.11	0.07	0.11	0.16	0.08	0.50
0.080	0	0.08	0.32	0.06	0.08	0.12	0.08	0.12	0.19	0.10	0.56
0.090	0	0.09	0.36	0.06	0.09	0.16	0.09	0.18	0.21	0.12	0.63
0.100	0	0.10	0.40	0.06	0.10	0.18	0.10	0.20	0.24	0.14	0.70
0.112	0	0.11	0.44	0.06	0.12	0.22	0.11	0.22	0.26	0.16	0.78
1.125	0	0.13	0.50	0.06	0.12	0.25	0.13	0.25	0.28	0.20	0.88
0.140	0	0.14	0.56	0.06	0.14	0.28	0.14	0.28	0.33	0.24	0.98
0.160	0	0.16	0.64	0.06	0.16	0.32	0.16	0.32	0.38	0.26	1.12
0.180	0	0.18	0.72	0.06	0.18	0.36	0.18	0.36	0.45	0.28	1.25
0.190	0	0.19	0.75	0.06	0.20	0.38	0.19	0.38	0.50	0.32	1.38
0.200	0	0.20	1.00	0.06	0.20	0.40	0.20	0.40	0.75	0.40	1.50
0.224	0	0.22	1.12	0.06	0.22	0.44	0.22	0.50	1.00	0.50	2.00
0.225 and over	1T	1T	3T	1T	1T	2T	1T	2T	5T	2.5T	10T

Notes

1. All dimensions are in inches.
2. *T* = material thickness.
3. Sheet-metal thicknesses are in accordance with MIL-STD 204A, except for 0.190-in. which is included because material is still supplied in this gauge.
4. All gauges less than 0.016 in. shall have the same bend radius as 0.016-in. sheet.
5. Decimal equivalents of fractions are in accordance with MIL-STD-8.

Table 4-4. Bend Radii for Magnesium Alloys.

	AZ318				HK 31A H-24			
	COND-O		H-26					
Thickness	Room	550°F	Room	325°F	Room	600°F	650°F	700°F
0.016	0.09	0.06	0.19	0.09	0.16	0.09	0.06	0.06
0.018	0.09	0.06	0.19	0.09	0.16	0.09	0.06	0.06
0.020	0.09	0.06	0.19	0.09	0.16	0.09	0.06	0.06
0.022	0.16	0.06	0.25	0.12	0.25	0.12	0.09	0.06
0.025	0.16	0.06	0.25	0.12	0.25	0.12	0.09	0.06
0.028	0.19	0.06	0.31	0.16	0.28	0.16	0.09	0.06
0.032	0.19	0.06	0.31	0.16	0.28	0.16	0.12	0.09
0.036	0.22	0.09	0.38	0.19	0.31	0.16	0.12	0.09
0.040	0.22	0.09	0.44	0.19	0.38	0.16	0.12	0.09
0.045	0.25	0.09	0.50	0.25	0.38	0.19	0.16	0.09
0.050	0.25	0.09	0.50	0.25	0.44	0.25	0.16	0.12
0.056	0.28	0.12	0.56	0.28	0.50	0.25	0.19	0.12
0.063	0.31	0.16	0.62	0.31	0.56	0.28	0.19	0.16
0.071	0.38	0.16	0.81	0.38	0.62	0.31	0.25	0.16
0.080	0.44	0.19	0.81	0.44	0.69	0.34	0.25	0.16
0.090	0.44	0.19	1.00	0.44	0.75	0.38	0.31	0.19
0.100	0.50	0.25	1.00	0.50	0.81	0.44	0.31	0.25
0.112	0.56	0.25	1.12	0.56	1.00	0.50	0.34	0.25
0.125	0.62	0.25	1.25	0.62	1.00	0.50	0.38	0.25
0.140	0.69	0.34	1.50	0.69	1.12	0.56	0.44	0.31
0.160	0.75	0.38	1.62	0.75	1.50	0.69	0.50	0.34
0.180	1.00	0.38	1.88	1.00	1.50	0.75	0.56	0.38
0.190	1.00	0.38	1.88	1.00	–	–	–	–
0.200	1.00	0.50	1.88	1.00	–	–	–	–
0.224	1.00	0.50	1.88	1.00	–	–	–	–
225 & over	5T	2.5T	8T	5T	8T	4T	3T	2.5T

Notes

1. All dimensions are in inches.
2. T = material thickness.
3. Sheet-metal thicknesses are in accordance with MIL-STD-204A, except for 0.190-in., which is included because material is still supplied in this gauge.
4. All gauges less than 0.016 in. shall have the same bend radius as 0.016-in. sheet.
5. Decimal equivalents of fractions are in accordance with MIL-STD-8.

Table 4-5. Bend Radii for Carbon Steels.

	SAE 1010		SAE 1020 SAE 1025 SAE 1095		AISI 4130 (Min Bend R for 4137, 4340, and 8630 steel are the same as for 4130 steel) Normalized		AISI 4130 Annealed	
Thickness	90°	180°	90°	180°	90°	180°	90°	180°
0.018	0.03	0.06	0.03	0.06	–	–	–	–
0.020	0.03	0.06	0.03	0.06	0.06	0.09	–	–
0.025	0.03	0.06	0.06	0.09	0.06	0.09	–	–
0.032	–	–	0.06	0.09	0.12	0.19	–	–
0.036	0.06	0.12	0.06	0.12	0.12	0.19	–	–
0.040	0.06	0.12	–	0.12	0.12	0.19	–	–
0.050	0.06	0.12	0.09	0.12	0.12	0.19	–	–
0.056	0.06	0.12	0.09	0.12	0.12	0.19	–	–
0.063	–	–	0.09	0.12	0.12	0.19	–	–
0.071	0.09	0.19	0.12	0.19	–	–	–	–
0.080	–	–	0.12	0.19	0.16	0.25	–	–
0.090	–	–	–	–	0.19	0.38	0.09	0.16
0.095	–	–	0.19	0.25	–	–	–	–
0.100	0.12	0.25	0.19	0.25	–	–	–	–
0.112	–	–	0.19	0.25	0.22	0.44	0.12	0.19
0.125	0.16	0.31	0.22	0.31	0.25	0.50	0.12	0.19
0.140	–	–	0.25	0.38	0.31	0.62	–	–
0.160	–	–	0.25	0.38	0.31	0.69	–	–
0.190	–	–	0.31	0.38	0.38	0.75	0.19	0.28
0.250	–	–	0.38	0.50	0.50	1.00	0.25	0.38
Over 0.250	–	–	$1\frac{1}{2}$T	2T	$1\frac{1}{2}$T	5T	$1\frac{1}{2}$T	2T

Notes

1. All dimensions are in inches.
2. T = material thickness.
3. Sheet-metal thicknesses are in accordance with MIL-STD-204A, except for 0.190-in., which is included because material is still supplied in this gauge.
4. All gauges less than 0.016 in. shall have the same bend radius as 0.016-in. sheet.
5. Decimal equivalents of fractions are in accordance with MIL-STD-8.

2. For aluminum alloys, machinability is roughly proportional to the tensile strength. (The higher the tensile strength, the easier it is to machine.)
3. Brasses are easier to machine than bronze.
4. All magnesium alloys are easy to machine.

Table 4-6. Bend Radii for Corrosion-Resistant Steel and Titanium.

	Corrosion-Resistant Steel and Super Alloys									Titanium		
	301 CRES 302 CRES 321 CRES 347 CRES		301 CRES 302 CRES		17-7PH CRES AMS5528		AM350 CRES 17-7PH CRES AMS5328 INCONEL X750	19-9DL 19-9DX	19-9DL 19-9DX	TI75A MIL-T-9046 Class 6	T15A12-5SN MIL-T-9046 Class 3	TI-64I 4V MIL-T-9046 Class 2
	Annealed		½ Hard		Strain Hardened		Annealed		Stress Rel			
Thickness	90°	180°	90°	180°	90°	180°	90°	90°	90°	90°	90°	90°
0.010	–	–	0.03	0.06	–	–	–	–	–	–	–	–
0.012	0.03	0.03	0.03	0.06	–	–	–	–	–	–	–	–
0.016	0.03	0.03	0.06	0.06	–	–	–	–	–	0.06	–	–
0.018	–	–	–	–	–	–	–	–	–	0.06	–	–
0.020	0.03	0.06	0.06	0.09	0.06	0.06	0.06	0.06	0.06	0.09	0.09	–
0.025	0.03	0.06	0.06	0.09	–	–	0.06	0.06	0.06	0.09	0.12	0.16
0.028	–	–	–	–	0.06	0.09	–	–	–	0.12	0.12	0.16
0.032	0.03	0.06	0.06	0.09	0.06	0.09	0.06	0.06	0.09	0.12	0.16	0.19
0.036	0.06	0.09	0.12	0.19	0.09	0.12	0.09	0.09	0.09	0.12	0.19	0.22
0.040	0.06	0.09	0.12	0.19	0.09	0.12	0.09	0.09	0.12	0.16	0.19	0.22
0.045	0.06	0.12	0.12	0.19	0.09	0.16	–	–	–	0.16	0.22	0.23
0.050	0.06	0.12	0.12	0.19	0.12	0.16	0.12	0.12	0.12	0.19	0.23	0.28
0.056	–	–	–	–	0.12	0.19	–	–	–	0.22	0.25	0.31
0.063	0.06	0.12	0.12	0.19	0.12	0.19	0.12	0.12	0.16	0.22	0.28	0.34
0.071	–	–	–	–	0.16	0.22	–	–	–	0.28	0.34	0.44
0.080	0.09	0.19	0.19	0.31	0.25	0.25	0.19	0.25	0.28	0.34	0.41	0.50
0.090	0.09	0.19	0.19	0.31	0.28	0.28	0.19	0.28	0.31	0.38	0.47	0.56
0.100	–	–	–	–	0.31	0.31	–	–	–	0.41	0.50	0.62
0.112	0.12	0.25	0.25	0.38	0.34	0.34	0.25	0.38	0.41	0.47	0.56	0.69
0.125	0.12	0.25	0.25	0.38	0.38	0.38	0.25	0.38	0.41	0.50	0.62	0.75
0.140	–	–	–	–	0.44	0.44	–	–	–	0.56	0.72	0.84
0.150	–	–	–	–	0.47	0.47	–	–	–	–	–	–
0.160	–	–	–	–	0.50	0.50	–	–	–	0.66	0.81	0.97
0.180	–	–	–	–	0.56	0.56	–	–	–	0.72	0.91	1.09
0.190	–	–	–	–	0.56	0.56	0.41	0.62	0.69	0.84	0.97	1.16
0.200	0.22	0.44	–	–	–	–	–	–	–	–	–	–
0.250	0.23	0.50	–	–	–	–	0.50	0.75	0.88	1.00	1.25	1.30
Over 0.250	1T	2T	–	–	–	–	–	–	–	4T	3T	6T

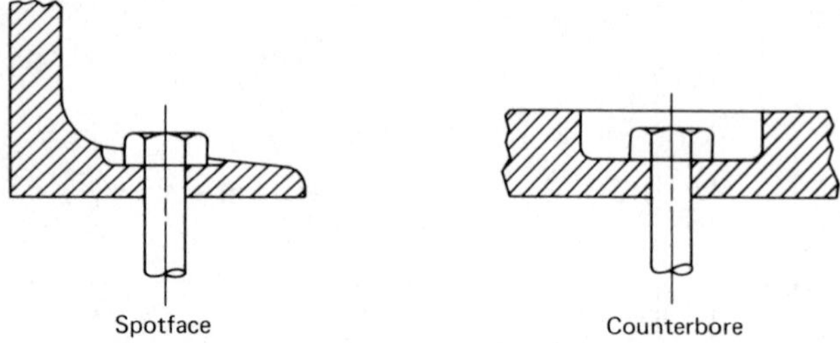

Figure 4-5. Examples of spotface and couterbore depth.

COUNTERBORING AND SPOT-FACING

Counterboring and spot-facing involve the use of a cutter, provided with a pilot, which is guided by a hole in a part that has been drilled (Figures 4-5 and 4-6). Although these operations may be performed by any practical method of machining, they are usually performed on a standard tool known as a counterbore. The counterbore is provided with interchangeable pilots in various sizes and can be used with any drilling device. Counterboring and spot-facing are similar operations. They differ in average depth and manner of use. Where the counterbore is used to recess a bolt head into a part, the spot-face is a means of providing a flat surface to seat the bolt head (see Figures 4-5 and 4-6).

COUNTERSINKING

Countersinking is the operation of removing material from a hole through the use of a fluted, conical, rotary cutting tool known as a countersink. The countersink is usually used as a seat for a flush-head screw or rivet. Occasionally, a

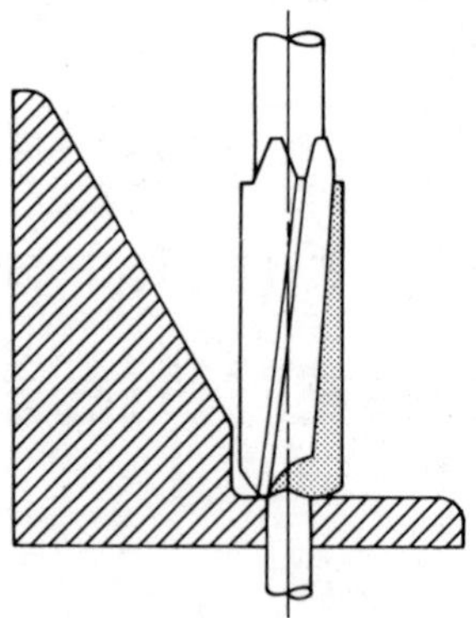

Figure 4-6. Spotfacing.

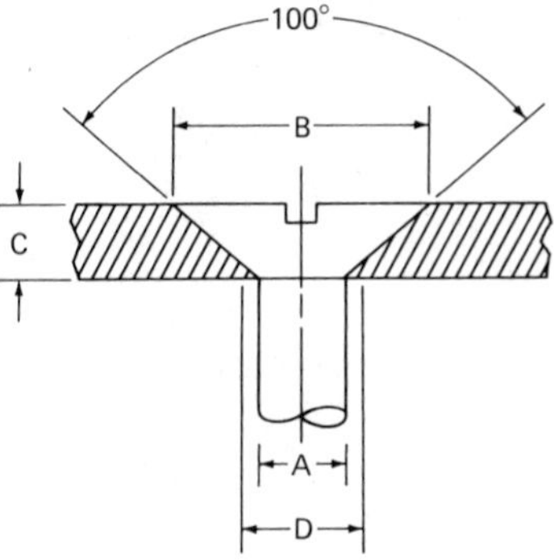

Countersinking Dimensions

Dimension	Screw Size							
	No. 2	No. 4	No. 6	No. 8	No. 18	1/4	5/16	3/8
A: Diameter (max)	0.086	0.112	0.138	0.164	0.190	0.250	0.313	0.375
B: Countersink diameter	0.19	0.23	0.29	0.34	0.40	0.52	0.65	0.79
D: Hole clearance	0.096	0.120	0.147	0.173	0.199	0.201	0.328	0.397
C: Recommended thickness[a]	0.062	0.071	0.090	0.122	0.125	0.160	–	–

[a]Minimum sheet thickness for maximum strength (equal to $1\frac{1}{2}$ times head height).

Figure 4-7. Countersinking.

countersink is used to provide easy entry into a hole for bolts or studs (Figure 4-7).

Two methods of fabrication can be employed to obtain a flush seating for the assembly of sheet-metal parts using screws: countersinking and dimpling. Countersinking is used where the surface under the metal to be countersunk is rigid enough to prevent buckling or tearing when pressure is applied to the screw. Countersinking into material less than 1.5 times the thickness of the screw head is not recommended. Dimpling is used for thin gauges and highly stressed parts.

DRILLING

Drilling is the operation of cutting round holes with a twist drill—a helically or straight fluted, rotary end-cutting tool. Drilling is normally classified as a roughing operation and should not be depended on for close tolerance holes. Where accuracy is required, as in close fitting shafting, a complementary operation such as reaming is required.

Figure 4-8. Plane milling cutters.

MILLING

Milling is a machining process in which shaping is accomplished by a rotating toothed cutter. The surface may be a plane, or a regular or irregular profile. In most applications, the cutter is stationary and rotating at high speed while the work is moved past it. Depending on the cutter configuration, milling can produce a wide variety of shapes. Whenever possible, designs should specify standard milling cutters to maintain an economic advantage (Figures 4-8 through 4-15). Nonstandard cutters are expensive and should be used only in special applications.

REAMING

Reaming is the machining operation used to improve the diametral accuracy and surface finish of existing holes when production quantities do not warrant the use of higher cost broaching. Careful consideration should be given to the depth of the drilled blind hole because the amount of material to be removed by the reaming operation is important in preventing reamer breakage. The drilled hole should extend well beyond the reamed depth to prevent bottoming of the reamer.

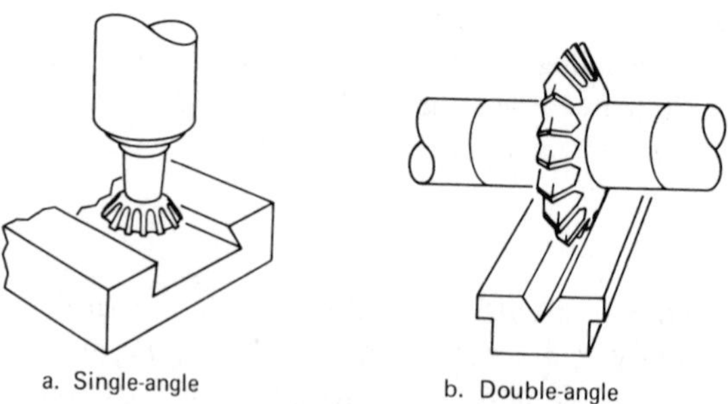

Figure 4-9. Angle milling cutters.

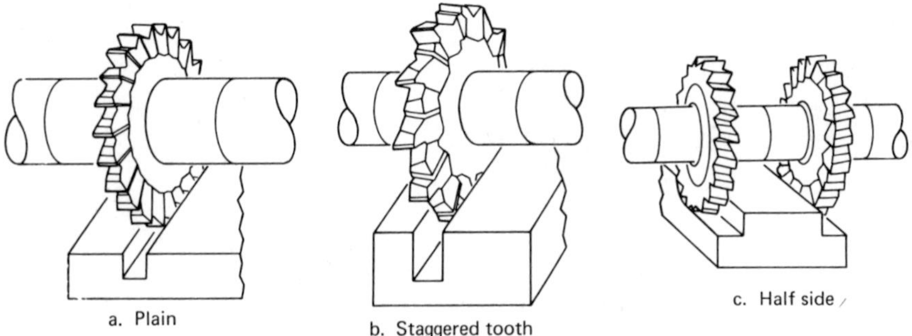

Figure 4-10. Side milling cutters.

CASTING METAL PARTS

General Design Requirements

Judgment must be exercised when designing castings. The various factors pertaining to pattern-making and diemaking, molding, and machining must be considered before deciding on the final design. Knowledge of the limitations and advantages of the various casting methods is essential to selecting the process best suited for the product (Tables 4-8 through 4-10).

Sand Castings

Sand castings have the lowest initial cost and provide flexibility in design, choice of alloy, and design changes during production. Sand casting is used to make a small number of pieces, or a moderate number requiring quick delivery. A sand-cast part will not be as uniform in dimensions as one produced by other methods, and therefore, greater tolerances and machining allowances must be provided.

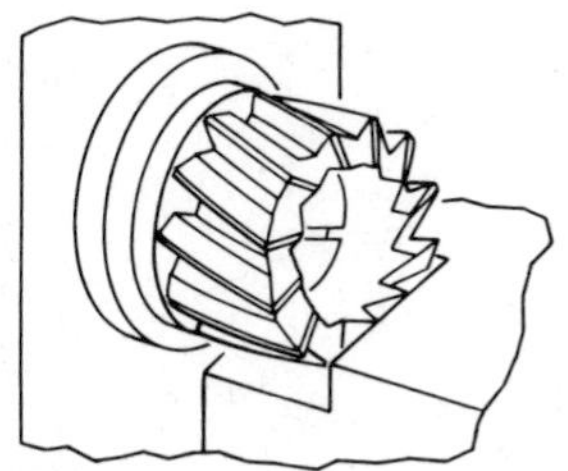

Figure 4-11. Shell-type end mills.

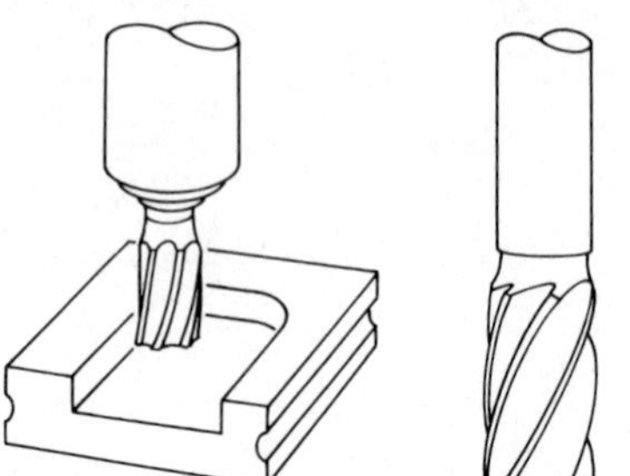

Figure 4-12. Standard end mills.

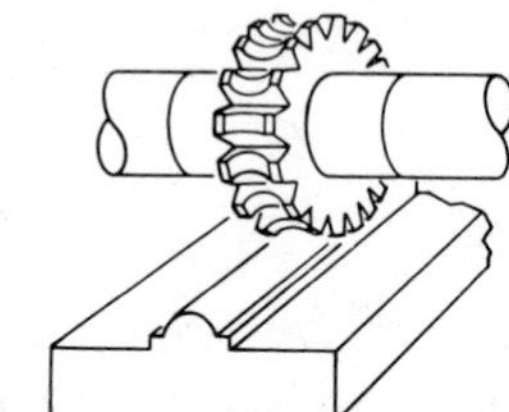

Figure 4-13. Convex milling cutters.

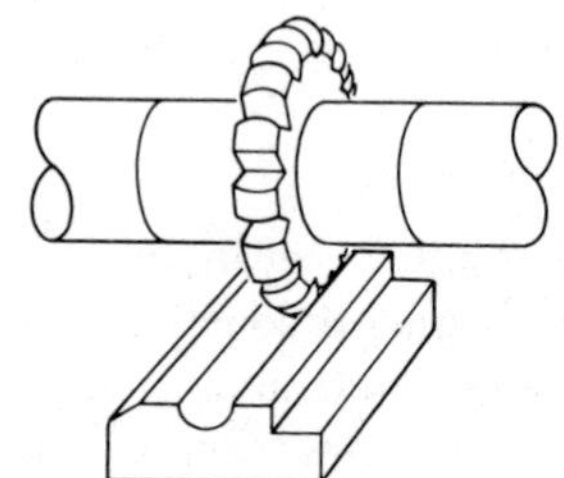

Figure 4-14. Concave milling cutters.

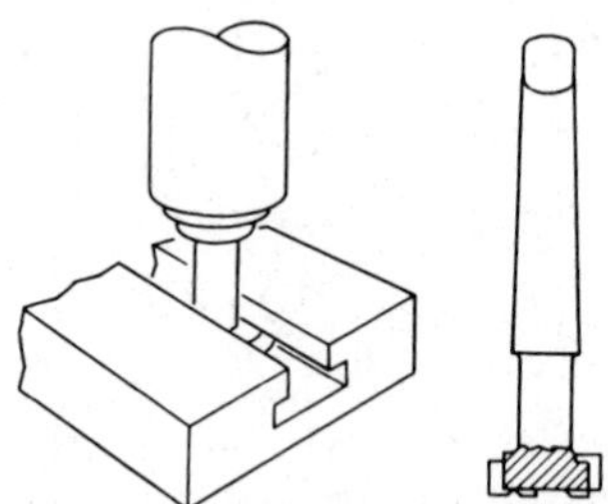

Figure 4-15. T-slot milling cutters.

Table 4-7. Notch Depth Dimensions.[a]

Material Thickness (T)	Bend Radius (R_b)							
	0.03	0.06	0.09	0.12	0.16	0.19	0.22	0.25
0.06	0.12	0.16	0.19	0.22	0.25	0.28	0.31	0.34
0.020	0.12	0.16	0.19	0.22	0.25	0.28	0.31	0.34
0.025	0.12	0.16	0.19	0.22	0.25	0.28	0.31	0.34
0.032	0.16	0.19	0.22	0.25	0.28	0.31	0.34	0.38
0.040	b	0.19	0.22	0.25	0.28	0.31	0.34	0.38
0.050	b	0.19	0.22	0.25	0.28	0.31	0.34	0.38
0.063	b	0.22	0.25	0.28	0.31	0.34	0.38	0.41
0.071	b	b	0.25	0.28	0.31	0.34	0.38	0.41
0.080	b	b	0.25	0.28	0.31	0.34	0.38	0.41
0.090	b	b	0.28	0.31	0.34	0.38	0.41	0.44
0.100	b	b	b	0.31	0.34	0.38	0.41	0.44
0.125	b	b	b	0.31	0.34	0.38	0.41	0.44
0.160	b	b	b	b	0.38	0.41	0.44	0.50
0.190	b	b	b	b	b	b	0.50	0.50
0.250	b	b	b	b	b	b	b	b

[a]Values given are for *ND* when R_n is equal to 0.16 for combinations of thickness and bend radii shown. (Other values may be calculated from Equation 4.1.)
[b]Not recommended.

Wall Thickness

For A356 and 43 aluminum alloy, it is recommended that the nominal wall thickness be not less than 0.125 in. For magnesium alloy, 0.156 in. is the minimum allowable thickness on structural castings, and therefore it is recommended that drawings specify 0.188 in. to allow for the usual 0.030 in. tolerance. For steel, 0.250 in. is the recommended minimum nominal; however, 0.188 may be possible in some cases. The designer must consider many variables (e.g., section area, stress, cast height, etc.) if thicknesses other than those given here are used.

Fillet and Corner Radii

Generous fillet radii avoid the weakening of a casting at the corners. However, too large a fillet may cause shrinkage at the corners due to the increased mass of metal, and too small a fillet may cause a weakened section to develop. The designer should attempt to use a common fillet radius throughout the design of any particular casting. Fillet radii should be in multiples of $\frac{1}{16}$ in.

Table 4-8. Casting Characteristics and Restrictions.

Sand Casting (Rough)	Sand Casting (Close Tolerance)	Permanent and Semipermanent Mold Casting	Close Tolerance Permanent Mold Casting	Die Casting	Investment Casting (Solid Mold)
1. Least initial cost. 2. Usually fast initial delivery. 3. Good physical properties. 4. Least dimensional accuracy with machining required for both size and surface. 5. Typical roughness 250–500 microinch AA. 6. Strength-weight ratio not equal to that of wrought parts (typical of all castings).	1. Higher cost than rough castings. 2. Requires close coordination with vendor. 3. Surface roughness down to 125 microinch AA. 4. Smaller draft angles. 5. Some machining can be eliminated. 6. Tolerances are two-thirds that of conventional sand casting.	1. Large quantities necessary to absorb tool costs. 2. Delivery retarded by time required for making tools. 3. Good dimensional accuracy. 4. A variety of shapes and alloys available. 5. Size is limited compared with sand casting. 6. Cast-in inserts may be hard. 7. Typical surface roughness of 250-microinch AA).	1. Similar to permanent mold except better finish (120-microinch AA) and dimensional accuracy with resulting higher cost and longer delivery schedule. 2. Good mechanical properties.	1. Large quantities necessary to absorb high tool cost. 2. Adaptable to a variety and intricacy of shapes in nonferrous alloys. 3. Good dimensional accuracy and natural finish, 63-microinch rms obtainable. 4. Only fair physical properties. 5. Cast-in inserts possible. 6. Subject to extensive porosity in heavy sections. 7. Do not use where extensive machining is required. 8. Thin sections can be cast.	1. Very intrincate shapes possible. 2. Good dimensional accuracy obtainable on special dimensions. 3. Wide variety of alloys castable. 4. Good mechanical properties in steel. 5. Natural finish is 125 microinch AA and below. 6. Size limits approximately 20 × 10 × 10 maximum. 7. Weight reduction. 8. Fastest initial delivery of all methods using plastic patterns.

The practical minimum cast corner radii is 0.03 in. Radii of 0.06 in. and larger are recommended. Sharp corners can be produced if the parting line coincides with an edge of the casting. Corner radii on webs and ribs should not be large enough to produce a fully rounded edge (see Figure 4-16).

Plaster Mold Castings

Plaster mold castings are a refinement of sand castings in which the sand mold is replaced by a plaster mold. It produces a casting with smoother surfaces and greater dimensional accuracy than is possible with sand casting.

Due to the low thermal conductivity of plaster of paris, molten metal remains fluid longer in plaster molds than it does in sand or metal molds. This permits much thinner casting sections than does sand casting or even metal mold casting, and there is less danger of internal porosity. However, the structural strength of castings made in plaster-of-paris molds is likely to be somewhat lower than that of metal mold or sand castings, especially in those alloys (such as aluminum alloys) that are susceptible to grain growth.

Permanent Mold Castings

Permanent mold castings should be considered when the quantity required justifies the cost of mold equipment. Permanent mold castings provide a better surface finish, and the dimensional tolerances can be held closer than is possible in sand casting. The higher strength of these castings permits thinner sections than are required for sand casting.

Permanent mold castings are formed by pouring molten metal into a metal mold and around cores, which are usually metal, without external pressure. The high cost of the metal mold makes this type of casting feasible only when production is in sufficient quantity to amortize the cost of the mold or when difficult and expensive machining can be eliminated.

Permanent mold castings are dense and fine-grained; they can be made with smoother surfaces and closer tolerances than sand castings. The process stands between sand casting and die casting in respect to possible complexity, dimensional accuracy, and production rates. For many parts where the ultimate, either in complexity of one-piece construction or in close tolerances, need not be fully realized, it provides a desirable compromise.

The process may be used to produce some ferrous alloy castings, but due to rapid deterioration of the mold caused by the high pouring temperature of the alloy and the attendant high mold cost, the process is confined largely to the production of lower melting-point, nonferrous alloy castings.

Many types of nonferrous alloy castings, particularly those of aluminum and magnesium, may be produced by this method. It produces a smoother finish

Table 4-9. Casting Features.

			Casting Method and Material											
			Sand			Permanent Mold		Die		Plaster Mold		Investment		
			Al	Mag	Steel	Al	Mag	Al	Mag	Al	Mag	Al	Mag	Steel
Draft/side on surfaces that[a]	Free themselves from mold parts after shrinkage	Minimum	$\frac{1}{2}°$ or 0.031[d,e]			1°		$\frac{1}{2}°$		$\frac{1}{4}°$		0°		
		Normal	1° or 0.031[d]			3°		2°		$\frac{1}{2}°$		0°		
	Bind on mold parts after shrinkage	Minimum	$\frac{1}{2}°$ or 0.031[d]			2°		1°		$\frac{1}{2}°$		0°		
		Normal	1° or 0.031[d]			5°		4°		1°		0°		
Minimum cored hole diameter			0.50	–	0.62	0.25		0.09		–		0.02		0.05
Maximum cored blind hole depth			1D			10D		f		–		2D		$1\frac{1}{4}$D
Maximum cored through hole depth			–			–		–		–		3D		
Round cored hole draft[a]			–			$\frac{1}{2}°$ to 2°		i	2°	3°		0°		
Wall thickness		Minimum	0.12	0.12	0.19	0.09	g	0.05		h		0.03		0.04
		Normal	0.19	0.16	0.25	0.12	0.19	0.12		0.09		0.06 to 0.25		
		Maximum	–			–		–		–		0.50		
Minimum fillet radius[b]			0.16	0.19	–	0.12		0.03		0.03		0.02		

Stock allowance[c]	Casting size (in.)	Up to 6	0.094	0.125	–	0.062	0.031	0.062	0.062	0.094
		Over 6 to 15	0.125	0.188	–	0.094	0.047	0.094	0.094	0.125
		Over 15	0.188	0.281	–	0.125	–	–	–	–
Casting size		Normal	No restrictions				300 in.^{3}	11 × 17 × 3	5 Dia × 6; or 1 lb	
		Maximum					–	23 × 35 × 11	10 Dia × 20; 9 × 12 × 15; or 2500 in.^{3}	

[a] For use in special cases only.
[b] When the wall thickness is more than the minimum radius specified in the table, the radius should equal one-half the wall thickness.
[c] Values specified in the table are given for information only.
[d] Whichever is larger.
[e] Very small castings sometimes are made without draft.
[f] 5D for holes under 0.25 in. diameter; 8D for holes 0.25 to 1.00 in. diameter.
[g] Minimum wall thickness for permanent mold magnesium castings shall be 0.16 in. for areas up to 4 in. wide; 0.19 in. for areas over 4 in. wide, or over 30 in.^{2}
[h] Minimum wall thickness for plaster mold castings is as follows:

Area (in.^{2})	Minimum Wall Thickness
Up to 2	0.04
Over 2 to 4 inclusive	0.05
Over 4 to 6 inclusive	0.05
Over 6 to 30 inclusive	0.09

[i] Draft for cored holes on die castings is as follows:

Diameter Range (in.)	Total Draft per Inch of Depth
0.09 to 0.12 inclusive	0.020
Over 0.12 to 0.25 inclusive	0.016
Over 0.25 to 1.00 inclusive	0.012
Over 1.00	0.012 plus 0.002 per each additional inch

Table 4-10. Preferred Casting Tolerances.[a]

Condition			Sand		Permanent Mold[b]		Die		Plaster Mold		Investment	
			Normal	Minimum	Normal	Minimum	Normal	Minimum	Normal	Minimum	Normal	Minimum
Linear[c]	Per inch or fraction of dimension	Al and Mg	0.005	0.003	0.005	0.003	0.005	0.002	0.005	0.003	0.005	0.002–0.004
		Steel	0.008	0.005	–	–	–	–	–	–	0.005	0.005
	Minimum per dimension	Al and Mg	0.03	0.02	0.010	0.010	–	–	–	–	–	–
		Steel	0.06	0.03	–	–	–	–	–	–	–	–
Angular[d]			1°	$\frac{1}{2}$°	1°	$\frac{1}{4}$°	1°	$\frac{1}{4}$°	1°	$\frac{1}{4}$°	1°	$\frac{1}{4}$°
Surface texture microinches AA[e]			500		125		63		125		125	

*0° when acrylic patterns are used.
[a]Tolerances, except for surface texture, are plus and minus.
[b]Tolerances for cored holes formed by sand cores in semipermanet molds may be obtained using values for sand castings.
[c]Linear casting tolerances are obtained by multiplying the "per inch" figure by the length involved. If the product is less than the "minimum per dimension" value specified, then this value shall be used. If the product is greater than the "minimum per dimension" specified, then the product shall be used. This is applicable to sand and permanent mold castings only. For die, plaster mold and investment castings, the tolerances shall be based on the "per inch" figure only.
[d]Applies to all specified angular casting dimensions. For 90° angles not specified, these tolerances are effective only while they lie within the tolerance zone of the controlling linear dimension. When the angular tolerance extends beyond the linear tolerance zone, it no longer is effective, and the linear tolerance shall govern.
[e]For use in special cases only.

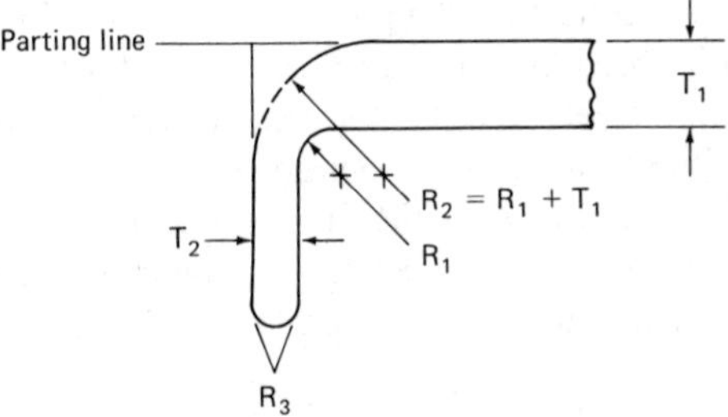

R_3 Should be less than 1/2 T_2 to provide flat edge on rib.
R_2 Is better design if parting is located elsewhere than shown.

Figure 4-16. Cast corner radii.

than the sand-casting method, improved mechanical properties, closer dimensional accuracy, and less production scrap.

Die Castings

Die castings afford extremely accurate dimensioning (more accurate than any other casting process except investment casting) and excellent surface finish. They require a minimum amount of machining, and afford weight saving due to the thin sections permissible. Die castings are produced in quantity at a cost lower than by the permanent mold method. Finer detail, which is not practical with other casting methods, can be produced by die casting. Die castings have poor resistance to shock loads and are subject to heavy corrosion.

In diecasting, the molten metal is forced under pressure into metal dies. Smooth surfaces, close tolerances, casting of insets, and the elimination of machining make it an economical method for quantity production.

On the other hand, die costs are high, and shapes are limited in complexity. Undercuts require movable pieces in the die. Die costs increase rapidly when the casting metal does not have a comparatively low melting point like that of aluminum or magnesium alloys. The most common nonferrous alloys used in die casting are zinc, aluminum, magnesium, and copper.

Investment Castings

Investment casting is a precision method of molding (1) all castable metals and alloys that are difficult to machine and (2) parts that, because of their intricate design or unusual contours, are too expensive to tool or to machine. Production rate is low, and the cost is relatively high because many steps are required for the process. In the investment casting process, parts composed of a number of elements, each of which formerly had to be machined and then assembled, may be cast in one piece.

Investment casting is practical for the production of parts that cannot be made satisfactorily on screw machines, by sand or die castings, or stampings. It produces parts for immediate use, in "as cast" condition, with good surface texture, sound structure, and close dimensional tolerance. It possesses the inherent characteristics necessary for the fabrication of never-before-cast precision parts, in unlimited quantities. Economy is one of its advantages; it eliminates machining, tooling, and assembly costs, and minimizes loss by reducing waste metal.

Die-Casting Design Recommendations

The cost of dies is naturally affected by the complexity of the design. Anything that will simplify the lines, eliminate sliding cores for undercuts, etc., will reduce the cost. The guidelines for good die-casting design can be summarized as follows:

1. Shapes should be kept as simple as requirements permit; projections should be avoided wherever possible.
2. Uniformity in sections is desirable. If varied, transition should be gradual to avoid concentration of stress.
3. A slight crown is better than a large flat surface. This is especially true on highly finished parts where surface blemishes will show up easily.
4. Undercuts that increase die or machining costs should be avoided.
5. Unless core pieces affect a considerable net saving or some other benefit, they should not be used.
6. Fillets should be provided on inside corners, and sharp outside corners should be avoided.
7. Sufficient draft must be allowed on cores and sidewalls to permit easy core extraction and casting ejection.
8. Coring very small holes or coring holes in walls less than 0.125 thick is undesirable.
9. Ejector pins should be located at points where surface marks are inconsequential. Enough pins to permit easy ejection without distortion of the casting should be provided.
10. Design should avoid sharp corners or small recesses, especially where these surfaces are later to be buffed or polished.
11. Die casting is a close-tolerance process. However, for economy, closer tolerances than actually are required should not be specified.
12. When permanent inserts are necessary, they should be provided with anchorage sufficient to secure them in the casting.

See Table 4-11.

DEFECTS IN CASTINGS

Castings are subject to certain defects that in a well-designed casting, are controllable by proper foundry technique, but are not wholly predictable or preventable. Most defects are hidden below the surface of the casting and therefore cannot be detected by visual inspection. The designer must either design to avoid or minimize such defects, or accept certain shortcomings and make allowances for them. Defects associated with the foundry practice include internal blowholes, slag inclusions, and centerline weakness (the junction of grains grown from opposite mold faces). The defects more directly attributable to poor design are hot tears and shrinkage cavities.

Hot tears can be prevented by eliminating sections that remain hot after others have cooled considerably and by eliminating stresses, particularly the concentrated stresses. Hot sections (or hot spots) may be formed by a difference in section thickness. The thinner section solidifies first and starts to contract even before the heavier section has solidified. The heavy section thus becomes a hot section. Occurrence of hot sections of this type can be reduced to a minimum if all sections of a casting are designed with uniform thickness. However, even with uniform thickness, hot spots will be formed at the junction of two or more sections because the ratio of cooling surface area to metal volume is less than for the straight sections. Therefore, these junctions cause a thermal gradient and accompanying stress. Since they also provide conditions for stress concentration, they are quite favorable to the formation of hot tears. The stress concentration can be reduced by using a fillet at sharp corners or junctions and by making all changes in sections as gradual as possible.

In some intricate castings, it is not possible to avoid hot spots and thermal gradients. With these designs, the foundry worker may reduce the stresses by using a mold-relieving technique.

WELDING

Methods of fastening are determined by the type of joint, strength required, materials, and the disassembly requirements. The tensile strength of a joint should be approximately that of the members to be joined, and the stresses should be distributed as evenly as possible. Defective mechanical connections are a major cause of defects in electronics equipment. Although some defects can be traced to improper design, most are the result of poor workmanship as evidenced by loose screws, improperly set rivets, and cracked welds. To minimize these defects and to improve reliability, good training programs, adequate supervision, and proper inspection and testing are required. One of the most effective methods of joining is the use of welding techniques.

Welding has many advantages over the other methods of joining parts such as high strength, simplicity, and uniform stress distribution. The cost of tooling

Table 4-11. Recommended Minimum Die-Casting Tolerances.

Example and Description of Dimension	Sum ($S_1 + S_2$)	Total Tolerance on Dimensions as Cast: Up to 1	1 to 2.5	2.5 to 4	4 to 5.5	5.5 to 7	7 to 10	10 to 14	14 to 18	18 to 28	28 to 36
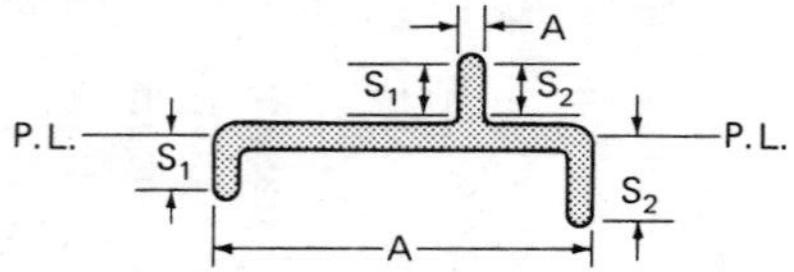 Dimension (A) between surfaces S_1 and S_2 which are parallel to direction of draw and in same portion of die	Up to 0.875	0.015	0.020	0.025	0.030	0.035	0.045	0.063	0.063	0.094	0.125
	0.875 to 1.188	0.020	0.025	0.030	0.035	0.040	0.063	0.063	0.094	0.094	0.125
	1.188 to 1.500	0.025	0.030	0.035	0.040	0.045	0.063	0.063	0.094	0.125	–
	1.500 to 1.813	0.030	0.035	0.040	0.045	0.050	0.063	0.094	0.094	0.125	–
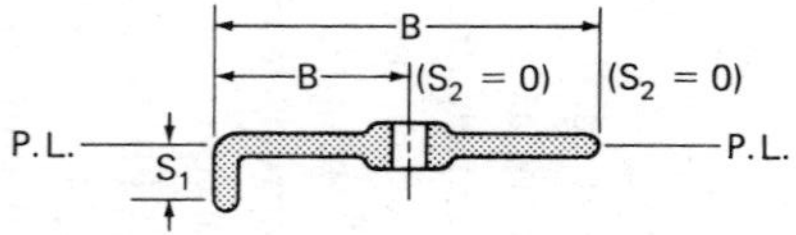 Dimension (B) between surface S_1 which is parallel to direction of draw and a point or center line in same portion of die	1.813 to 2.250	0.040	0.045	0.050	0.063	0.063	0.094	0.094	0.125	–	–
	2.250 to 3.000	0.050	0.063	0.063	0.063	0.094	0.094	0.094	0.125	–	–
	3.000 to 3.750	0.063	0.063	0.063	0.094	0.094	0.094	0.125	0.125	–	–
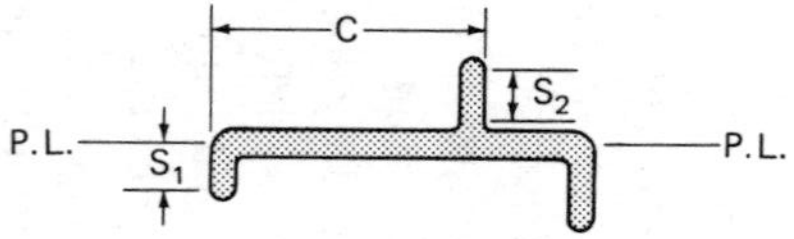 Dimension (C) between surfaces S_1 and S_2 which are parallel to direction of draw and in opposite portions of die	Up to 0.875	0.025	0.030	0.035	0.040	0.045	0.063	0.063	0.094	0.094	0.125
	0.875 to 1.188	0.030	0.035	0.040	0.045	0.050	0.063	0.094	0.094	0.125	0.125
	1.188 to 1.500	0.035	0.040	0.045	0.050	0.063	0.063	0.094	0.125	0.125	–
	1.500 to 1.813	0.040	0.045	0.050	0.063	0.063	0.094	0.094	0.125	–	–

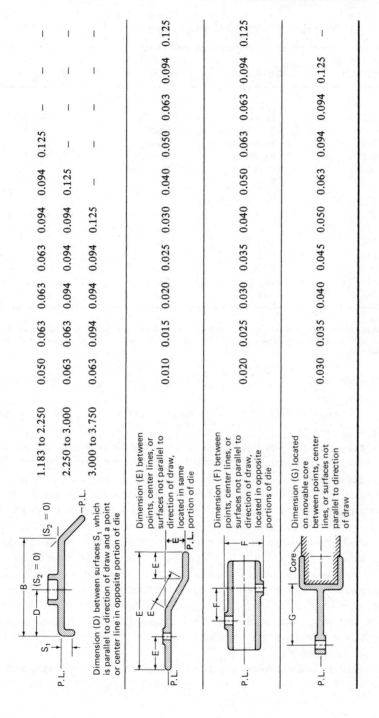

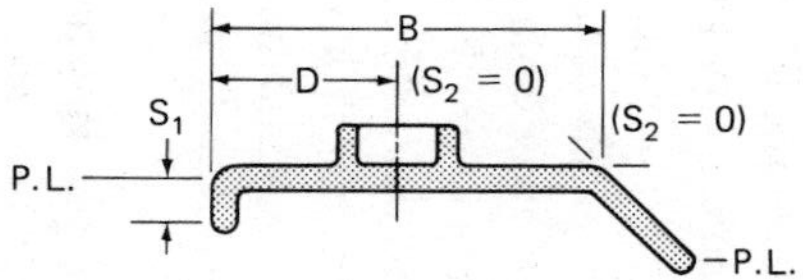 Dimension (D) between surfaces S_1 which is parallel to direction of draw and a point or center line in opposite portion of die	1.183 to 2.250	0.050	0.063	0.063	0.063	0.094	0.094	0.125	–	–	–
	2.250 to 3.000	0.063	0.063	0.094	0.094	0.094	0.125	–	–	–	–
	3.000 to 3.750	0.063	0.094	0.094	0.094	0.125	–	–	–	–	–
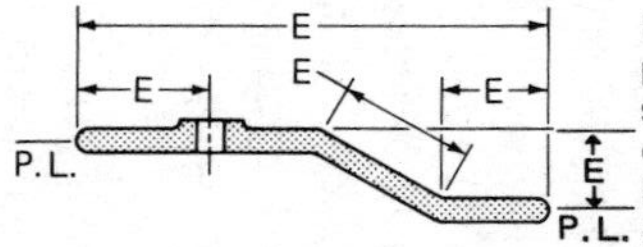 Dimension (E) between points, center lines, or surfaces not parallel to direction of draw, located in same portion of die		0.010	0.015	0.020	0.025	0.030	0.040	0.050	0.063	0.094	0.125
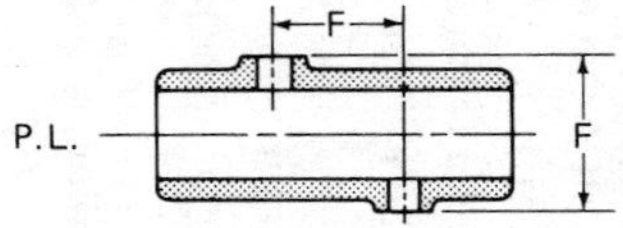 Dimension (F) between points, center lines, or surfaces not parallel to direction of draw, located in opposite portions of die		0.020	0.025	0.030	0.035	0.040	0.050	0.063	0.063	0.094	0.125
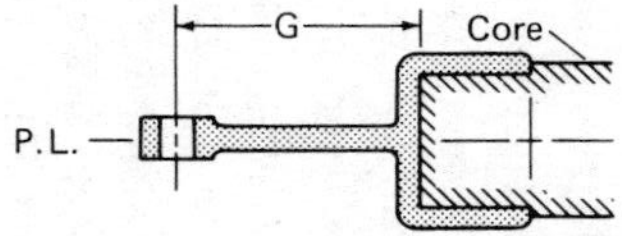 Dimension (G) located on movable core between points, center lines, or surfaces not parallel to direction of draw		0.030	0.035	0.040	0.045	0.050	0.063	0.094	0.094	0.125	–

Table 4-12. Methods of Welding Aluminum Alloys.

Aluminum Alloy Designation	Gas Welding	Arc Welding	Resistance Welding
Wrought			
3003	Excellent	Excellent	Good
2014	Poor	Good	Good
2024	Poor	Good	Good
5052	Excellent	Excellent	Good
6061	Excellent	Excellent	Excellent
7075	Poor	Poor	Good
Cast			
43	Excellent	Excellent	Excellent
122	Fair	Fair	Fair
142	Fair	Fair	Fair
195	Fair	Fair	Fair
220	Poor	Poor	Poor
356	Good	Excellent	Good
A612	Fair	Fair	Fair

Note:
Welding should never be used when partial or complete disassembly will be required. In selecting the type of welding joint to be used, cost and maximum effectiveness must be considered.

and labor is relatively low, and welded parts require a minimum of machining prior to assembly.

There are three principal types of welding: resistance, arc, and gas welding (Tables 4-12 through 4-14).

Resistance welding. When an electric current of sufficient magnitude is passed through two adjacent metal parts, heat is generated at the point of contact. This heat, together with the localized pressure exerted by the electrodes supplying the current, welds the two parts together. For resistance welding, the design should provide projections to localize the weld area and minimize distortion. The most common resistance welds are the seam weld and the spot weld. Spot welds should not be used in tension. In assemblies of thin metals containing long rows of spot welds that may be subjected to bending, it is advisable to use rivets at each end, and space a few along the row. For economy in spot welding, it is important to design the joint so that they are accessible to the welding electrodes. Zinc- and cadmium-plated parts can be spot-welded, but is is recommended that when maximum strength is required, another method be used. Extreme care should be used in welding cadmium parts since the fumes generated are a health hazard.

Arc welding. In arc welding, an electrode generally made of the same material that is being welded is used. The rod is coated with a flux, deoxidizer, and ionizer

Table 4-13. Welding Processes Based on Joint Design and Surface Buildup.

Welding Process	Butt Joint Section		Lap Joint Section		Tee Joint Section		Edge Joint Section		Surface Build-Up
	Light	Heavy	Light	Heavy	Light	Heavy	Light	Heavy	
Shielded metal-arc (coated electrode)	S	R	R	R	R	R	NR	R	S
Inert gas, tungsten-arc (nonconsumable electrodes)	S	S	R	S	R	S	R	S	R
Inert gas, metal-arc (consumable electrodes)	NR	R	NR	R	NR	R	NR	S	R
Spot welding	NA	NA	R	R	NA	NA	NA	NA	NA
Seam welding	NA	NA	R	R	NA	NA	R	R	NA
Gas welding (oxy-acetylene)	R	S	R	S	R	S	R	S	R

R = Recommended
S = Satisfactory
Light Section:
0.005 to 0.125

NR = Not recommended
NA = Not applicable
Heavy Section:
Over 0.115

Table 4-14. Welding Processes Based on Material.

Material		Welding Process					
		Shielded Metal-Arc (Coated Electrodes)	Inert Gas Tungsten-Arc	Inert Gas Metal-Arc	Gas Welding (Oxy-Acetylene)	Spot Welding	Seam Welding
Carbon steel SAE 1000, 1100 4100 and 4300 series		R	S	S	R	R	R
Stainless steel 300 series		R	R	R	S	R	R
High temperature steel 19.9DL, 16-25-6		R	S	S	S	S	S
Stainless steel 400 series		R	S	S	S	S	S
Aluminum and Al alloys		S	R	R	S	R	R
Nickel and Ni alloys		R	R	R	S	R	R
Copper and Cu alloys		NR	R	R	S	S	S
Cast and grey iron		S	S	NR	R	NA	NA
Magnesium and Mg alloys		NA	R	S	NR	S	S
Precious metals	Silver	NR	R	S	R	NR	NR
	Gold Platinum Iridium	NR	R	S	R	S	S
Rare metals	Titanium and Ti alloys	NA	R	NR	NA	S	S
	Uranium Molybdenum Vanadium Zirconium Tungsten	NA	R	NR	NR	S	S

R = Recommended NR = Not recommended S = Satisfactory NA = Not applicable to protect the filler when it is molten. Most metals, including stainless and high-alloy steels, can be satisfactorily welded using arc-welding methods. It is often necessary to perform some form of heat treatment after arc welding to obtain the desired mechanical properties. Aluminum alloys such as 1100, 3003, 5052, and 6061 can be arc-welded successfully.

Gas welding. A number of gas mixtures, such as air-acetylene, oxyacetylene, and oxyhydrogen, are commonly used in gas welding. In gas welding, the welding rod is puddled into the joint to form the weld. Gas welding, although not as fast as arc welding, permits greater control and prevents burn-in in thin ferrous metals. Gas welding is not as commonly used as the other methods. When objects of a large mass are gas-welded, distortion generally occurs, and cooling stresses may produce failures in ferrous metals of a high carbon content.

Basically, weldable metals are divided into four groups. The adaptability of combinations of these groups to welding is shown in Table 4-15.

The following information should be taken into consideration in the design of welded assemblies:

- Combinations of hard and soft materials should be avoided, especially where the softer material is also the thinner.
- The maximum number of members that may be welded in a joint is three for aluminum alloys, two for magnesium alloys, and four for corrosion-resistant steel alloys.
- The total joint thickness must not exceed four times the thickness of the thinnest outer member; it is further limited to 0.188 in. for aluminum and magnesium alloys, and to 0.375 in. for corrosion-resistant steel alloys.

Table 4-15. Adaptability of Metals to Welding.

Group[a]	Welded With Group	Weldability
1	1	Good
1	2	Not to be used
1	3	Not to be used
1	4	Not to be used
2	2	Good
2	3	Possible but not good practice
2	4	Not to be used
3	3	Good
3	4	Possible but not good practice
4	4	Possible but not good practice

[a]Group 1: Aluminum and aluminum alloys. Group 2: Brass, bronze, and copper. Group 3: Carbon and alloy steels. Group 4: Cast and malleable irons, cast steels.

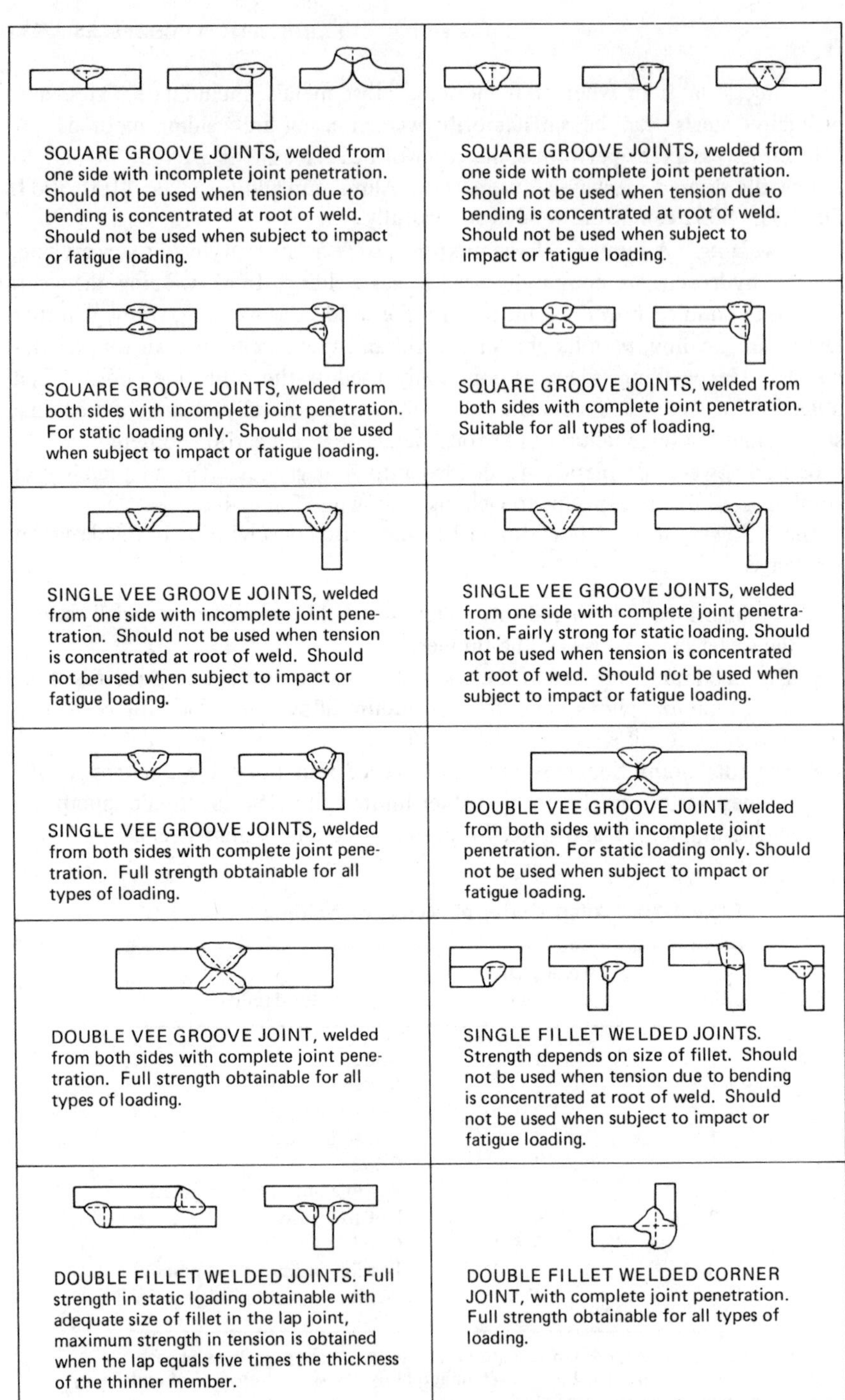

SQUARE GROOVE JOINTS, welded from one side with incomplete joint penetration. Should not be used when tension due to bending is concentrated at root of weld. Should not be used when subject to impact or fatigue loading.
SQUARE GROOVE JOINTS, welded from one side with complete joint penetration. Should not be used when tension due to bending is concentrated at root of weld. Should not be used when subject to impact or fatigue loading.
SQUARE GROOVE JOINTS, welded from both sides with incomplete joint penetration. For static loading only. Should not be used when subject to impact or fatigue loading.
SQUARE GROOVE JOINTS, welded from both sides with complete joint penetration. Suitable for all types of loading.
SINGLE VEE GROOVE JOINTS, welded from one side with incomplete joint penetration. Should not be used when tension is concentrated at root of weld. Should not be used when subject to impact or fatigue loading.
SINGLE VEE GROOVE JOINTS, welded from one side with complete joint penetration. Fairly strong for static loading. Should not be used when tension is concentrated at root of weld. Should not be used when subject to impact or fatigue loading.
SINGLE VEE GROOVE JOINTS, welded from both sides with complete joint penetration. Full strength obtainable for all types of loading.
DOUBLE VEE GROOVE JOINT, welded from both sides with incomplete joint penetration. For static loading only. Should not be used when subject to impact or fatigue loading.
DOUBLE VEE GROOVE JOINT, welded from both sides with complete joint penetration. Full strength obtainable for all types of loading.
SINGLE FILLET WELDED JOINTS. Strength depends on size of fillet. Should not be used when tension due to bending is concentrated at root of weld. Should not be used when subject to impact or fatigue loading.
DOUBLE FILLET WELDED JOINTS. Full strength in static loading obtainable with adequate size of fillet in the lap joint, maximum strength in tension is obtained when the lap equals five times the thickness of the thinner member.
DOUBLE FILLET WELDED CORNER JOINT, with complete joint penetration. Full strength obtainable for all types of loading.

Figure 4-17. Load limitations for welded joints.

- Where more than two members are used, the thinnest should be sandwiched between the outer members.
- Welding of formed parts should be considered carefully, since such parts are not usually consistent enough to provide the good contact necessary to produce satisfactory welds.

Figures 4-17 and 4-18 and Tables 4-16 through 4-18 provide the basic standard requirements for welding and are an aid in the selection of welding processes.

ADHESIVES AND CEMENTING

Metal fasteners, welding, and soldering are the most common methods of joining, but the use of adhesives is practical in many conditions. Adhesives have

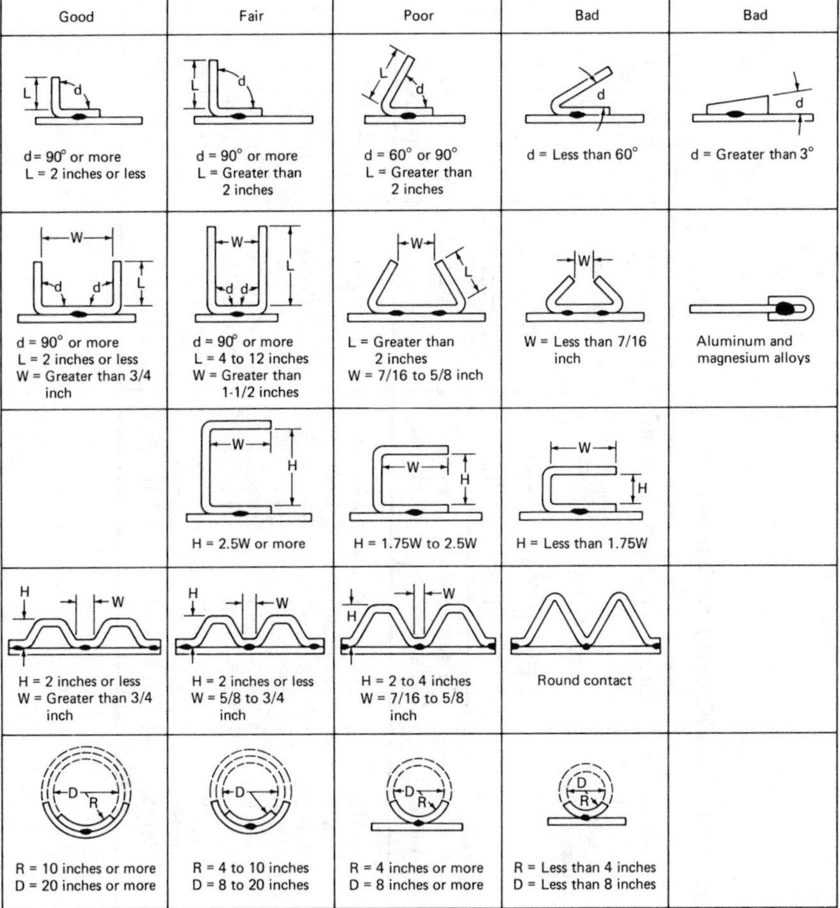

Good
Fair
Poor
Bad
Bad
d = 90° or more
L = 2 inches or less
d = 90° or more
L = Greater than 2 inches
d = 60° or 90°
L = Greater than 2 inches
d = Less than 60°
d = Greater than 3°
d = 90° or more
L = 2 inches or less
W = Greater than 3/4 inch
d = 90° or more
L = 4 to 12 inches
W = Greater than 1-1/2 inches
L = Greater than 2 inches
W = 7/16 to 5/8 inch
W = Less than 7/16 inch
Aluminum and magnesium alloys
H = 2.5W or more
H = 1.75W to 2.5W
H = Less than 1.75W
H = 2 inches or less
W = Greater than 3/4 inch
H = 2 inches or less
W = 5/8 to 3/4 inch
H = 2 to 4 inches
W = 7/16 to 5/8 inch
Round contact
R = 10 inches or more
D = 20 inches or more
R = 4 to 10 inches
D = 8 to 20 inches
R = 4 inches or more
D = 8 inches or more
R = Less than 4 inches
D = Less than 8 inches

Figure 4-18. Spot weld joint design.

Table 4-16. Spot-Welding Low-Carbon Steel.

Thickness "T" of Thinnest Outside Sheet	Electrode Diameter and Shape (20°; Either or)		Minimum Contacting Overlap	Minimum Spot Spacing	Diameter of Fused Zone	Minimum Weld Strength Per Spot (For Material Below 90000 Psi Ultimate Strength
Inches	D, Inches Minimum	d, Inches Maximum	(Inches)	(Inches)	(Inches, Approx)	(Lbs per Spot)
0.016	3/8 (0.38)	1/8 (0.12)	3/8 (0.38)	1/4 (0.25)	0.10	245
0.020	3/8 (0.38)	3/16 (0.18)	7/16 (0.44)	3/8 (0.38)	0.13	325
0.032	3/8 (0.38)	3/16 (0.18)	7/16 (0.44)	1/2 (0.50)	0.16	655
0.040	1/2 (0.50)	1/4 (0.25)	1/2 (0.50)	3/4 (0.75)	0.19	945
0.050	1/2 (0.50)	1/4 (0.25)	9/16 (0.56)	7/8 (0.88)	0.22	1390
0.063	1/2 (0.50)	1/4 (0.25)	5/8 (0.62)	1 (1.00)	0.25	2000
0.080	5/8 (0.62)	5/16 (0.31)	11/16 (0.69)	1-1/4 (1.25)	0.29	2750
0.090	5/8 (0.62)	5/16 (0.31)	3/4 (0.75)	1-1/2 (1.50)	0.31	3555
0.125	7/8 (0.88)	3/8 (0.38)	7/8 (0.88)	1-3/4 (1.75)	0.33	5000

Table 4-17. Preferred Spot-Weld Dimensions for Aluminum.*

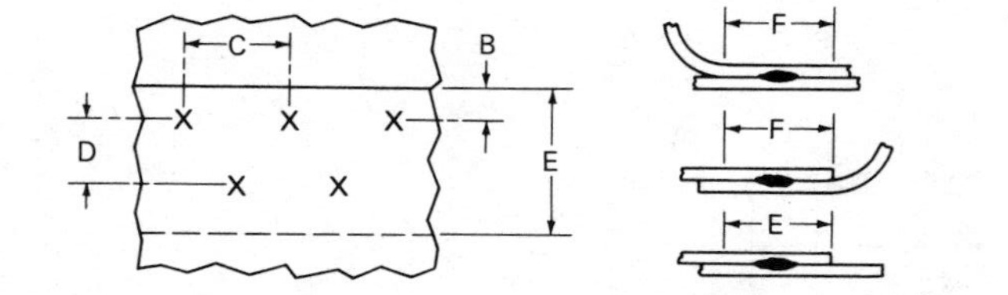

Dimension	Thickness of Pieces to be Joined									
	0.016	0.020	0.025	0.032	0.040	0.050	0.063	0.080	0.090	0.125
A Spot diameter	0.08	0.09	0.10	0.11	0.13	0.15	0.18	0.21	0.23	0.30
B Edge distance	0.19	0.19	0.22	0.25	0.25	0.31	0.38	0.38	0.44	0.50
C Spot spacing	0.31	0.38	0.38	0.38	0.44	0.50	0.50	0.63	0.62	1.00
D Distance between rows of staggered welds	0.25	0.25	0.31	0.31	0.31	0.38	0.38	0.50	0.50	0.62
E Overlap, flange width, or flat required	0.38	0.38	0.44	0.50	0.50	0.62	0.75	0.75	0.88	1.00
F Unobstructed area required to place weld (diameter)	0.44	0.56	0.56	0.69	0.69	0.69	0.69	0.69	0.94	0.94

*Based on equal thicknesses.

Table 4-18.

RESISTANCE WELDABILITY OF ALUMINUM ALLOY COMBINATIONS*

			Heat-Treatable						Non Heat-Treatable (Bore)										
Alloy and Temper		Parameter (See Legend)	Clad 7075 -T6	Bore 7075 -T6	Bore 6063 -T5 -T6	Bore 6061 -T4 -T6	Clad 2024 -T3 -T4	Bore 2024 -T3 -T4	Clad 2014 -T4 -T6	Bore 3004 -T4 -T6	5056 or 5154 -H32 -H34 -H36 -H38	5056 -0	5052 -H32 -H34 -H36 -H38	5052 -0	5050 -H32 -H34 -H36 -H38	5050 -0	3004 -H32 -H34 -H36 -H38	3304 -0	1100 -H12 -H14 -H16 -H18	1100 -0 3003 -0
Non Heat-Treatable (Bore)	1100-0 3003-0	Weldability			C	C							C	D	C	D	C	D	C	D
		Precleaning			B	B							B	B	A	A	A	A	A	A
		Corrosion			A	A							A	A	A	A	A	A	A	A
	1100, 3003 -H12, -H14, -H16, -H18	Weldability			A	A							A	C	A	A	A	C	A	
		Precleaning			A	B							B	B	B	B	A	B	A	
		Corrosion			A	A							A	A	A	A	A	A	A	
	3004-0	Weldability			C	C					C		C	D	C	D	C	D		
		Precleaning			B	B					B		B	B	B	B	A	A		
		Corrosion			A	A					A		A	A	A	A	A	A		
	3004-H32, -H34, -H36, -H38	Weldability			A	A					A		A	D	A	C	A			
		Precleaning			B	B					B		B	B	B	B	A			
		Corrosion			A	A					A		A	A	A	A	A			
	5050-0	Weldability			C	C							C	D	C	D				
		Precleaning			B	B							B	B	B	B				
		Corrosion			A	A							A	A	A	A				
	5050-H32, -H34, -H36, -H38	Weldability			A	A					A		A	C	A					
		Precleaning			B	B					B		B	B	B					
		Corrosion			A	A					A		A	A	A					
	5052-0	Weldability			C	C							C	D						
		Precleaning			B	B							B	B						
		Corrosion			A	A							A	A						
	5052-H32, -H34, -H36, -H38	Weldability	B	B	A	A	B	B	B	B	A		A							
		Precleaning	B	B	B	B	B	B	B	B	B		B							
		Corrosion	A	B	A	A	A	B	A	B	A		A							
	5056-0	Weldability			C	C					C	D								
		Precleaning			B	B					A	A								
		Corrosion			A	A					A	A								
	5056, 5154 -H32, -H34, -H36, -H38	Weldability			A	A					A									
		Precleaning			B	B					A									
		Corrosion			A	A					A									
Heat-Treatable	Bore 2014-T4, -T6	Weldability	B	B	B	B	B	B	B	B										
		Precleaning	B	B	B	B	B	B	B	B										
		Corrosion	B	B	B	B	B	B	A	B										
	Clad 2014-T4, -T6	Weldability	B	B	B	B	B	B	B											
		Precleaning	B	B	B	B	B	B	B											
		Corrosion	A	B	A	A	A	B	A											
	Bore 2024-T3, -T4	Weldability	B	B	B	B	B	B												
		Precleaning	B	B	B	B	B	B												
		Corrosion	B	B	B	B	B	B												
	Clad 2024-T3, -T4	Weldability	B	B	B	B	B													
		Precleaning	B	B	B	B	B													
		Corrosion	A	B	A	A	A													
	Bore 6061-T4, -T6	Weldability	B	B	A	B														
		Precleaning	B	B	B	B														
		Corrosion	A	B	A	A														
	Bore 6063-T5, -T6	Weldability	B	B	A															
		Precleaning	B	B	B															
		Corrosion	A	B	A															
	Bore 7075-T6	Weldability	B	B																
		Precleaning	B	B																
		Corrosion	B	B																
	Clad 7075-T6	Weldability	B																	
		Precleaning	B																	
		Corrosion	A																	

*Based on equal thicknesses

Legend

Weldability (based on equal thicknesses)
- A Good welds can be made over a wide range of machine settings.
- B Good welds but only over narrow range of machine settings.
- C Can be welded but material too soft to obtain consistent weld strength.
- D Difficult to weld, not recommended.

Precleaning
- A Easy to clean or frequently needs no cleaning.
- B Chemical or mechanical precleaning necessary to make good welds.

Resistance to corrosion
- A Corrosion resistance to weld zone equal to parent metal.
- B Corrosion resistance of weld zone not as good as parent metal.

such advantages as:

- The joining of heavy-gauge materials to thin sheets when other methods are impractical.
- Shearing and other stresses are evenly distributed.
- In dissimilar metals, the insulation effect prevents electrolytic corrosion.
- Lower weight resulting from the elimination of the added metal weight of welding and metallic fasteners.
- Smoother contours.
- Inconspicuous joints.
- Sealing of joints.
- More rapid assembly by the elimination of drilled or punched holes, jigs and fixtures, and care in mating parts.
- Elimination of stocking many sizes and types of fasteners.

Adhesives are generally classified by chemical type or physical properties:

- *Thermosetting* resins are permanently set by the application of heat and can be used over a wide range of temperatures. The thermosetting resins are also highly resistant to water and most solvents.
- *Thermoplastic* resins are affected by heat and should not be used in applications where the parts will be subjected to high temperatures. Also the thermoplastic resins have poor resistance to many solvents. The natural and synthetic rubber resins are used where a maximum of flexibility is required. Different types of rubber vary in resistance to solvents and temperatures.

The first and most important principle of adhesive-bond joint design is that joints must be specially designed for the use of adhesives. It is not an acceptable practice to design for another fastening method, alter the design slightly, and then use adhesive bonding.

An adhesive bond acts over an area, not at a single point. The joint, therefore, should be designed to minimize stress concentrations. In tension and shear, stresses are relatively uniform. In cleavage, however, stresses are concentrated on one side of the bond, while the remainder is virtually unstressed. In peel, only the edge of the bond is under load, and the remainder contributes nothing to the strength of the joint because the depth of the bond does not influence its strength in peel.

Bonded joints are rarely subjected to only a single type of stress. In the straight lap joint, which is one of the most commonly used designs, the bond is initially in shear. As the load increases, however, the lap begins to deflect, causing stresses

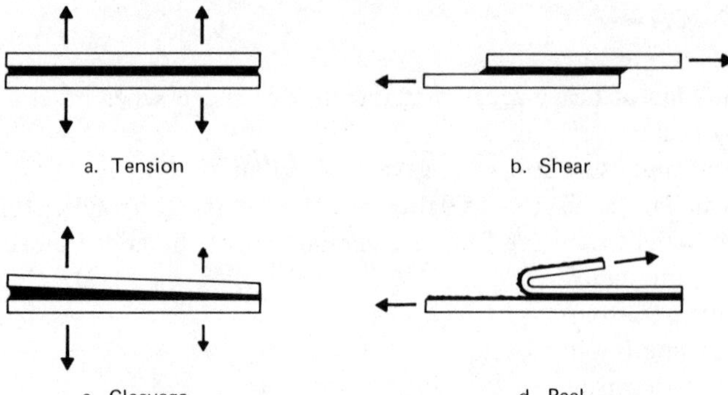

Figure 4-19. Stresses on bonded joints.

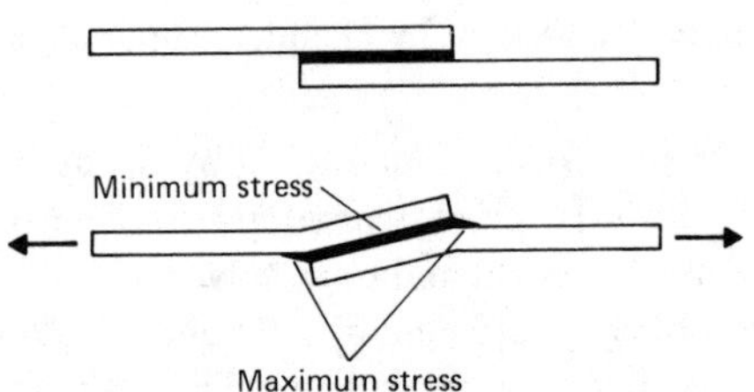

Figure 4-20. Stresses on straight lap joint.

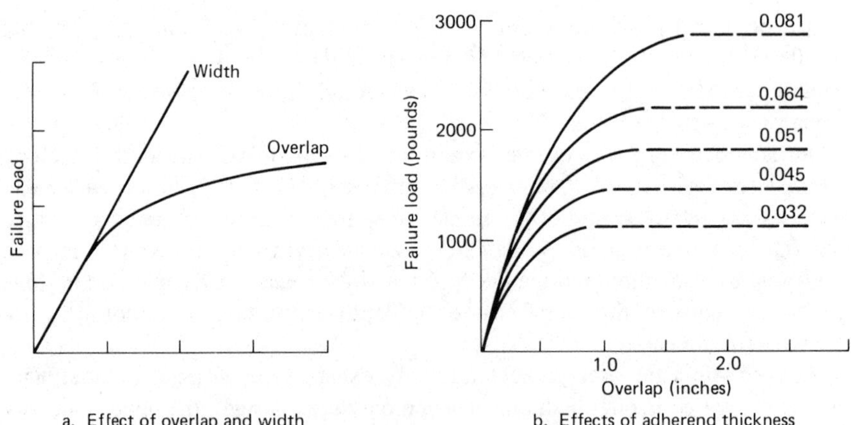

Figure 4-21. Strength analysis of straight lap joints.

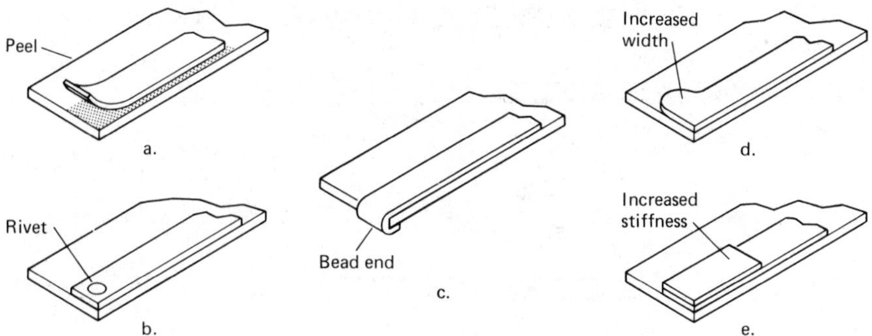

Figure 4-22. Bond designs for minimizing peel.

to concentrate at the ends of the lap. As bending continues, peel stresses are exerted on the bond, resulting in a sharp decreased in bond efficiency (Figure 4-19).

Since the adhesive bond acts over an area, the working area of the bond should

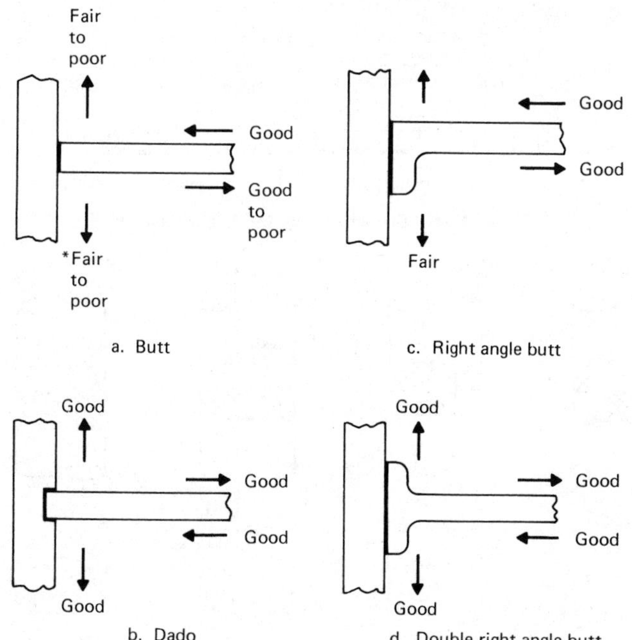

Figure 4-23. Angle joints.

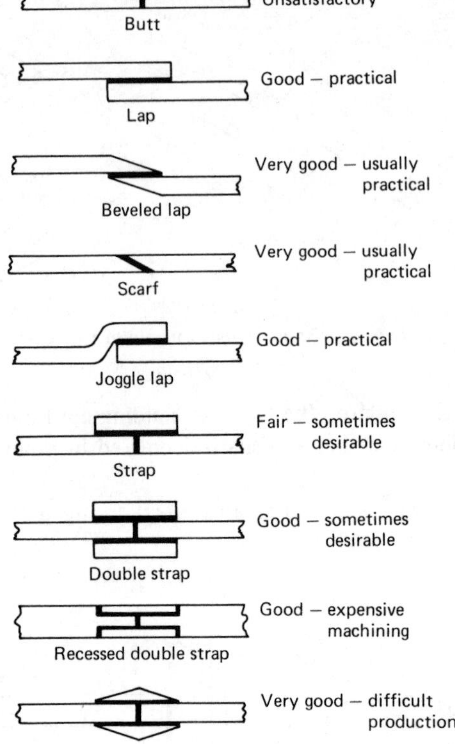

Figure 4-24. Straight joints for flat adherends.

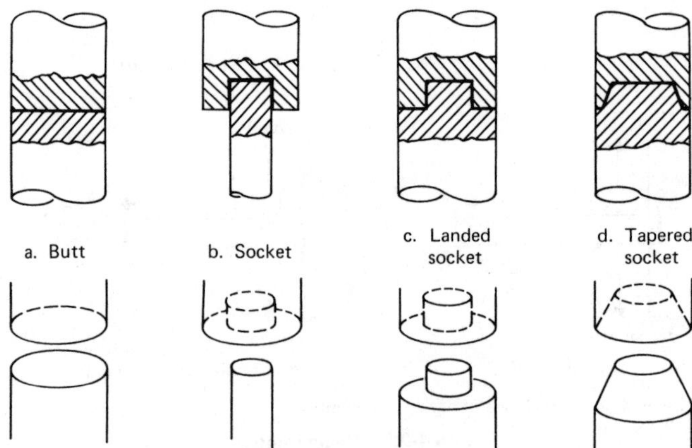

Figure 4-25. Straight joints for solid bars.

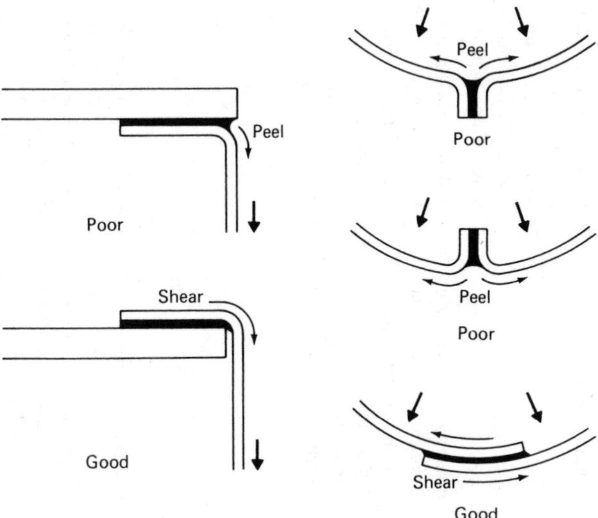

Figure 4-26. Joints for flexible materials.

be maximized. The strength of the straight lap joint, for example, is directly proportional to the width. Doubling the width, therefore, will always double the bond strength. Doubling the overlap, however, does not necessarily double the strength since, under load, peel stresses are applied to the bond, and bond depth does not influence peel strength. Therefore, increases in overlap are effective only within a certain limit, beyond which an increase in overlap results in no increase in bond strength. Thus, in a straight lap joint, a bond 10 in. deep and 1 in. wide will exhibit less strength than a bond 1 in. deep and 10 in. wide, although the area is the same (Figures 4-20 and 4-21).

Several methods for minimizing the effects of peel stresses on bonded joints are shown in Figure 4-22. Riveting may provide extra strength at the ends of the bond, but rivets often produce stress concentrations. Beading at the end of

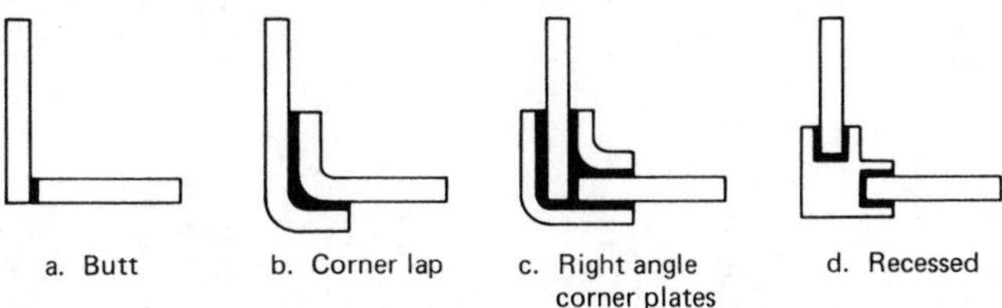

Figure 4-27. Corner joints.

Table 4-19. Properties of Synthetic Thermosetting Adhesives.

Type	Forms, Curing, Color	Strength	Application	Remarks
Allyl monomer	–	–	Optical cement and for laminating fibrous materials.	One form has superior optical qualities, good weathering resistance, and gives strong bond with glass. Partially thermoplastic. Insoluble. Water-clear.
Epoxy ethoxy	Solid or liquids; powder or stick. Solid cures at 390°F in 1 hr under slight pressure. Curing agent may or may not be included. Liquid uses separate curing agent; cures at 75° to 100°F under low pressure. Higher temperatures accelerate curing. Tan.	High tensile strength. Typical shear strength with aluminum strips, approx. 4500 psi.	Bonding wide variety of porous and nonporous materials; ceramics, glass, metal, glass cloth laminates. Bonding carbide tips to reamers.	Good adhesive qualities. Low shrinkage. Good "gap-filling" properties. Rigid.
Furane resins Furfuryl alcohol Furfuraldehyde	Cure at room temperature. Higher temperatures accelerate curing. Brown.	–	Bonding wide variety of materials. Recommended for thermosetting plastics, ceramics, and acrylics.	Unsuitable for bonding wood or other acid-vulnerable materials. Rigid.
Melamines	Curing requires little pressure.	–	Bonding imperfectly fitting parts, such as wood and plywood structures.	Higher resistance to moisture, boiling water, and acids than ureas, but more expensive.
Phenol formaldehyde	Film, powder, or liquid. Film placed between	–	Plywood gluing.	Excellent resistance to water, weathering,

	veneers, heat and pressure cured. Powder mixed with water, alcohol, acetone, or combination of all three. Brown.					chemical and bacterial action, and temperature variations. Good "gap-filling" properties. Rigid.
Resorcinol formaldehyde (resorcinol resin)	Liquid with separate curing agent. Cures at room temperature.	Adherend	Tensile psi	Shear psi	Boat-building and aircraft work.	Good resistance to deterioration. Good "gap-filling" properties. Rigid.
		Paper-phenolic laminate	820	1370		
		Birchwood	1180	1940		
		Hard rubber	1340	590		
Resorcinol-nylon	Cures at room temperature.		–		Bonding thermosetting and thermoplastic materials.	Produces relatively nonporous bond. Reduces strains caused by shrinkage or temperature differentials.
Silicones	Aromatic hydrocarbon solution. Requires 390°F and high pressure to cure. Amber.		–		Bonding silicone and other types of rubber to nonporous materials.	Highest heat resistance. Rigid.
Urea formaldehyde	Water-soluble powder, with curing agent combined or separate; syrupy water solutions with curing agents separate. Light color.		–		Replacing casein in plywood manufacture.	Less resistant to chemicals, heat, and weathering than phenolics. Film brittle and tends to craze, unless cellulose filler and thin glue line are used. Suitable up to 160°F. Rigid.

Table 4-20. Properties of Synthetic Thermoplastic Adhesives.

Type	Forms, Curing, Color	Strength	Application	Remarks
Acrylates	Colorless.	–	For nonporous materials: glass and metals.	Pliable, tough. Good aging properties.
Methacrylates	Ester, aromatic hydrocarbon, or ketone solvents. Cures by solvent release. Separate catalyst or activator added before use. Colorless.	–	Similar to acrylates.	Film is water-clear. Good weathering and aging properties. Higher softening points and tougher than acrylates. Flexible.
Polystyrene	Ester, aromatic hydrocarbon, or ketone solvents. Separate catalyst or activator added before use. Colorless.	–	Electrical and electronic parts.	High dielectric strength. Good water resistance. Rigid.
Polyvinyl chloride	Same as polyvinyl acetate, above.	–	–	Brittle film. Less adhesive than acetate. Good moisture resistance. High chemical resistance. Brittleness can be modified by plasticizers, but these impair adhesive properties.

Polyvinyl-chloride acetate	Same as polyvinyl acetate, above.	–	–	Combines chemical and moisture resistance of chloride with toughness of acetate.
Polyvinyl acetate	All vinyls can be bonded cold or hot. Cold bonding not suitable for nonporous surfaces, which tend to trap solvent and retard evaporation. Fabrication must be rapid, before solvent evaporates. Hot bonding: applied as hot melt or solution, cures in a few minutes at 225° to 250°F; can be bonded at later date by use of heat and low pressure. Colorless.	Adherend / Tensile psi / Shear psi Stainless steel / 3600 / 2960 Aluminum alloy / 3270 / 3560 Paper-phenolic laminate / 1060 / 2480 Glass / 2430 / 2310 Birchwood / 960 / 1990 Hard rubber / 400 / 630	Bonding of glass and metals.	Approximately same refractive index as glass. Good resistance to discoloration due to aging. Tough, elastic joint.
Polyvinyl alcohol		–	Bonding practically any material, porous or nonporous.	Clear, tough. Resists fats, many organic solvents. Stable and impermeable to gases. Soluble in water at 160°F.
Polyvinyl butyral		–	Bonding safety glass, and flexible mica sheets. Compounded with other resins to provide special adhesives.	Does not haze.

Table 4-21. Properties of Cellulose-Base Adhesives.

Type	Forms, Curing, Color	Strength			Application	Remarks
		Adherend	Tensile psi	Shear psi		
Cellulose acetate	Ester or ketone solvents. Cures by release of solvent. Colorless.	–			Cementing glass, metals, cloth, ceramics.	Superior to cellulose nitrate in aging properties and resistance to fire, but inferior in moisture resistance and adhesive qualities. More expensive. Flexible.
Cellulose acetate butyrate	–	–			Similar to cellulose acetate.	Greater durability and moisture resistance then cellulose acetate.
Cellulose nitrate	Ester or ketone solvents. Cures at moderate temperature by release of solvent. Colorless.	Stainless steel	2180	1580	Glass, leather, cloth, ceramics.	Care must be used in handling. Flexible. Replaced by polyvinyl resins because it discolors in sunlight and is flammable.
		Aluminum alloy	1500	1360		
		Paper-phenolic laminate	860	1680		
		Glass	1080	1680		
		Birchwood	1100	1390		
		Hard rubber	590	1000		
Ethyl cellulose	Solution or hot melt.	–			Bonding paper, cloth, metal foil.	More resistant to chemical action than other cellulose derivatives.

Table 4-22. Properties of Rubber-Base Adhesives.

Type	Forms, Curing, Color	Strength	Application	Remarks
Butadiene acrylonitrile	Solution.	–	–	Very resistant to oils and fuels. Specially compounded to resist deformation.
Butyl (GR-I)	Petroleum is best solvent.	Fair tensile strength.	–	Poor aliphatic resistance. Best resistance of rubber group to natural aging; resistance to high-temperature aging. Cold flows under low loads.
Butadiene-styrene (GR-S)	Petroleum best solvent. Available as solution or latex.	Fair tensile strength and elongation.	–	Poor aliphatic resistance. Fair resistance to natural or high-temperature aging. Lacks tack.
Natural rubber	Best solvents are aliphatic or straight chain hydrocarbons. Air drying at room temperature. Vulcanizing types may be cured under pressure and high temperature for greatest strength.	From benzene solvent, with high temperature-pressure cure: Adherend: Stainless steel; Tensile psi: 260; Shear psi: 270	Bonding cloth, felt, and rubber to metal.	Has poor aliphatic resistance. Ages rapidly in sunlight. Has good tack and tack retention. Flexible.

Table 4-22. *(Continued)*

Type	Forms, Curing, Color	Strength	Application	Remarks
		<table><tr><th>Adherend</th><th>Tensile psi</th><th>Shear psi</th></tr><tr><td>Aluminum alloy</td><td>390</td><td>250</td></tr><tr><td>Paper-phenolic laminate</td><td>160</td><td>130</td></tr><tr><td>Glass</td><td>34</td><td>43</td></tr><tr><td>Birchwood</td><td>170</td><td>160</td></tr><tr><td>Hard rubber</td><td>130</td><td>190</td></tr></table> Air-drying type is much lower in strength.		
Neoprene	Available as solution or latex. Ester or ketone solvents. Tan.	In tensile strength and elasticity, compares well with natural rubber. <table><tr><th>Adherend</th><th>Tensile psi</th><th>Shear psi</th></tr><tr><td>Stainless steel</td><td>170</td><td>90</td></tr><tr><td>Aluminum alloy</td><td>290</td><td>130</td></tr></table>	Most popular of all synthetic rubber adhesives.	Gives best resistance of all rubber group to organic solvents, sunlight, heat, oxidation, and chemicals. Continuous exposure to high temperatures liberates acid and may corrode metals. Compares well with natu-

						ral rubber in tensile strength and elasticity.
		Paper-phenolic laminate	170	250		
		Glass	90	100		
		Birchwood	340	180		
		Hard rubber	240	230		
Nitrile (acrylonitrilebutadiene, Buna N)	Available as liquid or tape. Ketone solvents. Tan.	Fair elongation. Moderately good tensile strength. Fair bond.			–	Do not weld readily to themselves and must be bonded with considerable solvent. Good aliphatic resistance. Fair aging resistance.
Polysulphide rubber (thiokol)	Chlorinated hydrocarbon solvents.	Low tensile strength and elongation.			Where bonds must be oil- and solvent-resistant.	Excellent resistance to aging. Best of all rubbers in resistance to solvents, greases, and oils. Low permeability to gases.
Reclaimed rubber	Reclaimed from discarded vulcanized rubber products. Commonly black, but red and gray available.	–			Bonding cloth, felt, and rubber to metal.	Lower in cost than natural rubber. Almost as good as natural rubber adhesives.

Table 4-23. Properties of Rubber-Resin Adhesives.

Type	Forms, Curing, Color	Strength	Application	Remarks
Liquid phenol formaldehyde plus resin.	Cures in 15 min. at 300°F, at 300 to 500 psi.	Shear strength with aluminum strips: 5000 psi.	Structural bonding, wood to aluminum alloys, stainless steel, magnesium, and chrome steel.	Not good for nickel, its alloys, lead, or copper.
Phenolic-elastomers	Film with liquid primer for metals; liquid two-component powder and liquid combination. Require high temperature and pressure for bonding.	Combines adhesive properties and strength of phenolic resins with flexibility of rubber.	Joining of stainless steel and aluminum. Good for all metals, and nonmetals not affected by curing heat and pressure.	Possesses characteristic phenolic resistance to corroding conditions.
Phenolic-synthetic rubber	Cures in 30 min. at 325°F, at 50 psi.	Shear strength with aluminum strips: 3000 psi.	Waffle-type sandwich-type constructions and attaching reinforcing stringers to sheets.	–
Rubber-resin (phenolic or nylon)	Cures at 325°F, bonds in 15 to 20 min. at 200 psi.	Shear strength with aluminum strips: 4000 psi.	Sandwich construction.	Must be protected from saltwater spray. Good oil resistance.

Table 4-24. Adhesives for Various Materials.

Material	Adhesive
Rubber	Phenol formaldehyde
	Resorcinol formaldehyde
	Epoxy
	Neoprene
	Nitrile rubber
	Phenol formaldehyde-vinyl acetal
Ceramics	Vinyl acetate
	Vinyl butyral
	Furfuryl alcohol
	Epoxy
	Polyurethane
	Neoprene
	Nitrile rubber
	Phenol formaldehyde-vinyl acetal
Acrylics	Melamine formaldehyde
	Resorcinol formaldehyde
	Nitrile rubber
	Phenol formaldehyde-nitrile rubber
Vinyls	Vinyl acetate
	Neoprene
	Phenol formaldehyde-nitrile rubber
	Resorcinol polyamide
Vinylidene chloride	Vinyl acetate
	Neoprene
Polyester	Silicone
	Epoxy
	Nitrile rubber
Polyamide	Resorcinol formaldehyde
	Neoprene
	Nitrile rubber
	Resorcinol polyamide
Polystyrene	Methyl methacrylate
	Styrene
	Alkyd
	Nitrile rubber
Thermosets: phenolics, ureas melamines	Phenol formaldehyde
	Urea formaldehyde
	Resorcinol formaldehyde
	Furfuryl alcohol
	Silicone
	Epoxy
	Neoprene
	Nitrile rubber

Table 4-24. *(Continued)*

Material	Adhesive
	Phenol formaldehyde-nitrile rubber
	Phenol formaldehyde-vinyl acetal
	Phenol formaldehyde-vinyl butyral
Steel	Vinyl acetate
	Phenol formaldehyde
	Epoxy
	Polyurethane
	Reclaimed rubber
	Neoprene
	Phenol formaldehyde-nitrile rubber
	Phenol formaldehyde-vinyl acetal
	Phenol formaldehyde-vinyl butyral
Aluminum	Vinyl acetate
	Polyisobutylene
	Phenol formaldehyde
	Epoxy
	Polyurethane
	Reclaimed rubber
	Neoprene
	Phenol formaldehyde-nitrile rubber
	Phenol formaldehyde-vinyl acetal
	Phenol formaldehyde-vinyl butyral
Magnesium	Silicone
	Epoxy
	Polyurethane
	Phenol formaldehyde-nitrile rubber
	Phenol formaldehyde-vinyl acetal
Zinc	Neoprene
	Nitrile rubber
	Phenol formaldehyde-nitrile rubber
	Phenol formaldehyde-vinyl butyral
Copper, brass, bronze	Vinyl acetate
	Silicone
	Epoxy
	Phenol formaldehyde-nitrile rubber
	Phenol formaldehyde-vinyl butyral
Nickel	Epoxy
	Neoprene
	Nitrile rubber
Tin	Epoxy
	Nitrile rubber
Silicones	Silicone
	Epoxy
Fluorocarbons	Silicone
	Epoxy

Table 4-24. *(Continued)*

Material	Adhesive
Leather	Vinyl acetate Vinyl butyral Resorcinol formaldehyde Epoxy Neoprene Phenol formaldehyde-nitrile rubber Phenol formaldehyde-vinyl acetal Resorcinol polyamide
Paper, textiles, felt, cork	Cellulose acetate Methyl methacrylate Vinyl acetate Vinyl butyral Phenol formaldehyde Urea formaldehyde Melamine formaldehyde Resorcinol formaldehyde Epoxy Neoprene Phenol formaldehyde-nitrile rubber Phenol formaldehyde-vinyl acetal Resorcinol polyamide
Cellulose derivatives (nylon-rayon)	Cellulose nitrate Cellulose acetate Resorcinol formaldehyde Polyurethane Nitrile rubber
Wood	Vinyl acetate Phenol formaldehyde Urea formaldehyde Melamine formaldehyde Resorcinol formaldehyde Furfuryl alcohol Neoprene Phenol formaldehyde-nitrile rubber Phenol formaldehyde-vinyl acetal
Teflon (polytetrafluoroethylene	Epoxy Silicone Phenol formaldehyde synthetic rubbers

Table 4-25. Adhesives for Various Service Conditions.

Service Condition	Adhesives
Static loads	Phenol formaldehyde Urea formaldehyde Melamine formaldehyde Resorcinol formaldehyde Furfuryl alcohol Alkyd Silicone Epoxy Phenol formaldehyde-nitrile rubber Phenol formaldehyde-vinyl acetal Phenol formaldehyde-vinyl butyral Resorcinol polyamide.
Impact or vibration	Cellulose nitrate Cellulose acetate Methyl methacrylate Vinyl acetate Vinyl butyral Polyisobutylene Natural rubber Reclaimed rubber Neoprene Nitrile rubber Phenol formaldehyde-nitrile rubber Phenol formaldehyde-vinyl acetal Phenol formaldehyde-vinyl butyral Resorcinol polyamide
Presence of water	Vinyl acetate Styrene Polyisobutylene Phenol formaldehyde Melamine formaldehyde Resorcinol formaldehyde Silicone Epoxy Neoprene Phenol formaldehyde-nitrile rubber Phenol formaldehyde-vinyl acetal Phenol formaldehyde-vinyl butyral Resorcinol polyamide
Presence of solvent	Phenol formaldehyde-nitrile rubber Phenol-formaldehyde

Table 4-25. *(Continued)*

Service Condition	Adhesives
Presence of solvent (*Continued*)	Urea formaldehyde
	Melamine formaldehyde
	Resorcinol formaldehyde
	Furfuryl alcohol
	Epoxy
	Polyurethane
	Phenol formaldehyde-nitrile rubber
	Phenol formaldehyde-vinyl acetal
	Phenol formaldehyde-vinyl butyral
	Resorcinol butyral
High temperature	Phenol formaldehyde
	Melamine formaldehyde
	Resorcinol formaldehyde
	Furfuryl alcohol
	Silicone
	Epoxy
	Phenol formaldehyde-nitrile rubber
	Phenol formaldehyde-vinyl acetal
	Phenol formaldehyde-vinyl butryal
	Resorcinol polyamide
Low temperature	Phenol formaldehyde
	Urea formaldehyde
	Melamine formaldehyde
	Resorcinol formaldehyde
	Silicone
	Epoxy
	Phenol formaldehyde-nitrile rubber
Differential expansion	Cellulose nitrate
	Cellulose acetate
	Methyl methacrylate
	Vinyl acetate
	Vinyl butyral
	Natural rubber
	Reclaimed rubber
	Neoprene
	Nitrile rubber
	Phenol formaldehyde-nitrile rubber
	Phenol formaldehyde-vinyl acetal
	Phenol formaldehyde-vinyl butyral
	Resorcinol polyamide

Table 4-26. Adhesives for Rubber.

				Resistance to					
Type	Strength	Static Load	Tack	Water	Oil	Gasoline	Heat	Cold	Aging
Nitrile rubber	E	F	P–G	E	E	E	G	G	G
Butyl rubber	F	P	G	E	P	P	P	G	E
GR-S	F	P	F	E	P	P	F	G	F
Natural rubber	G	P	E	E	P	P	P	G	F
Neoprene	E	G	P–G	E	G	G	G	G	G–E
Reclaimed rubber	G	P	G	E	P	P	P	G	F
Thiokol	F	P	F	E	E	E	F	E	E
Cyclized rubber	G	F	G	E	P	P	F	M	G
Chlorinated rubber	G	F	G	E	G	G	M	M	G

Code: E, excellent; G, good; M, moderate; F, fair; P, poor.

the joint is satisfactory, but not always compatible with other design criteria. Increasing the width of the end of the bond will increase its peel strength. Increasing the stiffness of the adherends is often effective since the stiffer adherends result in a smaller deflection of the joint for a given force and, consequently, reduce the peel stresses. One method for increasing stiffness is to increase adherend thickness.

Unfortunately, no adequate method for calculating stresses present in bonded joints has yet been developed and proven. Therefore, empirical strength data for various bonding surface dimensions and adhesive film thicknesses should be accumulated before the final design is prepared.

Designing joints for adhesive bonding is not always simple. Careful planning is essential if the engineer is to achieve the best possible design. Such planning consists of four basic steps:

1. The design should be selected based on the stresses and service conditions that the joint is expected to encounter. Whenever possible, the design shall be such that the bond is stressed in shear, tension, or compression, minimizing peel and cleavage (Figures 4-23 through 4-27).

Table 4-27. Adhesives for Porous Materials.

Adhesive Type	Supplied as	Mix with	Pot life (hr)	Sets at °F	Resistance to Water	Weather	Fungus	Heat	Solvents	Suitable for
Melamine	Powder	Water	24 to 48	240 to 280	E	G	E	E	E	Exterior
Phenolic, acid catalyst	Liquid	Hardener	1 to 6	70 to 210	E	E	E	E	.E	Exterior
Phenolic, hot set	Liquid Film Powder	– – Water	Indefinite – Indefinite	250 to 300	E	E	E	E	E	Exterior
Resorcinol	Liquid	Hardener	1 to 6	70 to 210	E	E	E	E	E	Exterior
Urea	Powder or Liquid	Water or Hardener	1 to 24	70 to 210	G	M	E	M	E	Interior
Vinyl acetate	Liquid	–	Indefinite	70	F–M	P	G	P	F	Interior

Code: E, excellent; G, good; M, moderate; F, fair; P, poor.

Table 4-28. Adhesives for Structural Purposes.

Adhesive	Supplied as	Mix with	Requires Drying	Curing Temperature (°F)	Resistance to					
					Water	Oil	Gasoline	Glycol	Heat	Cold
Epoxy	Liquid	Hardener	No	70 to 250	F–G	E	E	G	F	G
	Powder	—	No	250 to 500	F–G	E	E	G	G	G
	Rod	—	No	250 to 500	F–G	E	E	G	G	G
Polyester	Liquid	Hardener	No	70 to 220	G	E	E	E	G	M
Phenolic-vinyl	Liquid	—	Yes	240 to 500	E	E	E	E	G	G
	Film	—	No	240 to 500						
Phenolic-nitrile rubber	Liquid	—	Yes	325 to 500	E	E	E	E	E	P–G
	Film	—	No	325 to 500						
Phenolic-neoprene	Liquid	—	Yes	325 to 500	E	E	E	E	G	E
	Film	—	No	325 to 500						
Phenolic-nylon	Liquid	—	Yes	325 to 500	G	E	E	G	F–M	G

Code: E, excellent; G, good; M, moderate; F, fair; P, poor.

2. A rough estimate should be made of the strength required of the adhesive on the basis of the available bond area, predicted load on the joint, and the required safety factor. (This should never be used as the final design figure since the relationship between bond strength and bond area is rarely linear.)
3. The proper adhesive should be selected (Tables 4-19 through 4-28).
4. Test specimens should be prepared to determine the optimum bond dimensions and adhesive film thickness. These specimens should simulate the final assembly as closely as possible with respect to adherends, bonding technique, and type of stress.

5

MECHANICAL FASTENERS

Reliability is one of the prime requirements in the design of electronics equipment. Reliability is the ability of a product to perform as designed for the specified life of the equipment.

Since mechanical fasteners are a major component in every design, it is extremely important to select the right fastener in order to insure maximum reliability. Fortunately, the designer's choice of the right parts, materials, and finishes is simplified due to the wide range of hardware available. Basically, the selection of the right parts is dependent on the proper materials with the right physical properties for the mechanical and environmental applications.

Packaging engineers must have a knowledge of the properties of the available materials at their fingertips. The following section reviews the physical, mechanical, and environmental properties that must be considered in selecting the proper hardware for electronics packaging.

Strength. The strength characteristics of fastener materials are important since they must satisfy the conditions of load, vibration, and shock.

Size. Not every type of fastener is readily available in a full range of sizes. Size availability may require a review in the strength due to any change in size.

Wear and abrasion. The harder the material, the better the wear resistance. Methods such as case hardening can be used to minimize costs. This is especially important when fasteners are to be used on moving assemblies.

Weight. In many instaces, weight is a major factor. Aluminum and titanium offer good weight-to-strength ratios. Plastics, when properly used, can also provide substantial weight savings.

Electrical conductivity. Materials can be selected to provide both good electrical conductivity or good insulation properties. For example, copper and aluminum offer exceptional conductivity, whereas plastics provide the best electrical insulation.

Thermal conductivity. The rate of heat transfer (thermal control) depends primarily on the materials selected. When it is desired to obtain the maximum in heat transfer, copper or aluminum should be used.

Thermal expansion. Fasteners used in applications where different metals are joined must be carefully selected to minimize or prevent mechanical misalignment or tearing caused by various rates of expansion.

Appearance. Good appearance is often an advantage in selling a product. There are many different decorative finishes available for commercially available fasteners. Design engineers should have knowledge of the various finishes and configurations available to them.

Corrosion resistance. The selection of materials and finishes is important in preventing or minimizing corrosion in a product. The fundamentals involved in corrosion reactions are important. The following considerations should be taken:

- Select fastener materials that will most likely resist the corrosive environment to which they will be subjected.
- Use metal combinations that are close together in the galvanic series of dissimilar metals (Table 5-1). Avoid combinations that are farther apart in the galvanic series where the less noble material is relatively small.
- Paint, chemically coat, or insulate metals.
- Never use dissimilar metals without insulation, and never use dissimilar metals in threaded connections.

The nonmetallic fasteners have many advantages of the metals. For example:

- Corrosion resistance. Special coatings and finishes are not required.
- Colorable. Color coding and color matching are simply achieved.
- Good thermal and electrical insulation.
- Lightweight.
- Strength. For example, nylon has a higher strength-to-weight ratio than steel.
- Bearing qualities. Plastic hardware is used in moving assemblies where the screw heads act as a glide.
- Heat resistance.

Nylon offers good insulating properties plus resistance to shock, vibration, and chemical solvents. Nylon also possesses lightness, elasticity, and high torque strength. Care should be taken when using nylon in applications where exposure to direct sunlight will occur, since it has a tendency to oxidize and embrittle. Polycarbonate is recommended for high-impact applications. Both nails and rivets have been manufactured from polycarbonate.

Table 5-1. Galvanic Series of Dissimilar Metals.

(Least Noble)
Corroded End-Anode
Magnesium
Magnesium Alloys
Zinc
Aluminum 2002S
Cadmium
Aluminum 2021S-T4
Steel or Iron
Cast Iron
Chromium-Iron (Active)
Ni-Resist
Type 304 Stainless (active)
Type 316 Stainless (active)
Lead-Tin Solders
Lead
Tin
Nickel (active)
Iconel (active)
Brasses
Copper
Bronzes
Copper-Nickel Alloys
Monel
Silver Solder
Nickel (passive)
Iconel (passive)
Chromium-Iron (passive)
Type 304 Stainless (passive)
Type 316 Stainless (passive)
Silver
Graphite
Gold
Platinum
(Most Noble)
Protected End–Cathode

FINISHES AND COATINGS

Coatings and finishes are applied to fasteners to improve their corrosion resistance and to enhance their appearance. Frequently, a standard part may be selected; by using the proper finish, the desired properties may be obtained at a lower cost than if special materials are specified. Many decorative finishes, however, provide little or no protection from severely corrosive environments. Table 5-2 gives the properties of frequently used finishes and coatings. In gen-

Table 5.2. Fastener Finishes and Coatings.

Coating or Finish For Fasteners	Used On	Corrosion Resistance	Characteristics
Anodizing	Aluminum	Excellent	Frosty-etched appearance.
Black oxide, blued	Steel	Indoor satisfactory, outdoor poor.	Black, can be waxes or oiled.
Black chromate	Zinc-plated or cadmium-plated steel	Added corrosion protection on plated surfaces.	Black, semilustrous. Used for decorative outdoor purposes. Can be lacquered.
Blueing	Steel	Indoor satisfactory, outdoor poor.	Decorative use. Blue to black, can be waxed or oiled.
Brass plate, lacquered	Steel, usually	Fair	Brass finish which is lacquered. For indoor decorative use.
Bronze plate, lacquered	Steel, usually	Fair	Color similar to 80% copper, 20% zinc alloy. Lacquered finish. Recommended for indoor decorative use.
Cadmium plate	Most metals	Excellent	Bright silver-gray, dull gray, or black finish. For decoration and corrosion protection (especially applications).
Clear chromate finish	Cadmium and zinc-plated parts	Very good to excellent	Clear bright or iridescent chemical coating for added corrosion protection, coloring, and paint bonding. Colored coatings usually provide greater corrosion resistance than the clear.
Dichromate	Cadmium and zinc-plated parts	Very good to excellent	Yellow, brown, green, or iridescent colored coating same as clear chromate.
Olive drab, gold, or bronze chromate	Cadmium and zinc-plated parts	Very good to excellent	Green, gold, or bronze tones same as clear chromate.
Chromium plate	Most metals	Good (improves with increased copper and nickel undercoats)	Bright blue-white, lustrous finish. Relatively hard surface used for decorative purposes.

Table 5.2. *(Continued)*

Coating or Finish For Fasteners	Used On	Corrosion Resistance	Characteristics
Copper plate	Most metals	Fair	Used for nickel and chromium-plate undercoat. Can be blackened and relieved to obtain Antique Statuary, and Venetian finishes.
Copper, brass, bronze, miscellaneous finishes[a]	Most metals	Indoor, very good	Decorative finishes. Applied to copper, brass, bronze-plated parts to match colors. Tones vary from black to almost the original color. Finish names are: Antique, Black Oxide, Statuary, Old English, Venetian, Copper Oxidized.
Lacquering, clear or color-matched	All metals	Improves corrosion resistance. Some types suitable for humid or other severe applications.	Decorative finishes. Clear or colored to match mating color or luster.
Lead-tin	Steel, usually	Fair to good	Silver-gray, dull coating. Gives good lubrication to tapping screws.
Bright nickel	Most metals	Indoor excellent. Outdoor good if thickness is at least 0.0005 in.	Silver finish used for applicances, hardware.
Dull nickel	Most metals	Same as bright nickel	Whitish cast.
Passivating	Stainless steel	Excellent	Removes iron partciels and produces a passive surface.
Phosphate Bearing Surfaces, Army 57-0-2, Type II, Class A	Steel	Good	Antichafing properties used on sliding or bearing surfaces.
Phosphate Rust Preventive, Army 57-0-2, Type II, Class B	Steel	Fair to good	Rustproofs steel. Plain grayish surface. Rust-preventive oils can be applied over it. Can be dyed black.

Phosphate Paint-base Preparations, Army 57-0-2, Type II, Class C	Steel, aluminum, zinc plate	Good, after paint or lacquer applications	Plain gray. Prepares steel, aluminum, and zinc-plated parts for painting or lacquering increases bond between metal and coating.
Colored phosphate coatings	Steel	Superior to regular phosphated and oiled surfaces	Increases corrosion resistance. Available in green, red, purple, blue, black, etc.
Rust preventives	All metals	Varies with function of oil	Various colors. Usually applied to phosphate and black oxide finishes to protect parts in transit or prolonged storage.
Silver plate	All metals	Excellent	Decorative, expensive. Excellent electrical conductor.
Electroplated tin	All metals	Excellent	Silver-gray color. Excellent corrosion protec- in contact with food.
Hot-dip tin	All metals	Excellent	Silver-gray. Thickness hard to control, especially on fine-thread parts.
Electroplated zinc	All metals	Very good	Bright-blue-white gray coating.
Electrogalvanized zinc	All metals	Very good	Dull grayish color used where bright appearance is not wanted.
Hot-dip zinc	All metals	Very good	For maximum corrosion protection. Dull grayish color. Use where coating thickness not important.
Hot-dip aluminum	Steel	Very good	For maximum corrosion protection. Dull grayish color. Use where coating thickness not important.

[a] All of these finishes require a coating of lacquer to prevent color change.

eral, finishes and coatings should be used only in mildly corrosive applications. The following criteria should be considered in deciding where to use finishes and coatings.

- Degree of corrosion resistance required
- Physical properties required of the base material
- Cost
- Finish process availability
- Intended use of the finish, decorative or protective
- Type of fastener used

THREADED FASTENERS

Despite the advances in welding and adhesives, threaded fasteners are the basic method of assembly. A *bolt* is defined as a threaded fastener intended to be mated with a nut. A *screw* is a threaded fastener that engages into performed or self-made internal threads. Following are various types of commercially available threaded fasteners.

Common Standard Bolt Styles

Square. A square bolt is supplied in two strength grades and in popular sizes from 0.25 to 1.5.

Bent. A bent bolt is a threaded rod with an end formed to meet special requirements such as an eye or right-angle bend.

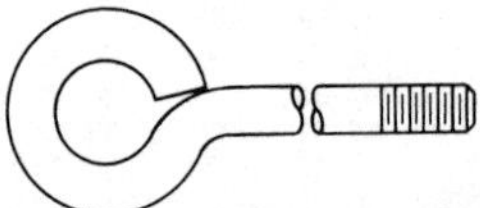

Elevator. The large-diameter flathead elevator bolt provides a nearly flush surface and large bearing area for use in softer materials. Its square neck prevents rotation.

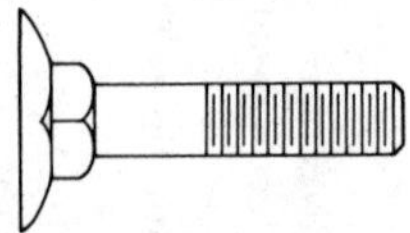

Round. A round bolt has a smooth, attractive external appearance and is tightened by torquing the mated nut.

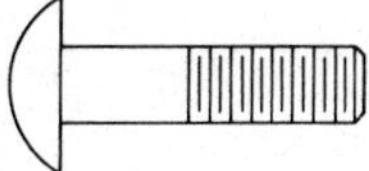

Hex. A hex bolt is the most commonly used standard fastener. It is supplied in three basic strength grades and in popular shank diameters ranging from 0.25 to 4 in. The hex head design offers greater strength, easier torque input, and larger area for manufacturer's identification than does the square head.

Countersunk. A countersunk bolt is used for flush mounting of assemblies. When supplied with a machine screw, it is called a *stove bolt*.

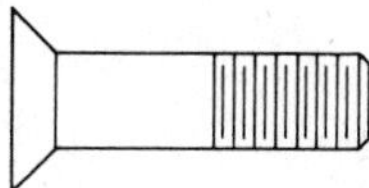

Aircraft. Aircraft bolt is becoming a generic term for any high-strength fastener, but officially it is any bolt that conforms to the Air Force-Navy Aeronautics Standard Group and National Aerospace Standards Committee specifications.

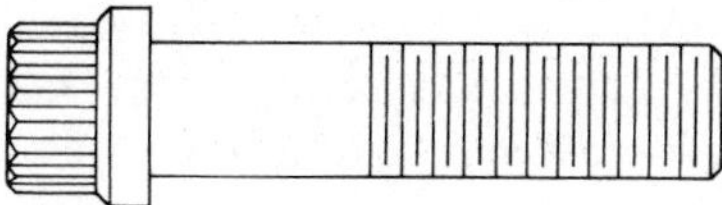

Carriage. A carriage bolt, normally made with a round head for an attractive external appearance, has ribs or flats on the shank to prevent turning when the bolt is tightened. Some versions require a prepunched square hole; others are pressed into place.

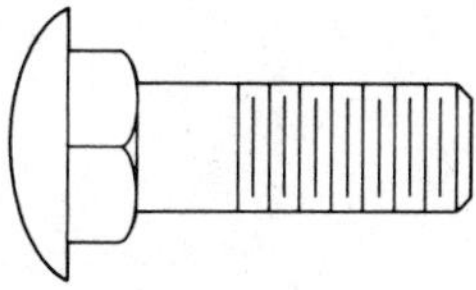

Plow. Usually made for flush mounting, the plow bolt has a square countersunk head, sometimes also including a key, to prevent rotation.

Track. A track bolt is one of the family of bolts designed specifically for use with railroad tracks. This version has an elliptical head to prevent rotation.

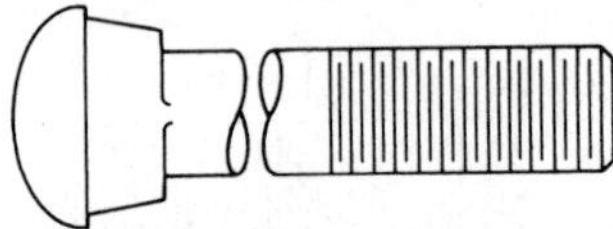

Drive-head bolts. Several head configurations other than hex, square, and slotted exist. When selecting a drive configuration, fastener size, materials, torque requirements, economics, and frequency of assembly and disassembly should be considered. Common drive-head designs include:

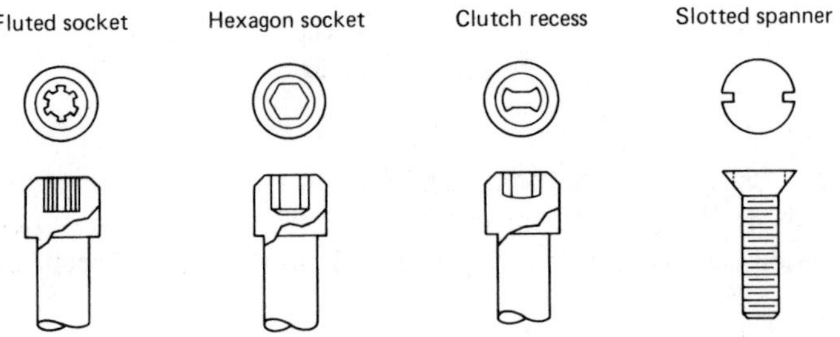

Tamperproof heads. A number of special-purpose fastener heads are available for making a product theft-proof and tamperproof. Most require special-purpose tools.

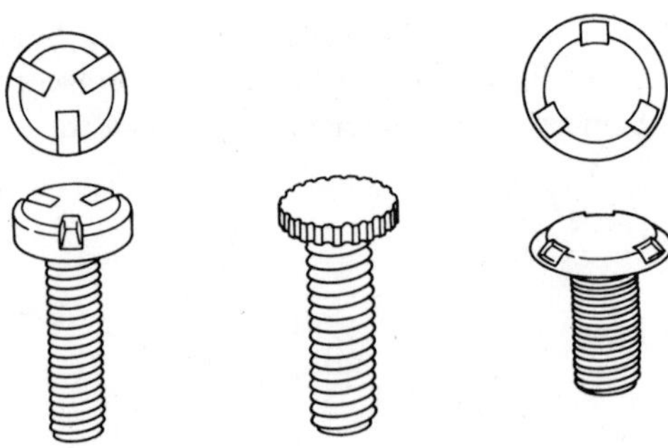

Thread Types

Thread configurations are covered by a number of standard specifications, most notable of which is ANSI 1974 Unified Inch Screw Threads (UN and UNR thread form). The long-term trend toward simplification of fastener standards has led to the use of Unified in most applications. Ultimately, after an extensive transition period, these too will be phased out as metric standards become more prevalent.

Head Styles

Which head configuration to specify depends on the type of driving equipment used (e.g., screw-driver, socket wrench), the type of joint load, and the external appearance desired. The head styles shown herein can be used for both bolts and screws; however, they are most commonly identified with the "machine screw" or "cap screw" fastener category.

Hex and square. Hex and square heads have the same advantages as the corresponding bolts.

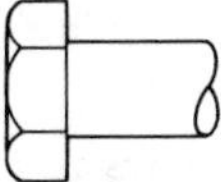

Binding. A binding head is commonly used in electrical connections because the undercut prevents fraying of stranded wire.

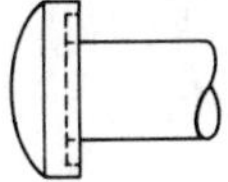

Washer. Also called flanged head, a washer head eliminates the need for a separate assembly step when a washer is required, increases the bearing areas of the head, and protects the material finish during assembly.

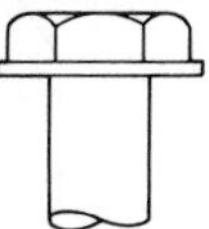

Oval. The oval head, similar to a flat head, is sometimes preferred because of its neat appearance.

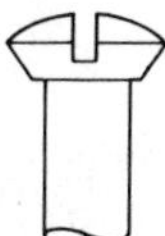

Fillister. The fillister's deep slot and small head allow a high torque to be applied during assembly.

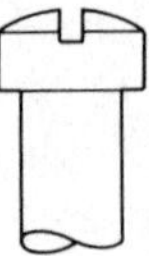

Truss. The truss head covers a large area. It is used where extra holding power is needed, holes are oversize, or material is soft.

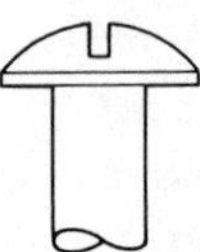

Pan. The pan head combines the qualities of the truss, binding, and round heads.

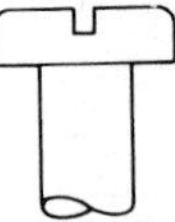

12-point. The twelve-sided 12-point head is normally used on aircraft grade fasteners. Multiple sides allow sure grip and high torque during assembly.

Flat. The flat head is available with various head angles; the fastener centers well and provides a flush surface.

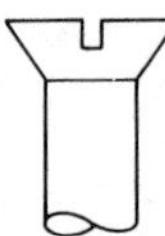

Tapping Screw Guide

Thread-Forming Screws

Type AB. For sheet metal up to 18 gauge, resin-impregnated plywood, wood, and asbestos compositions. Used in pierced or punched holes where a sharp point for starting is needed. Joint strength of easily deformed materials can be increased with pilot holes less than root diameter of screw. Fast driving.

Type B. For heavy-gauge sheet metal and nonferrous castings. Used in assembling easily deformed materials where pilot hole is larger than root diameter of screw. Fast driving.

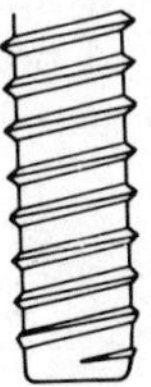

Type BP. Used for locating and aligning holes, or piercing soft materials.

Type C. Used where a machine screw thread is preferable to the spaced-thread form. Makes a chip-free assembly. May require extremely high driving torques due to long thread engagement. Resists loosening by vibration. Provides tighter clamping action than that of Type B for equivalent driving torques.

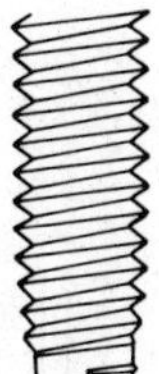

Type U. For permanent fastenings in metals and plastics. Should not be used in materials less than one screw-diameter thick.

Thread-Cutting Screws

Type D. Requires less driving torque than Type C and has longer length of thread engagement. Good for low-strength metals and plastics, high-strength brittle metals, and rethreading clogged pretapped holes. Easy starting. Gives highest clamping force for a given torque of any tapping screw.

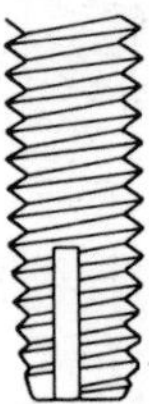

Type F. Used in a wide range of materials. Fast driving. Resists vibration.

Type G. Recommended for same general use as Type C, but requires less driving torque. Good for low-strength metals.

Type T. Same as Type D but has more chip clearance and cuts easier.

Type BF. Cutting grooves remove only a small part of material, thus mtaintaining maximum shear strength in threaded hole wall. Wall thickness should be 1.5 times the major diameter of the screw. Reduces stripping in brittle plastics and die castings. Good for long thread engagement, especially in blind holes. Faster driving than fine thread types.

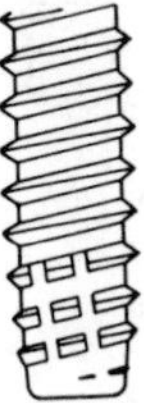

Type BT. Similar to Type BF except for single wide flute, which provides room for twisted, curly chips so that binding or reaming of hole is avoided.

Thread Rolling Screws

Type SF. Four-point contact provides straight driving and permits low driving torque. Recommended for thin and heavy-gauge materials.

Type SW. Recommended for sheet metal, structural steel, zinc, and aluminum castings, and steel, brass, and bronze forgings.

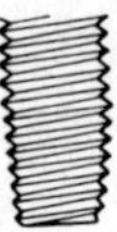

Type TT. Trilobular cross section, providing slight radial relief for its full length. Recommended for heavy materials and structural applications to give deep engagement with low driving torque.

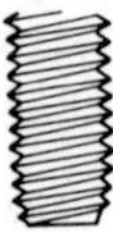

Standard Points

Cup. Most widely used where cutting-in action of point is not objectionable. Heat-treated screws of Rockwell C 45 hardness or greater can be used on shafts with surface hardness up to Rockwell C 35 without deforming the point.

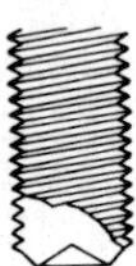

Half dog. Normally applied where permanent location of one part in relation to another is desired. It is spotted in a shaft hole. Drilled hole must match the point diameter to prevent side play. Recommended for use with hardened members and on hollow tubing, provided some locking device holds the screw in place.

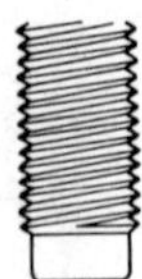

Cone. Used for permanent location of parts. It develops greatest axial and torsional holding power when bearing against material of Rockwell C 15 hardness or greater. Usually spotted in a hole to half its length.

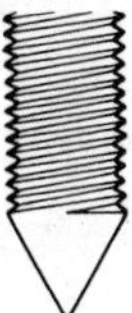

Flat. Used when frequent resetting of a machine part is required. Particularly suited for use against hardened steel shafts. Can also be used as adjusting screws for fine linear adjustments. A flat is usually ground on the shaft for better contact. This point is preferred where walls are thin or threaded member is a soft metal.

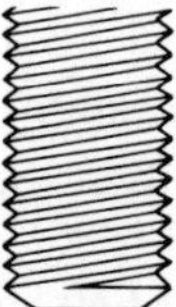

Oval. Used when frequent adjustment is necessary or for seating against angular surfaces. In some applications, shaft is spotted to receive the point. It has the lowest axial or torsional holding power.

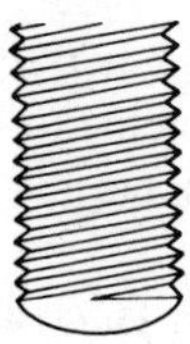

CHEMICAL THREAD LOCKING

Epoxy and anaerobic adhesives are the most commonly used thread-locking compounds. Anaerobics are sold in liquid form and are applied directly to the threads just prior to assembly and cure-in-place. Both anaerobic and epoxy adhesives are available as preapplied adhesives, which are applied to the fastener by the manufacturer. In specifying the thread-locking adhesive, consideration must be given to matching the strength of the adhesive to the application. The packaging engineer must be familiar with:

- Self-life of the materials
- Sealing of effectiveness in the presence of chemicals
- Effect of the fastener finish on the adhesive holding strength
- Reusability of the fastener
- Performance advantages of mechanical locking devices

NUTS

Nuts are internally threaded fastener elements designed to mate with a bolt. Hex and square nuts, also referred to as full nuts, are the most commonly used. Square nuts are normally used for lighter duty than hex nuts. Flanged nuts are available with integral washers and may be used to bridge oversize holes. In many assembly and loading conditions, the standard full nuts cannot maintain a reliably tight joint. There are several standard methods of preventing nuts from working loose.

- Peening the end of the bolt extending through the nut
- Staking or deforming the threads of the nut
- Using lockwashers
- Providing special threads on the bolt and nut
- Using a chemical locking compound, doping the threads

Locknuts should be considered if the joint will be subjected to:

- Vibration or cyclic motions
- An added safety factor for unknown service conditions
- Accurate positioning and adjustment—for example, spacer applications and when parts must be free to rotate.

Several nut types use nonmetallic (such as nylon) or soft metal plugs, strips, or collars, which are retained in the nut. These locking elements are plastically deformed by the bolt threads, which produces an interference-fit action to resist rotation on the bolt threads.

Jam Nuts

A standardized fastener, the jam nut (Figure 5-1) is a thin nut normally used under a full nut to develop locking action. It is recommended that the jam nut be torqued to seat only, and then a full nut assembled on top of jam nut. The same effect can also be achieved with two nuts if preload must be developed when the first nut is tightened into position.

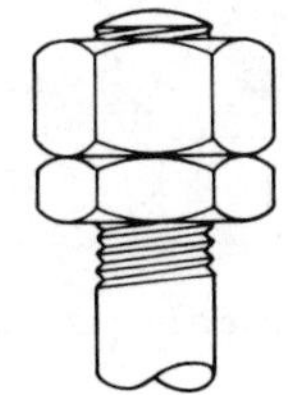

Figure 5-1. Jam nuts.

Castle and Slotted Nuts

Both castle and slotted nuts (Figure 5-2) are slotted to receive a cotter pin or wire that passes through a drilled hole in the bolt and serves as the locking member. These nuts are used primarily as safety nuts.

Prevailing-Torque Locknuts

The prevailing-torque nut has a built-in locking feature that develops full locking action as soon as it is engaged with the bolt threads and then must be wrenched to final seated position (Figure 5-3). A common design for the nut is a noncircular shape that grips the bolt threads to produce a tight friction clamp. In other designs, the upper portion of the nut is slotted and pressed inward. These locking fingers provide a frictional grip on the bolt thread.

Ideally, prevailing-torque nuts should be used with the minimum thread engagement necessary to develop the holding strength of the nut and the locking action. Torquing these nuts over a long thread travel under loaded conditions could damage the locking feature. Prevailing-torque locknuts can be used as spacer nuts or stepnuts where components must be free to rotate without end play.

Captive Nuts

Captive or self-retained nuts (Figure 5-4) provide a permanent, strong fastener when used on thin materials and especially in blind locations. Captive nuts can be attached in a variety of ways, and before selecting a fastener, designers should consider what assembly tools will be needed.

There are four general categories of captive nuts.

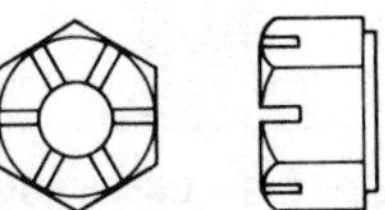

Figure 5-2. Castle and slotted nuts.

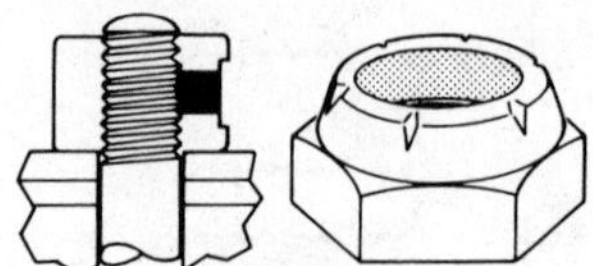

Figure 5-3. Prevailing-torque locknuts.

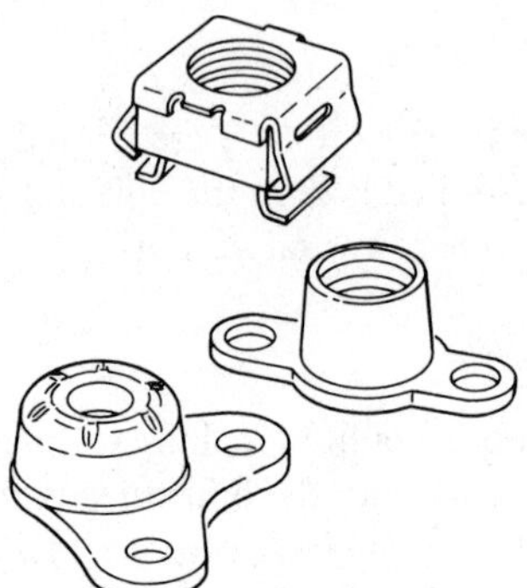

Figure 5-4. Captive nuts.

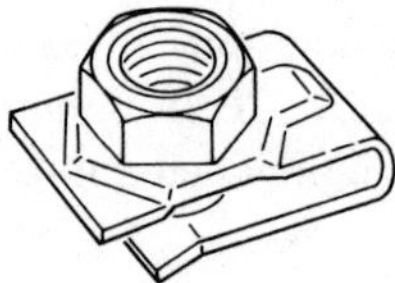

Figure 5-5. Nut retainer design for a panel edges.

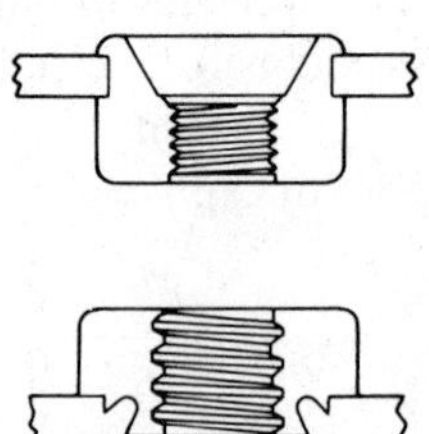

Figure 5-6. Clinch nuts.

Plate or Anchor Nuts

Plate or anchor nuts have mounting lugs, which are riveted, welded, or screwed to the part. Plate nuts provide permanent attachment for threaded fasteners at inaccessible or blind locations. They assure positive positioning of the mating bolt and are self-wrenching. Plate nuts are the preferred nut type of stressed-skin applications because they do not introduce additional stresses around the bolt hole. Riveting is the most common method of attaching plate nuts.

Caged Nuts

Caged nuts use a spring-steel cage that retains a standard nut. It may snap into a holder or clip over a panel edge (Figure 5-5). A full cage retainer has lugs designed for use with thin sheet metal.

Clinch Nuts

A clinch nut is a solid nut that has a pilot or other feature that is inserted into a preformed hole and permanently "clinched" to the parent material (see Figure 5-6). Clinch nuts provide a strong, multiple-threaded fastener in materials too thin to be extruded or tapped. They are suitable for blind assembly. Two basic methods are used to attach clinch nuts:

1. The nut may be squeezed into a punched or drilled hole by pressure. Parent material flows into angular slots in the nut, which holds the nut in place. These nuts are sometimes called self-clinching.
2. The pilot may be peened over (clinched), staked, or expanded to retain the nut and prevent rotation.

Clinch nuts fit into prepared holes, so locating jigs or fixtures are not necessary. They can be used in metals that are unsuitable for welding and in plastics and fiberglass. In addition, clinch nuts can be installed after parts are painted, plated, or coated.

Self-Piercing Nuts

A self-piercing nut is an internally threaded, precision, work-hardened steel nut with external undercuts on two sides. Used in sheet-metal panels, it is installed by automatic machinery (Figure 5-7). They are permanently attached, threaded fasteners that can be installed in most open or blind locations. Self-piercing nuts have high resistance to torque, vibration, tension, and shear loads—and to a combination of such forces.

Figure 5-7. Universal pierce nut.

In the assembly operation, pierce nuts are fed continuously under a plunger. With each press stroke, the nut pilot is driven through the metal panel, piercing the mounting hole for the fastener. Clinching die ribs cause a cold-flow of the panel metal into the undercut, securing the nut to the panel.

Nut installation may proceed simultaneously with other forming, blanking, and piercing operations at high speed in both single and progressive die setups. The nut can be installed progressively in strip or coil stock, as multiples in one panel, or individually.

Standard self-piercing nuts may be installed in metal panels up to 0.145-in. thick. The work-hardened steel nut has a pilot-head hardness greater than Rockwell B 85. Proper malleability is retained in the threaded area.

Self-piercing nuts for flush mounting are available where thin metals and high stresses are factors. Application of the high-stress nut is generally limited to metal thickness of 0.03 to 0.09 in.

Universal pierce nut is the most economic of the self-piercing nuts. This nut is also available as a strip pierce nut in which strands of wire hold nuts together in a continuous strip (Figure 5-8). The nut strip simplifies automatic assembly by eliminating feed hoppers and nut orientation stations. High-strength pierce nuts are used where greater stresses are encountered.

Figure 5-9 defines some commercially available single-thread nuts typically used in low-stress applications. Most single-thread nuts do not require lockwashers since the design provides a springback force of the thread against the screw thread.

Steels are the most common fastener materials and cover a broad range of physical properties. The most readily available steels include the carbon steels and the stainless steels (Table 5-3). Brass (composed of copper, zinc, and a

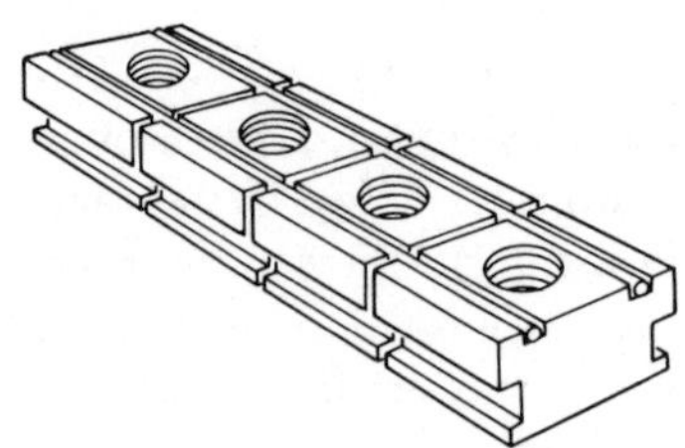

Figure 5-8. Strip pierce nut.

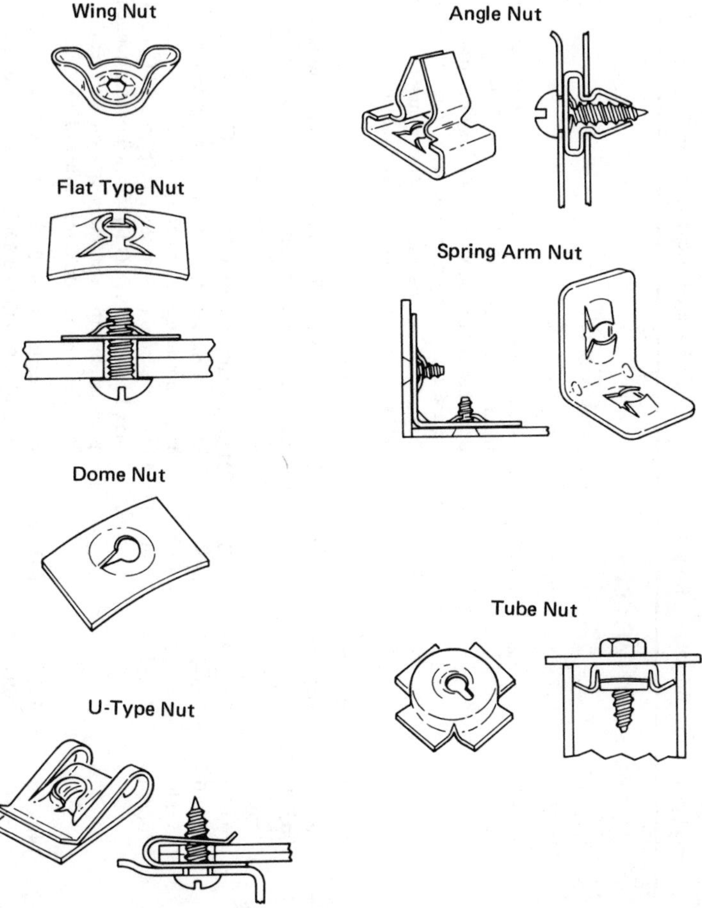

Figure 5-9. Typical single-thread nuts.

hardening agent) has many uses in electronics packaging. Brass cannot be hardened or tempered by heat treatment. The tensile strength and hardness can only be improved by cold-working. In many cases, cold-working and stress-relieving will improve corrosion resistance. Cold-drawn brass has a greater tensile strength than the mild carbon steels (Table 5-4).

Aluminum weighs about one-third as much as mild steel, and can equal or exceed the tensile strength of mild steel. Aluminum is nonmagnetic and can be highly finished, it also possesses high thermal and electrical conductivity properties. Aluminum is ideal for applications where strength-to-weight ratio, durability, corrosion resistance, thermal conductivity, and appearance are required (Table 5-4).

Table 5-3. Physical and Mechanical Requirements for Threaded Fasteners.

Grade	Bolt Size Diameter (in.)	Proof Load (psi)	Tensile Strength, Min (psi)	Hardness		Material–Heat Treatment
				Brinell	Rockwell	
			BOLTS AND CAPSCREWS			
0	All sizes	–	–	–	–	No requirements.
1	All sizes	–	55,000	207 max	95 B max	Commercial steel.
2	Up to $\frac{1}{2}$ incl.	55,000	69,000	241 max	100 B max	This is intended to be a cold-headed product made from low-carbon steel; 0.28 max C, 0.04 max P, 0.05 max S. Lengths over 6 in. may be hot-heated from medium-carbon steel; 0.55 max C. Deviation from specified chemistry may be made by agreement between producer and consumer.
	Over $\frac{1}{2}$ to $\frac{3}{4}$ incl.	52,000	64,000	–	–	
	Over $\frac{3}{4}$ to $1\frac{1}{2}$ incl.	28,000	55,000	207 max	–	Commercial steel.
3	Up to $\frac{1}{2}$ incl.	85,000	110,000	207–269	95–104 B	Produced by the cold heading process, up to and including 6 in. in length from medium carbon steel; 0.28 to 0.55 C, 0.04 max P, 0.05 max S.
	Over $\frac{1}{2}$ to $\frac{5}{8}$ incl.	80,000	100,000	–	–	
5	Up to $\frac{3}{4}$ incl.	85,000	120,000	241–302	23–32 C	Medium-carbon steel; 0.28 to 0.55 C, 0.04 max P, and 0.05 max S. Quenched and tempered at a minimum temperature of 800°F.
	Over $\frac{3}{4}$ to 1 incl.	78,000	115,000	235–302	22–32 C	
	Over 1 to $1\frac{1}{2}$ incl.	74,000	105,000	223–285	19–30 C	
5.1	Up to $\frac{3}{8}$ incl.	85,000	120,000	241–375	23–40 C	Low or medium carbon steel; 0.15 to 0.30 C; 0.04 max P, 0.05 max S. Oil quenched (water quench permissible with customer approval) and tempered at a minimum temperature of 650°F.
6	Up to $\frac{5}{8}$ incl.	110,000	140,000	285–331	30–36 C	Medium-carbon steel; 0.28 to 0.55 C, 0.04 max P, 0.05 max S. Oil quenched and tempered at a minimum temperature of 800°F.
	Over $\frac{5}{8}$ to $\frac{3}{4}$ incl.	105,000	133,000	269–331	28–36 C	
7	Up to $1\frac{1}{2}$ incl.	105,000	133,000	269–321	28–34 C	Medium-carbon fine-grain alloy steel; 0.28 to 0.55 C, 0.04 max, P. 0.05 max S, providing sufficient hardenability to have a minimum oil-quenched

						hardness of 47 Rc at the center of the threaded section, one diameter from the end of the bolt. Oil-quenched and tempered at a minimum temperature of 800°F. Roll threaded after heat treatment.
8	Up to 1½ incl.	120,000	150,000	302–352	32–38 C	Medium-carbon fine-grain alloy steel; 0.28 to 0.55 C, 0.04 max P, 0.05 max S, providing sufficient hardenability to have a minimum oil-quenched hardness of 47 Rc at the center of the threaded section, one diameter from the end of the bolt. Oil-quenched and tempered at a minimum temperature of 800°F.
				STUDS		
1	Up to 1½ incl.	–	55,000	241 max	100 B max	Commercial steel.
2	Up to ½ incl.	55,000	69,000	241 max	100 B max	Any SAE steel, 0.55 max C, 0.12 max P, 0.33 max S.
	Over ½ to ¾ incl.	52,000	64,000	–	–	
	Over ¾ to 1½ incl.	28,000	55,000	–	–	
4	Up to 2 incl.	–	115,000	–	22–32 C	Medium-carbon, cold-drawn SAE steel, 0.55 max C, 0.04 max P, 0.33 max S.
		–	100,000	–	–	
5	Up to ¾ incl.	85,000	120,000	241–302	23–32 C	Any SAE steel, 0.28 to 0.55 C, 0.04 max P, 0.13 max S. Quenched and tempered at a minimum temperature of 800°F.
	Over ¾ to 1 incl.	78,000	115,000	235–302	22–32 C	
	Over 1 to 1½ incl.	74,000	105,000	223–285	19–30 C	
8	Up to 1½ incl.	120,000	150,000	302–352	32–38 C	Any SAE fine-grain alloy steel, 0.28 to 0.55 C, 0.04 max P, 0.04 max S, providing sufficient hardenability to have a minimum oil-quenched hardness of 47 Rc at the center of the threaded section, one diameter from the end of the bolt. Oil quenched and tempered at a minimum temperature of 800°F.
8.1	Up to 1½ incl.	120,000	150,000	302–352	32–38 C	Medium carbon alloy or SAE 1041 modified. Elevated temperature drawn (EDT) steel.

Table 5-4. Nonferrous Fastener Materials.

Material	Tensile Strength (psi)	Yield Strength, 0.5% Elongation (psi)	Rockwell Hardness	Electrical Conductivity[a]
Aluminum				
2024-T4	55,000 min 60,000 avg	50,000	B 54–63	30
2011-T3	55,000 avg	48,000	B 50–60	40
1100	13,000 min 16,000 avg	5,000	F 20–25	57
Brasses				
Yellow brass	60,000 min 72,000 avg	43,000	B 57–65	27
Free-cutting brass	50,000 min 55,000 avg	25,000	B 53–59	26
Commercial bronze	45,000 min 50,000 avg	40,000	B 40–50	44
Naval bronze, composition A	55,000 min 65,000 avg	33,000	B 45–50	26
Naval bronze, composition B	50,000 min 60,000 avg	31,000	B 45–50	26
Copper	35,000 to 45,000	10,000 to 37,000	F 40–80	101
Silicon Bronze				
High-silicon, Type A	70,000 min 80,000 avg	38,000	B 74–80	7
Low-silicon, Type B	70,000 min 76,000 avg	35,000	B 67–75	12
Silicon-aluminum	80,000 min 85,000 avg	42,000	B 76–82	7
Nickel and High Nickel				
Monel	82,000 min 97,000 avg	60,000	B 90	–
Nickel	68,000 to 82,000	20,000 to 65,000	B 75–86	–
Inconel	80,000 to 120,000	25,000 to 70,000	B 72 to C 24	–
Stainless Steel (AISI)				
Type 302	90,000 to 124,000	35,000 to 116,000	B 89 to C 25	–
Type 303	90,000 to 124,000	35,000 to 116,000	B 89 to C 25	–
Type 304	85,000 to 112,000	30,000 to 92,000	B 87 to B 98	–
Type 305	80,000 to 110,000	30,000 to 95,000	B 84 to B 97	–

Table 5-4. *(Continued)*

Material	Tensile Strength (psi)	Yield Strength, 0.5% Elongation (psi)	Rockwell Hardness	Electrical Conductivity[a]
Stainless Steel (AISI) *(Continued)*				
Type 309	100,000 to 120,000	40,000 to 100,000	B 94 to C 23	–
Type 310	100,000 to 120,000	40,000 to 100,000	B 94 to C 23	–
Type 316	90,000 to 115,000	30,000 to 95,000	B 89 to B 99	–
Type 317	90,000 to 115,000	30,000 to 95,000	B 89 to B 99	–
Type 321	90,000 to 112,000	30,000 to 92,000	B 89 to B 98	–
Type 347	90,000 to 112,000	35,000 to 95,000	B 89 to B 98	–
Type 410	75,000 to 190,000	40,000 to 140,000	B 81 to C 42	–
Type 416	75,000 to 190,000	40,000 to 140,000	B 81 to C 42	–
Type 430	70,000 to 90,000	40,000 to 183,000	B 77 to B 90	–

[a]Approximately the percentage of conductivity of international annealed copper standard at 68°F. (Actual conductivity of this copper standard is 101.)

WASHERS

Washers are added to bolts and screws to serve one or more of these functions:

- Distribute the load
- Lock the fastener
- Protect the surface
- Insulate
- Span oversize holes
- Seal
- Provide electrical connections
- Provide spring tension

Flat washers, also called plain washers, primarily provide a bearing surface for a nut or screw head, cover large clearance holes, and distribute fastener loads over a large area, particularly on soft materials such as aluminum or wood.

Tooth lockwashers are used with screw and nuts to add spring takeup to the screw elongation and to increase the frictional resistance under the screw head or nut face. They bite into both the head of the screw and the work surface to provide an interference lock. Even at zero tension the tooth lockwasher will provide frictional resistance to loosening.

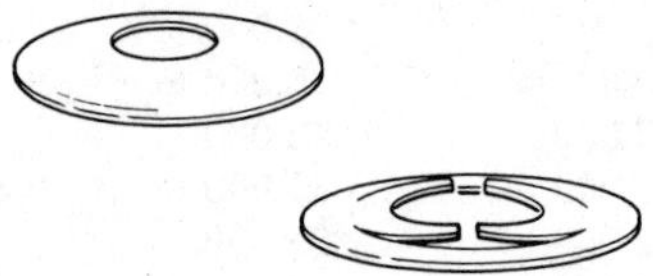

Spring washers have no industry standards except shapes. Except for the cone shape, they usually do not have a high value of spring reactance, but can be designed to have greater distance of spring action.

Spring washers are generally made of spring steel, but spring bronze or any other resilient material can be used.

6

HEAT TRANSFER AND THERMAL CONTROL

The field of high temperature control in Electronic Assemblies is, of course, too broad to be completely covered in this text. Nevertheless, this chapter is intended to establish a basic approach to a thermal study for any unit under consideration, and to obviate the minor problems often encountered in a thermal design.

When designing a mechanical package, where is the most practical point to begin? A logical study of each major requirement (thermal, shock, vibration, etc.) and a little common sense can delete many of the problems that are normally encountered in the later stages of design. Heat transfer, for example, can be greatly simplified by closely planning the thermal paths and "hot spot" locations throughout the unit without regard to thermal size and conductivities or the use of long strings of equations and exotic mathematics. As the design progresses, heat-dissipating devices must be developed based on a firm mathematical foundation; however, at this stage of the design, the flow paths and "hot spots" have already been logically placed. A critical problem in high-density packaging is the removal of excessive heat generated within the unit. Since heat has a deleterious effect upon electronic components (tending to decrease the mean-time-between-failures), new techniques must be developed for temperature control. "Paint it black and drill lotsa holes" can no longer be accepted. Such items as draft, proper dissipation surface, and short thermal paths to external equipment surfaces must be carefully considered.

The black-body concept is of great advantage in many cases; however, it must be remembered that a good thermal radiator is also a good thermal absorber. This technique requires careful planning, and thermal reflective devices should be considered. The major obstacle in removing excessive heat from present-day and future equipment is space. A systems approach, rather than a component approach, offers greater capability in solving the problem of heat removal. Flow

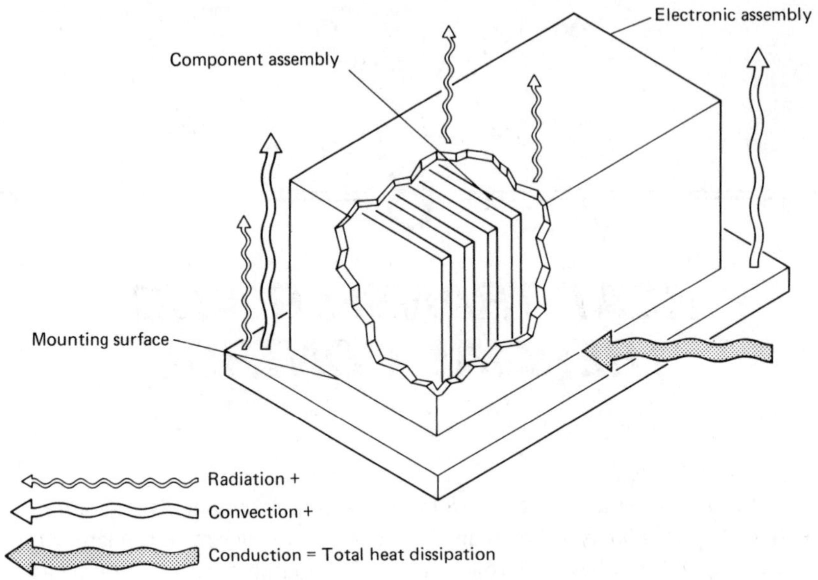

Figure 6-1. Basic modes of heat transfer.

paths for the entire equipment, rather than for individual components, must be considered. Figure 6-1 is a broad pictorial definition of the basic modes of heat transfer.

The engineer who must design a package that will insure cool operation of components can approach the problem in various ways. The engineer can, of course, make a detailed heat-balance analysis–assuming that such an extensive study is within the scope of the design project. However, using an empirical approach as outlined in this chapter, the engineer can design a package that will come reasonably close to meeting the requirements. Corrective measures and a little common sense will eliminate the problems normally encountered in the later stages of design. For best results, a systems approach rather than a component approach should be used. After making a logical study of major requirements, thermal flow paths for the entire unit can be planned, and "hot spots" located to greatest advantage. As the design progresses, it will be necessary to develop heat-dissipating techniques based on a firm mathematical foundation.

PROCEDURE

It is best to begin a thermal study by organizing all available facts and establishing some concept of the magnitude of the problem. The product description issued by the customer is an excellent source of information. Specifications covering the size, configuration, thermal environment, and installation of electronic equipment are normally covered in this document.

Early in the development program, the circuit designer can estimate the power dissipation and the maximum operating temperature of circuit components. The thermal design can then proceed as follows:

Step 1

Assuming the package to be in free air with all sides exposed to radiation and convection, determine the unit dissipation, designated as Q_u. This is an approximate measure of the amount of heat that can be effectively dissipated from the surface of the enclosure, and is expressed in watts per square inch (W/in.^{2}). Estimate the power dissipation, and compute the area of the external dissipating surfaces. Apply to Equation 6.1:

$$\text{Unit Dissipation } Q_u = \frac{\text{Estimated Power Dissipation}}{\text{Total Area of Dissipating Surfaces}} \tag{6.1}$$

Compare the results with the chart in Figure 6-2. This chart indicates the cooling technique required, based on the amount of heat that can be dissipated by the outside surfaces of the box. For example, natural cooling is adequate if unit dissipation is 0.5 W/in.^{2} or less. Enough heat will be dissipated by natural radiation and convection not to require the use of other techniques. However, if unit dissipation exceeds 0.5 W/in.^{2}, forced air or direct liquid cooling will be required to dissipate excessive heat (Figure 6-2).

To illustrate, electronic equipment is planned to be packaged in a box 4 X 5 X 8 in. The estimated power dissipation of the components is 50 W. Q_u is determined as follows:

$$\begin{aligned} Q_u &= \frac{50}{2(4 \times 5) + 2(4 \times 8) + 2(5 \times 8)} \\ &= \frac{50}{40 + 64 + 80} \\ &= \frac{50}{184} = 0.27 \text{ W/in.}^2 \end{aligned}$$

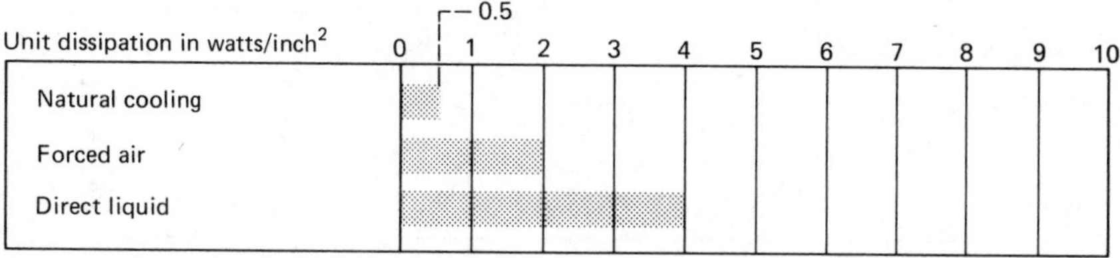

Figure 6-2. Maximum capabilities of external cooling techniques.

Referring to the chart in Figure 6-2, we see that 0.27 W/in.^{2} falls within the limits of natural cooling. In effect, enough heat will be removed by natural convection and radiation to insure that the equipment will operate at normal room temperature without the use of additional techniques or devices.

If forced air or direct liquid cooling has not been specified by the customer, the dissipation surfaces must be expanded to allow sufficient radiation and convection to maintain proper operating temperatures. This can be accomplished in a variety of ways–by the use of corrugated surfaces, for example, or by the addition of a finned sink, which will serve as a control surface to guide the flow of heat along the desired path. This type of sink offers maximum heat transference, because it exposes a large dissipating area in a minimum of space.

Step 2

Continuing the procedure of defining the package in free air, determine the heat concentration, designated as ϕ. This is an approximate measure of the amount of heat generated by the components packaged inside the enclosure and is expressed in watts per cubic inch (W/in.^{3}). Compute the internal volume, and apply to Equation 6.2.

$$\text{Heat Concentration } \phi = \frac{\text{Estimated Power Dissipation}}{\text{Internal Volume of Package}} \tag{6.2}$$

Compare the results with the chart in Figure 6-3. This chart indicates the internal cooling technique required to transfer excessive heat to the external surfaces for dissipation. Natural cooling will permit safe operation of the equipment if the heat concentration is 0.15 W/in.^{3} or below, but any excess must be transferred to the external surfaces for dissipation, and the designer will have to employ other cooling techniques. Selection will depend upon the results of Equation 6.2.

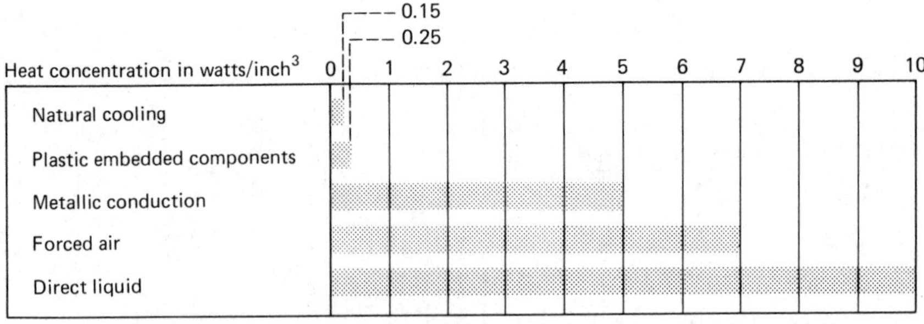

Figure 6-3. Maximum capabilities of internal coding techniques.

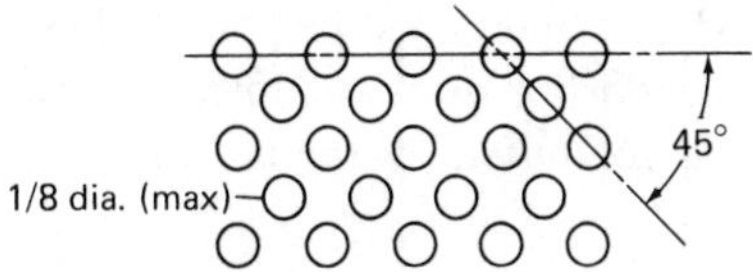

Figure 6-4. Staggered hole pattern for intake air turbulence reduction.

To illustrate, the volume of a package 4 X 5 X 8 in. is 160 in.^{3}. The estimated power dissipation is 50 W. Therefore:

$$\text{Heat Concentration } \phi = \frac{50}{160} = 0.31 \text{ W/in.}^3 \tag{6.3}$$

Referring to the chart in Figure 6-4, we find that 0.31 W/in.^{3} of heat concentration will require metallic conduction paths to transfer heat to the external surfaces where heat will be dissipated by natural convection.

If installation requirements permit, the external surfaces can be expanded by using a larger box, or by employing fins or corrugated construction. If some of the internal packaging space is sacrificed in expanding the external surfaces, this difference must be taken into consideration. It may be feasible to place the hottest components externally so that heat can be dissipated directly, thus allowing the equipment to operate at room temperature.

If the heat concentration of a specific package does not exceed 0.15 W/in.^{3}, the components can be cooled by natural convection. A convective current can be set up to draw cool air into the unit and force the warm air out. This is known as a chimney effect. Holes must be located strategically so that the air current will flow across heat-generating components, but not across components that would be adversely affected by heat from the warm air being forced out. Although it is not possible to completely eliminate turbulence, especially when high-density packaging is necessary, it can be reduced by staggering the holes (Figure 6-4).

The area of the intake and exhaust holes required to create the proper draft can be determined by Equation 6.4.*

$$A = 0.568 \frac{Q}{\rho(t_2 - t_1)^{3/2}} \text{ (approximate)} \tag{6.4}$$

where

Q = power dissipated (in watts)
ρ = density of the internal air = $22.1/t_2 + 273$ lb/ft^3
t_2 = maximum allowable internal temperature in °C
t_1 = external temperature in °C

*This equation does not apply when $t_2 - t_1$ exceeds 35°C and pressure is not constant.

The total area of the exhaust holes must be greater than that of the intake holes because air expands in proportion to its temperature rise. An approximation of the area required may be computed by the use of the Charles Law of Thermal Expansion.*

Step 3

Steps 1 and 2 have defined the thermal capability of a given package in free air using radiation and convection to transfer heat to the atmosphere. However, when the unit is installed in the system, heat will be conducted to the system structure through the fastening devices. To determine the heat concentration of the package in an operating condition, Equation 6.2, used in Step 2 applies, but the amount of heat conducted through the bolts must be computed and substracted from the estimated power dissipation of the box in free air.

To illustrate, assume the customer's description specifies four No. 10 aluminum bolts, 0.5 in. long. A No. 10 bolt has a diameter of 0.19 in.: the area of the cross section is 0.028 in.^{2}, or a total area of 0.112 for the four bolts. The heat flowing through a unit is expressed in Btu per hour (Btu/hr) by Equation 6.5.

$$Q = \frac{kA\Delta T}{L} \tag{6.5}$$

where

k = thermal conductivity of the substance = 100 (for aluminum)
L = length of the bolt = 0.5 in.
ΔT = temperature differential = 25 (ΔT must be kept below 25°C for heat-sensitive components)
A = total area of the cross section of the four bolts = 0.112 in.^{2}

Therefore, the conductivity of the bolts is computed as follows:

$$Q = \frac{100 \times 0.112 \times 25}{0.5} = 56 \text{ Btu/hr}$$

Converted to watts,

$$Q = \frac{56}{3.415} = 16.4 \text{ W}$$

The bolts are capable of dissipating 16.4 W, but due to the thermal resistance at the interface, it is advisable to use a 50% safety factor—or an estimated power dissipation of 8.2 W.

Since 8.2 W are dissipated by conduction through the bolts, this must be sub-

*The coefficient of expansion of air is equal to 1/273 of the volume at 0°C for each °C increase in temperature.

tracted from the estimated power dissipation of 50 W (as given in Equation 6.3). The heat concentration of the box in an operating condition is determined as follows:

$$\phi = \frac{50 - 8.2}{160} = \frac{41.8}{160} = 0.26 \text{ W/in.}^3$$

GENERAL CONSIDERATIONS

Having defined the package, the engineer must locate and mount components using the most effective cooling techniques or modes of transfer. The ideal way to remove heat is by natural cooling, but whatever method or combination of methods is used, thermal paths must be carefully planned to carry a maximum quantity of heat over the shortest possible distance.

High-heat-generating components should be located near the top of the case, and, when possible, mounted in direct contact with the external structure. High-temperature components should be arranged to form a bank of medium height. If vertical stacking is required, the parts should be staggered to provide convective paths for the removal of heat. (See Figure 6-5.)

Temperature-sensitive components should be mounted near the bottom. If they are located in close proximity to hot components, they must be protected from heat damage. A thin sheet of aluminum highly polished on both sides makes an excellent reflecting shield (Figure 6-6).

To transfer heat by metallic conduction, a direct metallic path with good thermal joints must be provided. Conductive members should be short with large cross section, and should be made of materials having good thermal conductivity. Conductive paths should offer the least possible resistance. To de-

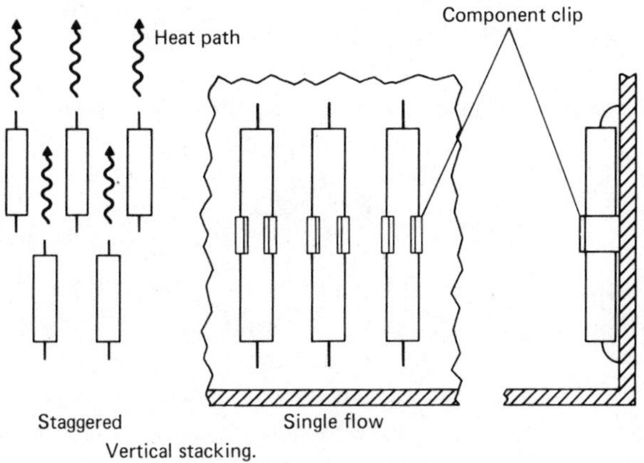

Figure 6-5. Vertical stacking.

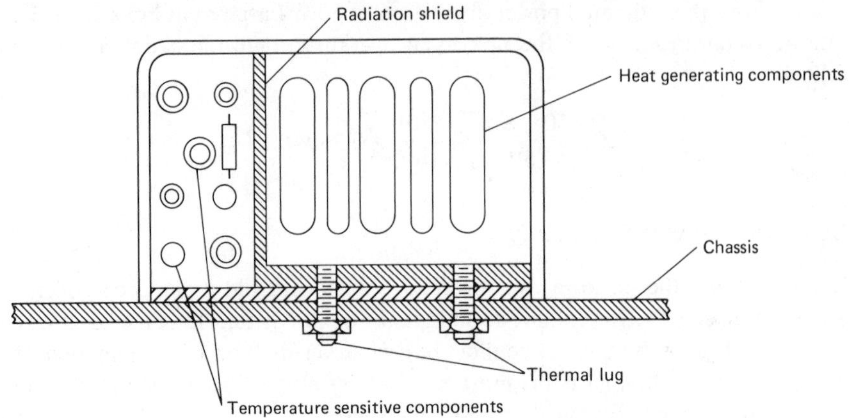

Figure 6-6. Isolating temperature sensitive components.

crease resistance, care must be taken to limit the number of thermal interfaces and to provide maximum contact surface area. Joints should be soldered, welded, or brazed to form a molecular bond. The best bond is obtained by dip brazing or furnace brazing, because the additive materials flow evenly with almost no air traps. If a heat sink is required for a high-heat-generating component or group of components, the design will be determined by the space available for installation and the amount of dissipating surface required.

Heat loss by convection is usually referred to as the film coefficient, and is measured in Btu/hr/ft²/°F/ft. The film coefficient of a particular type of heat sink varies according to its size, shape, and position. Heat loss, designated as H_c, is expressed by Equation 6.6.

$$H_c = 0.28\left(\frac{\Delta T}{L}\right)^{0.25} \quad \text{Btu/hr/ft}^2/^\circ\text{F/ft} \tag{6.6}$$

where L is the vertical height in feet when less than 2 ft. Converted to watts per square inch,

$$H_c = 5.698 \times 10^{-4}\left(\frac{12\Delta T}{L}\right)0.25 \text{ W/in.}^2 \tag{6.7}$$

The film coefficients of several types of heat sinks in common use are listed in Table 6-1. Whatever type of heat sink is selected, it should always be tested to verify that sufficient heat can be dissipated to insure proper thermal control.

In high-density assemblies where heat removal by convection is limited because of insufficient space to allow a flow of air, the designer can provide an intermediate sink to transfer heat by radiation or conduction.

Table 6-1. Film Coefficients.
c = Film coefficient for various shapes and positions.

Vertical plate	0.55
Horizontal cylinder	0.45
Long vertical cylinder	0.45 to 0.55
Horizontal plate, warm side down	0.35
Horizontal plate, warm side up	0.71
Sphere	0.63

In some cases, electrical insulation may be required when mounting semiconductors or installing heat-generating electronic devices with electrically hot cases. Various products and processes, including mica washers, silicon grease, thermal pads, and hard anodizing, have been used in the industry.

Mica washers have been used successfully for several years, and are often supplied by manufacturers of semiconductors as standard installation hardware. Mica is a good electrical insulator, but a poor thermal conductor. Silicon grease compound is comparable to mica as an insulating material. However, the compound has a tendency to rub off or break down if exposed to atmospheric contamination during handling, and it is difficult to control its film thickness. Although adequate for laboratory use, it is too unreliable for production equipment.

Durafilm is an aqueous dispersion of an acrylic polymer high-dielectric enamel that is applied by dip-coating or spraying. A nonflammable, nontoxic liquid, Durafilm is easy to handle. It adheres well to most materials, and has been used successfully as an insulation coating for semiconductor clips, heat sinks, and general-purpose equipment.

Hard anodizing is an electromechanical process for coating aluminum with an aluminum oxide film. Because of its superior *thermal qualities*, it is the best heat-sinking material to date. Tests have shown that Hardas Process Anodized Finishes increase in temperature at a rate of only 0.3°C/W/mil, as compared to 1.25°C for mica washers. The thermal conductivity and the dielectric strength of different insulating materials are given in Table 6-2.

It is commonly accepted by the industry to use the figure of 80 Btu/ft^2/°F/ft for the thermal conductivity of anodized aluminum. However, after considerable investigation of various electrical insulation films, the author believes that coating in the range of 0.002 through 0.0005 in. thick integrally bonded to the base material (normal for insulation coatings in electronic applications) will cause negligible resistance to thermal flow. To support this opinion, L/k is the thermal resistance in (hr $\times$ ft^2 $\times$ °F)/Btu, where L is the length of the thermal path and k the conductivity of the material or materials under consideration.

Table 6-2. Comparison of Electrical Insulators Possessing Good Heat Transfer Properties.

Material	Thermal Conductivity (Btu/hr/ft²/°F/ft)	Dielectric Strength (V/mil)
Silver	241	Conductor
Aluminum and aluminum alloys	124	Conductor
Lead and lead alloys	19.6	Conductor
Hardas process anodized finish	[a]	650
Durafilm	[b]	550
Epoxies (cast)	0.8	550
Mica	0.4	2000
Silicon grease	0.363[c]	250
Phenolic	0.39	425
Silicon rubber	0.12	400
Diallyl phthalate	Insulator	450
Fillers	[c]	–

[a] Thermal conductivity for hard anodized surfaces has not been established. However, one large manufacturer of semiconductor heat sinks claims the difference between the base material and the hard anodize finish is so minute it is difficult to measure.
[b] A 2.2°C rate of temperature increase, per watt per mil of Durafilm coating, has been claimed by users in component clip applications.
[c] An increase in thermal conductivity can be achieved by the addition of fillers.

It follows that the smaller the resultant number, the lesser the effect on heat transfer by conduction.

Consider a composite wall of 0.060-in. aluminum and an insulation coating of 0.0004-in. If the equation for conduction through a composite wall is:

$$Q = \frac{A(t_1 - t_3)}{R_1 + R_2}$$

then the smaller the sum of the resistances, the larger the flow Q.

Proof: L/k = Thermal resistance = R

$$\text{Aluminum, } R = \frac{0.060 \text{ in.}}{k} = \frac{0.005 \text{ ft}}{100} = 0.00005$$

$$\text{Anodized } (al_2O_1) = \frac{0.0004}{k} = \frac{0.000033}{1.8} = 0.000018$$

The same effect is encountered with Durafilm coatings. Durafilm is basically an acrylic polymer that bonds with the base material and has a thermal conductivity of 1.4 Btu/hr/ft²/°F/ft.

Then: The total resistance of an anodized plate to thermal flow is 0.000068 compared to the base material of 0.00005; a plate coated with Dura-

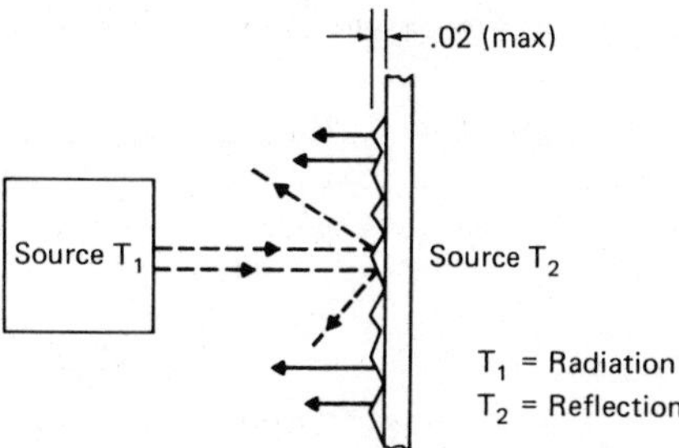

Figure 6-7. Finish cross section-laminated.

film offers a resistance of 0.00002, indicating that thin integral coatings have little effect on tranɔfer properties.

Radiation, in general, offers the least effective transference of heat of the three modes. However, it must be taken into consideration since it has noticeable effect on heat transfer, and care must be taken to protect low-temperature components from the damaging effect of radiant heat.

Since radiation occurs in the form of electromagnetic waves, a black-body finish may be applied to increase emittance and lower absorption. Application of a lamp black or other radiation finish may be controlled to reflect a large portion of the heat that would otherwise be absorbed. A heat wave reflects from a surface not normal to it, identical to light waves. Therefore, by applying a laminated or crackle surface, absorption is reduced (Figures 6-7 through 6-9).

CONCLUSION

Every package presents a special problem in heat control that must be resolved within the limits of the design specifications. The success of the package will

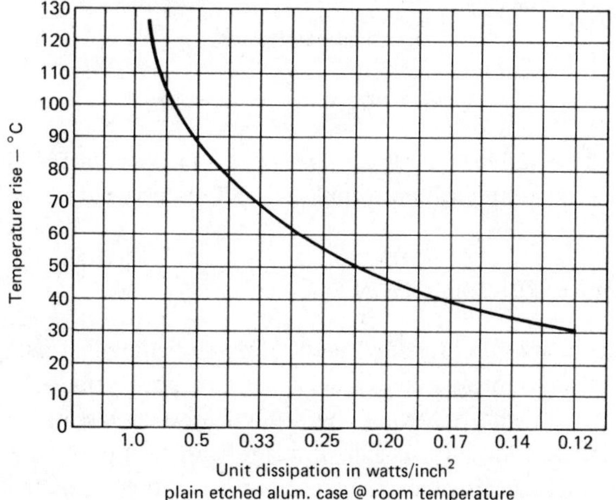

Figure 6-8. Unit dissipation-plain etched alum. case.

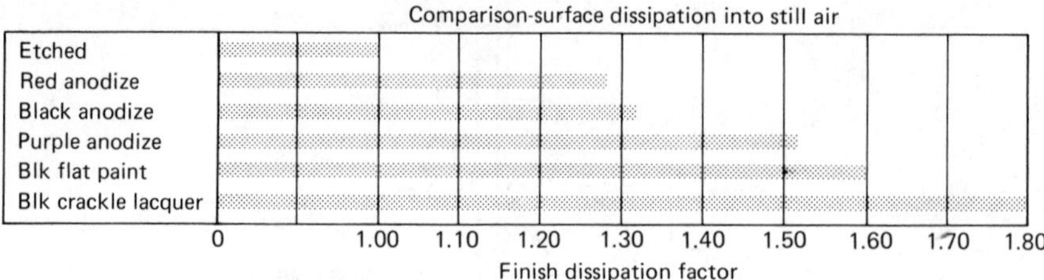

Figure 6-9. Comparison-surface dissipation into still air.

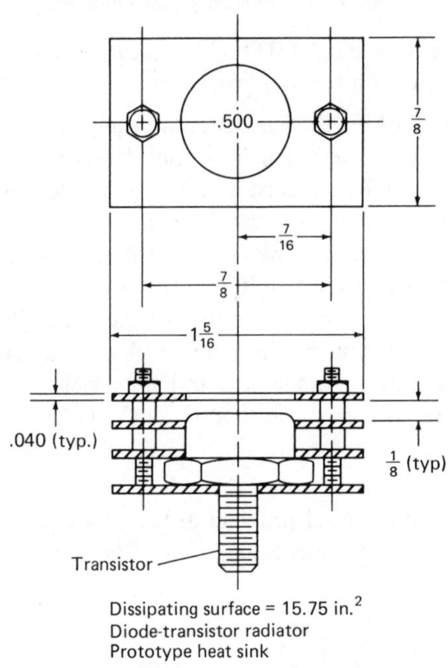

Figure 6-10. Prototype heat sink.

depend upon how well the designer defines the package in the first place, and how well the designer applies the techniques and materials at his or her disposal.

EXPERIMENTAL VERIFICATION

Based on the results of an investigation of commercial dissipators, an experimental sink was built and evaluated (Figure 6-10). A power transistor was used for the evaluation under the following conditions: A power dissipation of 36 W was generated in the test module, and an ambient temperature of 55°C (131°F) was maintained in the test chamber. The system was allowed to stabilize for 3 hr, and a reading of 97°C was taken on the transistor case. The results of this test

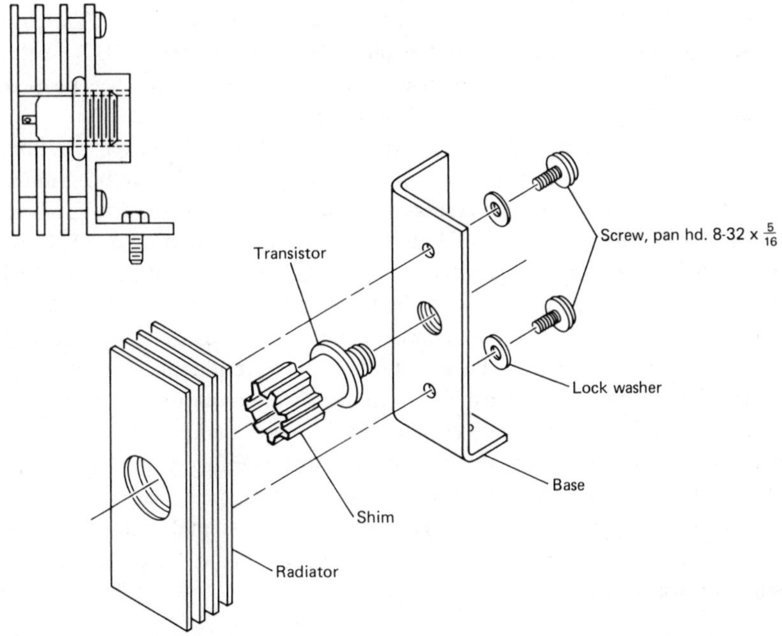

Figure 6-11. Diode-transistor radiator.

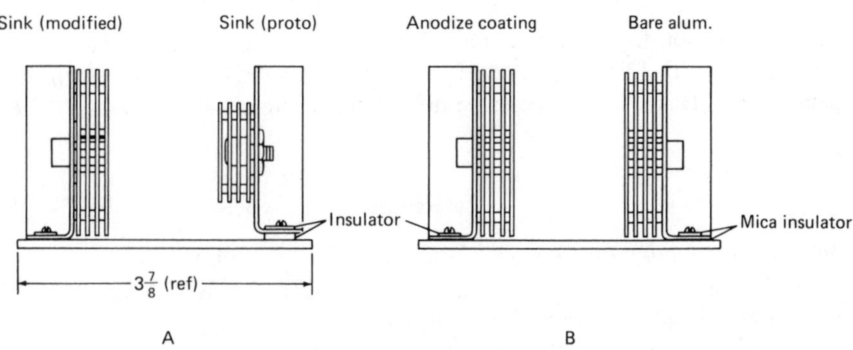

Figure 6-12. Comparison test set-up.

on the prototype were acceptable to the manufacturer's recommended operating conditions. A second dip-brazed sink was built (Figure 6-11), to improve conduction throughout the sink by removing dry joints (joints that may entrap still air). Due to the electrical characteristics of the transistor being used, the sink had to be insulated electrically; this was accomplished by application of a Hardas Process Anodized Finish.

A comparison test was made of the improved sink against the prototype. Both units were mounted on an aluminum plate, $\frac{1}{8}$ in. thick by 3-$\frac{7}{8}$ by 8-$\frac{5}{8}$, as shown in Figure 6-12a. An identical transistor was used in each assembly. The prototype was insulated from the aluminum plate by a mica washer 0.032 in. thick, while the modified unit was in direct contact. (The mica washer was needed to electrically insulate the sink from the aluminum plate.)

Power dissipation of each transistor was 20 W, and an ambient temperature of 25°C (77°F) was maintained in the test chamber. After allowing both units to stablize for 3 hr, the following readings were taken: the anodized and dip-brazed unit had a sink temperature of 83°C and a transistor temperature of 95°C, compared to the mechanically assembled and flat-black-painted prototype, which read 106°C on the sink and 122°C on the transistor, an improvement of 27°C.

A third evaluation was made to compare the Hardas Process Anodized Finish to a bare aluminum finish. The two units used were identical except for the finish (modified units, Figure 6-11). Both units were mounted on an aluminum plate $\frac{1}{8}$-inch thick by 3-$\frac{7}{8}$ by 8-$\frac{5}{8}$ (Figure 6-12b). The anodized unit was in direct contact with the aluminum plate (the oxide coating serving as an electrical insulation), while the bare unit was insulated by using a mica washer 0.032 inch thick and fiber shoulder washers. Transistors were assembled into both units which dissipated 20 W each. After stabilizing for 2 hr, at 25°C, the following information was recorded:

1. *Anodized Unit*
   Transistor: 110°C
   Sink: 95°C
2. *Bare Unit*
   Transistor: 141°C
   Sink: 126°C

Due to reflection from the bare unit, the temperature of the anodized assembly was increased by 12°C from the original test against the prototype, which was painted flat black. The temperature differential of transistor to sink was 15°C in both cases.

References

Air Transport Equipment Cases and Racking, ARINC Report No. 404.

Cornell Aeronautical Laboratory, Inc., Report No. HF-845-D-8, *Design Manual of Natural Methods of Cooling Electronic Equipment.*

Guidance for Designers of Airborne Electronic Equipment, ARINC Report No. 403.

Guide Manual of Cooling Methods for Electronic Equipment, NAVSHIPS 900, 190.

Matisoff, B. S. 1962. Cool electronic packages. *Product Engineering Magazine*, 33:22.

Study of Hard Coating for Aluminum Alloys, Dept. of Commerce, Bulletin No. PB111320.

Suggestions for Designers of Electronic Equipment, U.S. Navy Electronics Laboratory, San Diego, California.

7

SHOCK AND VIBRATION DESIGN

Shock and vibration start to become problems long before electronics equipment is installed. The designer must consider handling and transportation at the early stages of design in order to achieve factory-to-field dependability.

The most common vibration failures are due to the flexure of the component leads supporting the resistors and capacitors, and to the integrated circuits and transistors that loosen, fall out of the sockets, or short to the adjacent circuitry. The overall fragility of the entire system is best determined by analyzing the fragility of the components and subelements that make up the system (Figure 7-1).

In electronics equipment, vibration failures are four times as frequent as shock failures. Many components and systems can take up to 75 g of shock yet cannot perform under as little as 2 g of sustained vibration. Vibration causes the structural elements to build up to a natural resonance, and fatigue failures are brought on rapidly by the continuous energy input and the overstressing of structural parts. Whereas transient shock may cause large motions of structural elements, these motions diminish rapidly and no additional energy is applied. Electronics equipment chassis will normally fail under vibrations of 5 g up to 2000 Hz, but will usually withstand shocks of 50 g for 6 msec, although with some physical distortion.

In the interests of sound dynamic design, avoid cantilevers and sliding joints, and use stiff brackets and where appropriate, damped structures (see Table 7-1). Try to achieve high stiffness-to-mass ratios. Always firmly anchor large masses, such as transformers, to stiff structures, and keep large masses off panels and large flat surfaces. Where a printed circuit board is long or flexible, additional supports are required to prevent warpage and damage by vibration. Mounting holes should be located at least three times the diameter of the screws from the edge of printed circuit boards.

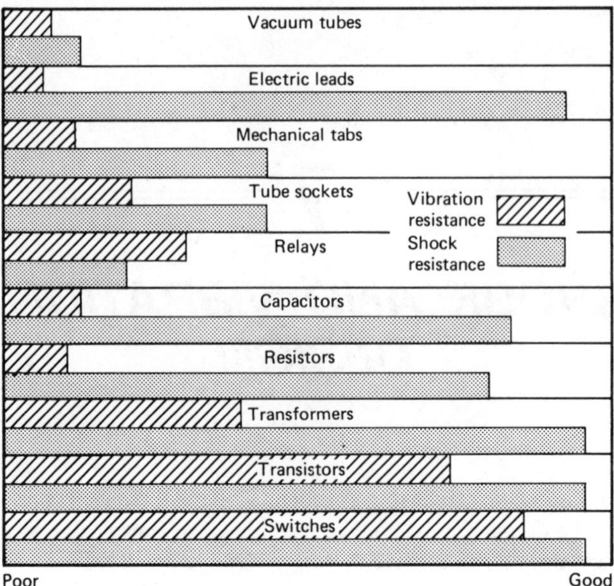

Figure 7-1. Relative dynamic resistance of common components.

Either rugged design, tuned to tolerate vibration, or a conventional design using shock or vibration isolators can provide dynamic resistance. Ruggedized units are sturdy, reliable, and self-contained. Their disadvantages include increased development time and cost, and often excessive weight.

Some basic characteristics of commonly used electronics components are shown in Figure 7-1. Care should be taken in both the selection of components and their location in the assembly. Heavy components should be positioned as

Table 7-1. Design for Maximum Dynamic Resistance.

Use	Avoid
Minimum spans	Cantilever elements
Short load paths	Sliding joints
High stiffness-to-mass ratios	Notches, corners, cracks, points, and other stress raisers
Box, Z-shaped, and channel sections for structural elements	High static stresses
Stress relief in sheet-metal parts	Mass concentrations
Positive locking and fastening devices	Attachment of masses to panels
Good clamp support for leads	Undersized fasteners

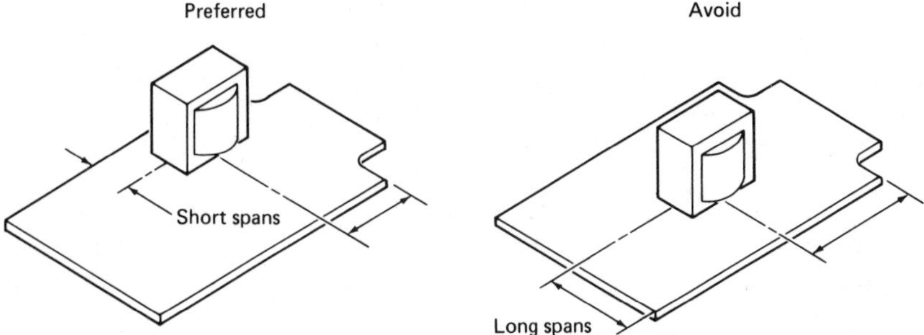

Figure 7-2. Mounting heavy mass components.

close to the edges of the printed circuit board or chassis as practical to maintain the shortest span to a mechanical support. Large flat surfaces are the most critical and should be carefully planned. (Figure 7-2.)

RELATIVE DYNAMIC RESISTANCE OF COMMON COMPONENTS

- Relays chatter and ultimately fail when subjected to 10 to 15 g vibration at more than 500 Hz. Special relays can take up to 100 g. Plunger and rotary types are the most rugged, while the clapper types are the most fragile.
- Transistors, capacitors, and resistors have low mass and good rigidity. Correctly supported and anchored, they can withstand as much as 15 g in vibration. Capacitors and resistors in particular usually fail from lead flexure.
- Terminals, switches, transformers, and circuit breakers are generally very rugged and stand up in severe dynamic environments. Shock and vibration failures are often the result of poor bracket and hold-down devices.
- Electric cables and connectors need to be well anchored. Care must be taken to prevent flexure, abrasion, and noise voltage due to flexing high-voltage coaxial cables and leads.

The best approach to rugged design is to contact the component vendors for the recommended dynamic tolerances of their products.

Axial lead parts weighing less than 0.5 oz may be mounted by the leads. In such cases, the distance between the body of the part and each terminal should be from 0.25 to 0.5 in. The leads should extend straight out, at least 0.125 in. beyond the body of the part, since most vibration failures occur at the junction of the lead and the body where the flexure stress is the greatest. This practice will also prevent breakage of the seal between the lead and the body (Figures 7-3 and 7-4).

Radial lead mounting is characterized by low mechanical resonant frequencies. It is not recommended unless the leads are rigid enough to raise the frequencies

Straight
Lead length
1/4" min. 1/2" max.

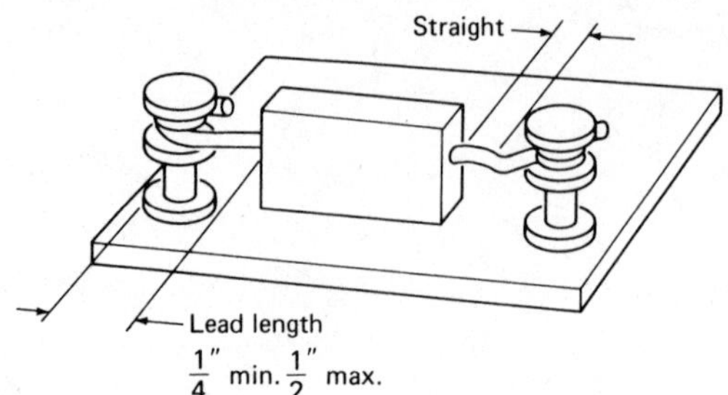

Figure 7-3. Mounting by axial leads.

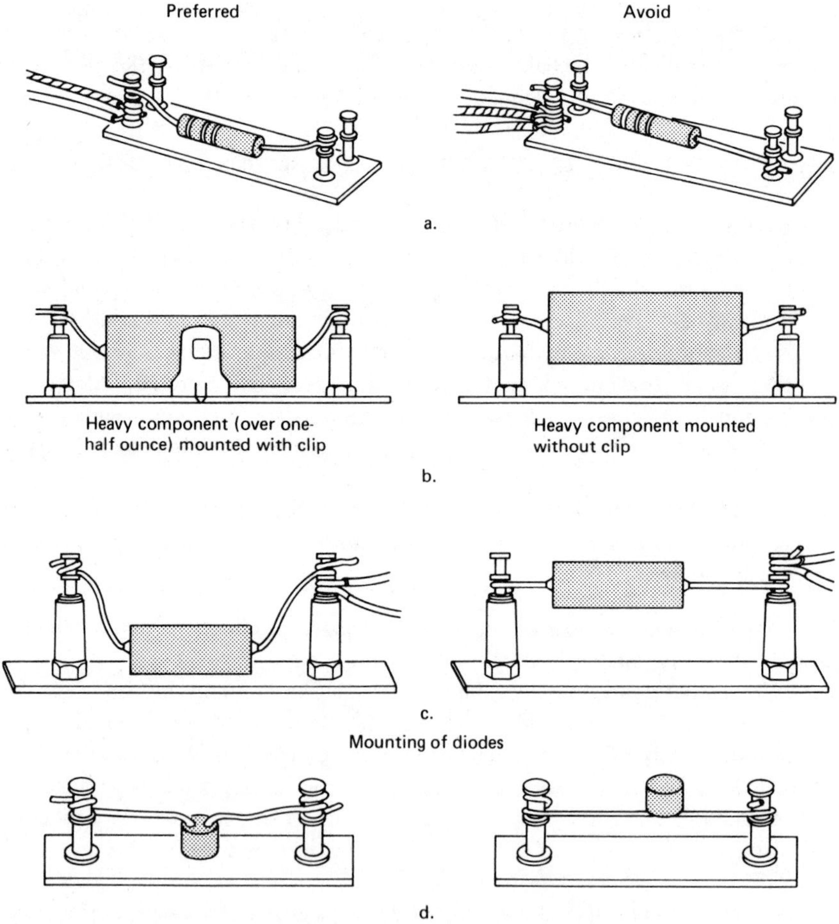

Preferred
Avoid
a.
Heavy component (over one-half ounce) mounted with clip
Heavy component mounted without clip
b.
c.
Mounting of diodes
d.

Figure 7-4. Mounting of resistors and tubular capacitors.

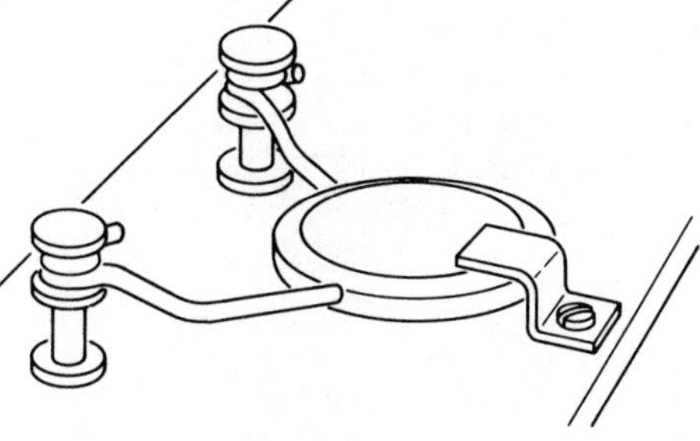

Clamp improves mounting

Figure 7-5. Clamping radial lead devices.

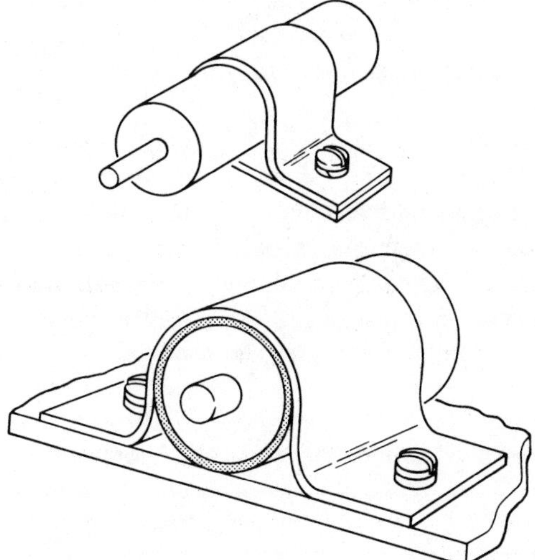

Figure 7-6. Clamp mountings.

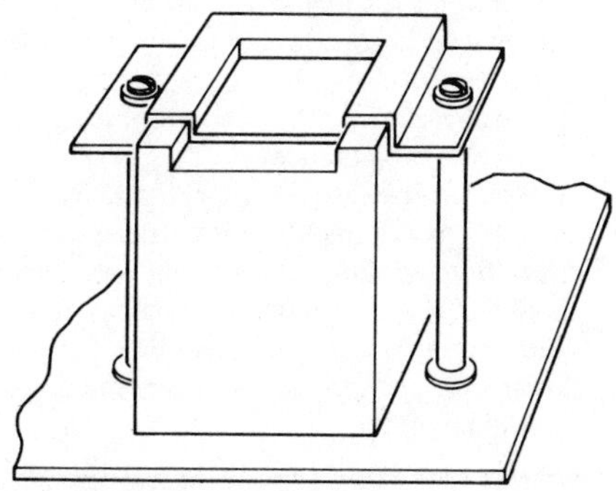

Figure 7-7. Plug-in unit with clamp.

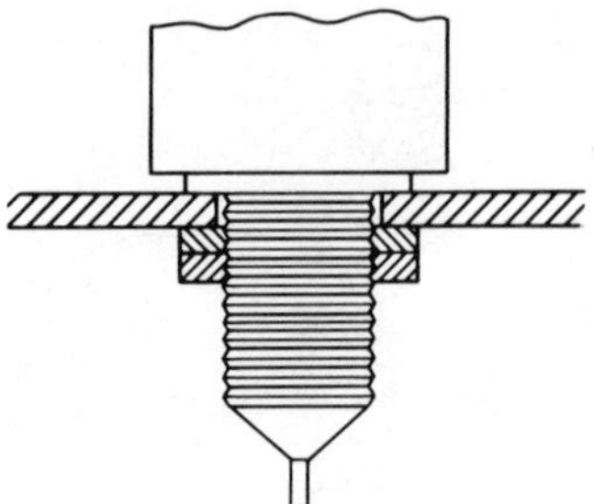

Figure 7-8. Single-hole mounting.

above 100 Hz. Certain miniature devices with short leads and small body mass will meet this requirement. Large parts such as wire-wound resistors and printed circuit assemblies of the radial lead configuration always require a mechanical clamp (Figure 7-5).

Axial lead or radial lead components that weigh more than 0.5 oz must be constrained by clamping (Figures 7-6 and 7-7). Single-hole mountings are adequate where the torsional stress is low, but long slender parts are subject to shock and vibration, and mountings should be kept above 100 Hz (Figure 7-8). Table 7-2 lists some of the common military specifications required for military equipment. These specifications provide an excellent guide for establishing the mechanical goals of a commercially designed product.

Table 7-2. Military Specifications for Shock and Vibration.

Specification	Agency	Description
Mil-C-172C	DOD	Cases bases, vibration mounts for electronic equipment in aircraft
Mil-S-4456	USAF	Method and apparatus for shock, variable duration
Mil-T-4807	USAF	Tests: vibration and shock, ground electronic equipment
Mil-E-4970	USAF	General specification for environmental testing, ground support equipment
Mil-E-5272C	ASG	General specification for environmental testing, aeronautical and associated equipment
Mil-E-5400D	ASG	General specification for electronic equipment, aircraft
Mil-T-5422E	ASG	Testing, environmental, aircraft electronic equipment
Mil-P-9024B	USAF	Specifications and general design requirements for packaging: guided missile weapon systems
Mil-G-9412	USAF	Engineering and procurement data for ground support equipment for weapons systems, support systems, subsystems, and equipment.

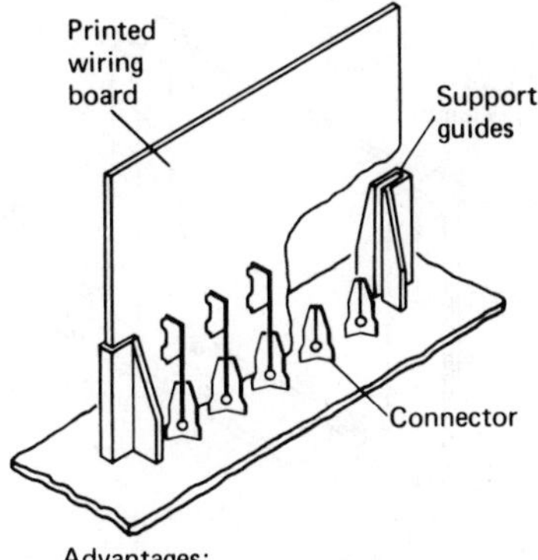

Advantages:
1. Guides board during insertion
2. No loose hardware
Disadvantage:
No positive retention

Figure 7-9. Sheet metal support guides.

PRINTED CIRCUITS

The following factors should be considered in selecting the method of mounting printed circuits.

1. board size
2. board configuration

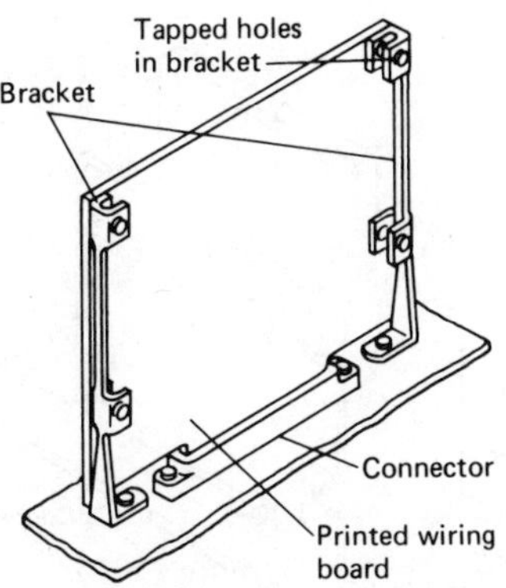

Advantages:
1. Commercially available in several sizes
2. Mounts single or double boards
Disadvantage:
Loose hardware

Figure 7-10. Vertical flanged bracket.

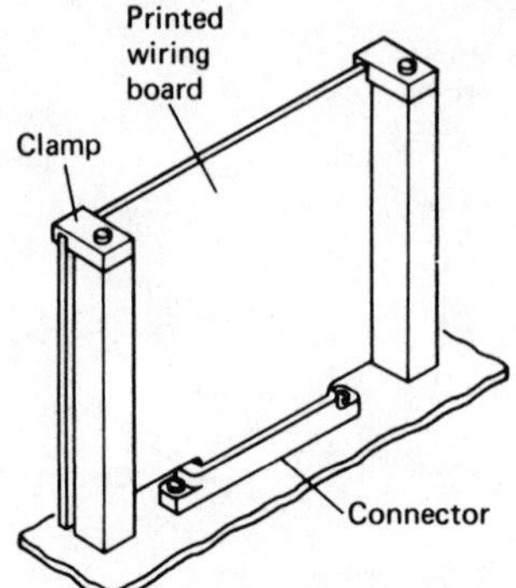

Advantages:
1. Guides board during insertion into connector
2. Can be made to accomodate any size board

Disadvantage:
Relatively high cost

Figure 7-11. Vertical post and grooved clamps.

3. type of connector
4. available space
5. vertical or horizontal mounting
6. support and retention
7. heat dissipation
8. relation with other nearby circuits

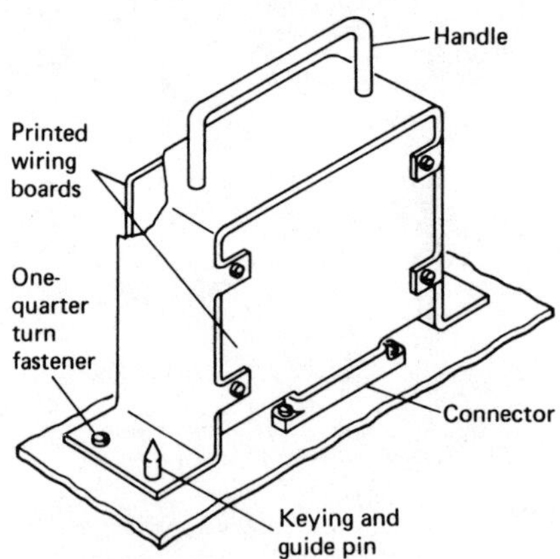

Advantages:
1. Positive retention
2. Boards can be used in pairs

Disadvantage:
Additional chassis area required for mounting

Figure 7-12. Dual mounting fixture.

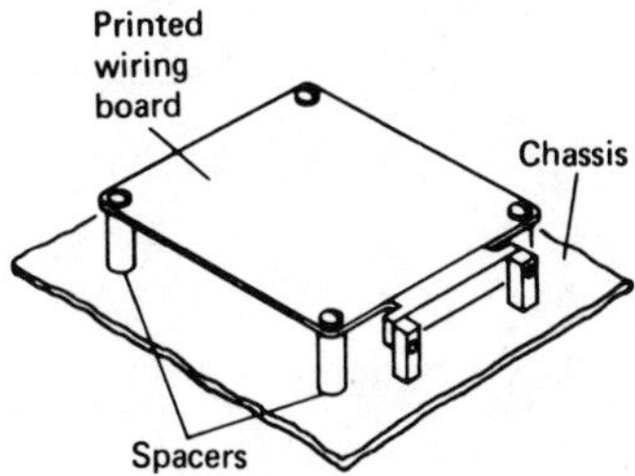

Advantages:
1. Accommodates any size of board
2. Simple inexpensive mounting

Disadvantages:
1. Loose hardware
2. Occupies large chassis area

Figure 7-13. Mount on tapped spacers.

Adequate support and retention must be provided for holding the printed circuit board since friction alone cannot be depended on. The friction connector should never be used as the sole means of mechanical support (Figures 7-9 through 7-15). Boards that are $\frac{1}{16}$ to $\frac{3}{32}$ in. thick should be supported at not more than 4.00-in intervals. Boards that are thicker than $\frac{3}{32}$ in. should be supported at not more than 5.00-in. intervals. The mechanical support for a printed circuit board should hold it in position and minimize its relative motion to the chassis.

Printed circuit boards usually employ four basic types of mounting: (1) cantilever, (2) clamped at each end, (3) hinged and clamped, and (4) hinged at each end (Figure 7-16). In practice, a form of the cantilever method is normally used. In most cases, this is the least desirable method of mechanically supporting a

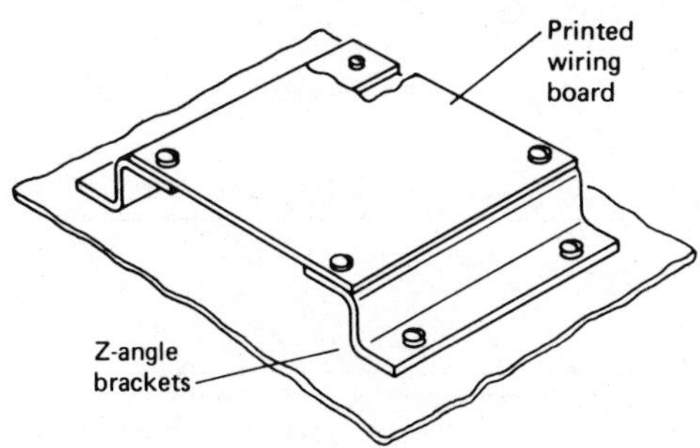

Advantages:
1. Suitable for any size board
2. Provides good heat dissipation to chassis

Disadvantages:
1. Loose hardware
2. Occupies large area on the chassis

Figure 7-14. Z-angle brackets.

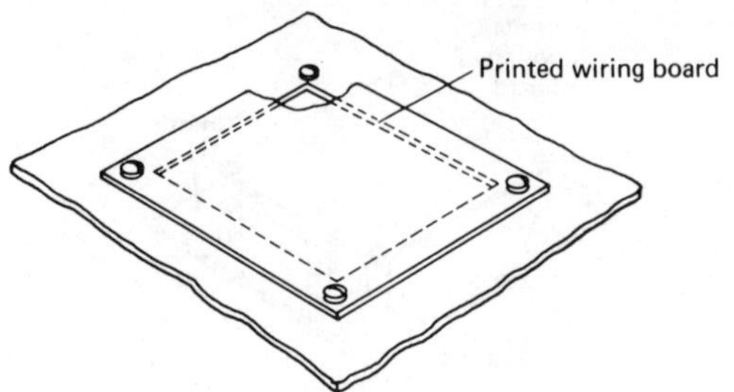

Advantages:
1. No mounting brackets required
2. Provides accessibility to underside of printed board

Disadvantage:
Loose hardware

Figure 7-15. Cut-out in chassis.

board. When a cantilever mounting is required, an additional bracket should be used at the top of the board to provide lateral rigidity.

The various printed circuit board parameters may be calculated using Equations 7.1 and 7.2.

$$A = \frac{1}{1 - (fa/fr)^2} \tag{7.1}$$

where $$fr = \frac{fa}{\sqrt{1 - (1/A)}} = \left(\frac{K}{2\pi}\right)\sqrt{\frac{EIg}{WL^3}} = \left(\frac{K}{2\pi}\right)\sqrt{\frac{Et^3}{12\,\mu L^4}}$$

where $W = bL$

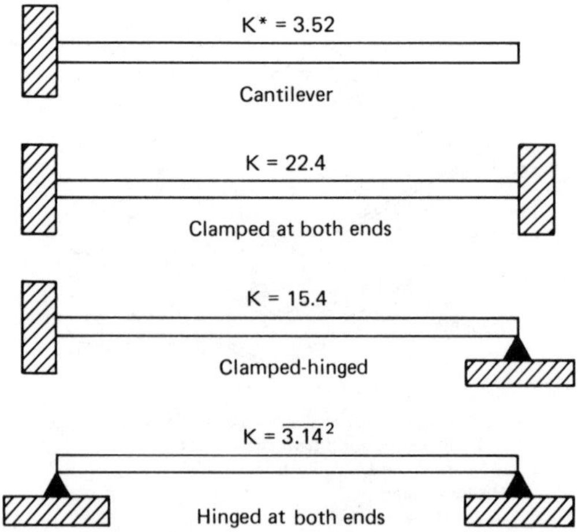

K* = 3.52

Cantilever

K = 22.4

Clamped at both ends

K = 15.4

Clamped-hinged

$K = \overline{3.14}^2$

Hinged at both ends

Figure 7-16. Types of mounting printed circuit boards.

Then the length of the board may be calculated from;

$$L = \left(\frac{K}{\pi fr}\right)^{1/2} \left(\frac{Et^3 g}{48\mu}\right) \tag{7.2}$$

A more refined method of a theoretical analysis of an electronics assembly using printed circuit boards is based on the following empirical data. As a printed circuit board vibrates up-and-down during a resonant condition, the component leads are forced to bend back-and-forth as shown in Figure 7-17, causing component damage. This is the most common mode of failure in electronics equipment. Therefore, the following method is used to establish the acceptable limits of displacement and frequencies.

To start, the maximum permissible displacement of the printed circuit board which will provide the longest fatigue life (10 million cycles or better) is computed (Figure 7-17).

$$Y_{max} = 0.003\,b \text{ (the maximum static displacement)} \qquad (7\text{-}3)$$

The natural frequency of the printed circuit board under evaluation, based on the maximum allowable static displacement of a rectangular board on the length of the shortest side, b, is computed by;

$$f_n = (9.8\,gA/0.003\,b)^{2/3}$$

(minimum natural frequency based on allowable displacement) (7.4)

where A is an empirical constant and is based on the natural frequency, f_n, and the acceleration inputs (Table 7-3). Start by assuming the value of A is 1.00 in equation 7.4. If the natural frequency is computed to be between 100 and 400 Hz., the assumed value of 1.00 is valid and you may continue on.

Using the computed natural frequency, the next step is to establish the approximate transmissibility;

$$Q = A(f_n)^{1/2} \text{ (transmissibility based on the minimum natural frequency)}$$

(7.5)

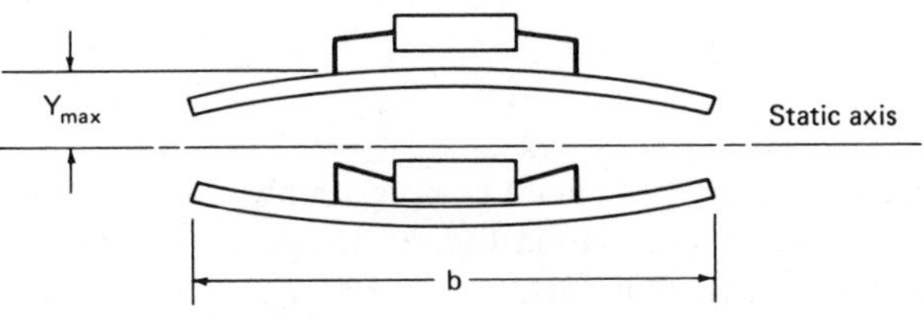

Figure 7-17. Circuit board displacement (bending).

Table 7-3. Values for Constant *A*.

g Input Acceleration	Resonant Frequencies	*A*
(Note)	50 to 100 Hz.	0.70
	100 to 400 Hz.	1.00
	400 to 700 Hz.	1.40

(values for input acceleration, for 3 to 10 g)

You have now defined all the parameters required to compute the dynamic single amplitude displacement of the printed circuit board using the equation;

$$Y = 9.8gQ/f_n \text{ (dynamic displacement)} \tag{7.6}$$

The maximum dynamic displacement during a resonant condition is computed using the transmissibility, Q, (the amplification factor of the input displacement to the output displacement) and the natural frequency of a printed circuit board based on the shortest dimension, b, and the maximum allowable displacement only. Now, the actual natural frequency of a specific printed circuit board using all the factors, size, weight, and the physical characteristics of the materials to be used is computed and compared to the natural frequency requirements computed above.

Actual natural frequency, f_n, of a specific printed circuit board.

$$f_n = \frac{\pi}{2}\left[\frac{K}{\rho}\right]^{1/2}\left[\frac{1}{a^2}+\frac{1}{b^2}\right] \text{ (natural frequency of the actual printed circuit board)} \tag{7.7}$$

$$K = Eh^3/12\,(1-\mu^2) \text{ (Stiffness factor)} \tag{7.8}$$

$$E = 2.0 \times 10^6 \text{ lbs/inch. (Modulus of elasticity for Epoxy Glass)} \tag{7.9}$$

h = printed circuit board thickness

$\mu = 0.12$ (Poisson's ratio for Epoxy Glass)

$$\rho = W/gab \tag{7.10}$$

where

W = the weight of the printed circuit board assembly
g = 386 inches/second (gravitational acceleration)
a = longest side dimension of board

b = shortest side dimension of board
A = amplification factor = yi/ya
ya = amplitude of the applied vibration (in.)
yi = amplitude of the induced vibration (in.)
fa = frequency of the applied vibration (Hz)
fr = frequency at resonance (Hz)
K = support constant (Figure 7-16)
E = modulus of elasticity of the board material (psi)
b = board width
t = board thickness
I = moment of inertia = $bt^3/12$
g = gravitational constant = 386 in./sec^2
W = unit loading (parts, board, and hardware, in psi)
L = maximum theoretical length of the board (in.)

The modulus of elasticity of the actual printed circuit board material is modified due to the assembly of the components onto the board increasing the stiffness. For a $\frac{1}{16}$-in. phenolic board, the increased stiffness is approximately 50%.

If the natural frequency of the actual printed circuit board falls under the computed acceptable limits in Equation 7.4, then the natural frequency must be raised by the addition of stiffeners or by increasing the thickness of the board material.

The printed circuit board is only one part of the vibrating system. The natural frequency of the chassis into which the printed circuit is to be installed must also be evaluated and should not exceed one-half of the natural frequency of the printed circuit board.

For a preliminary analysis of the chassis, we may consider the chassis as a uniformly loaded, simply supported beam.

$$f_n = \frac{1}{2\pi}\sqrt{\frac{384\,EIg}{5\,WL^3}} \tag{7.11}$$

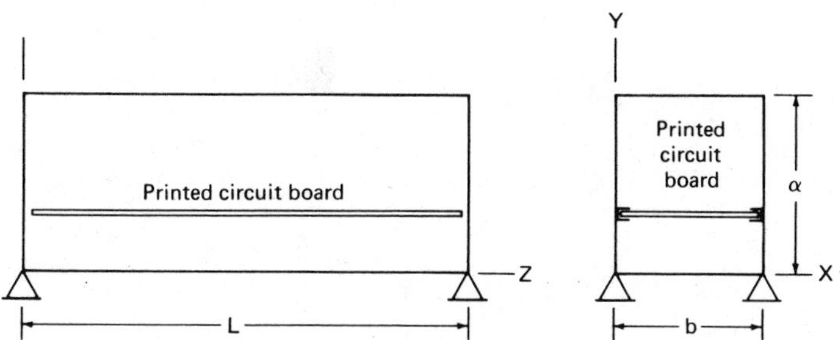

Figure 7-18. Uniformly loaded beam.

$$I = \frac{bd^3}{3} \text{ (moment of inertia)} \tag{7.12}$$

$E = 10 \times 10^6$ (modulus of elasticity for aluminum)
$g = 386$ in./sec (gravitational acceleration)
L = length of the beam (chassis)
W = weight of the chassis

8

SUBASSEMBLIES AND ASSEMBLIES

Experience has shown that the so-called scrambled parts and wiring layouts are unsatisfactory for ease of manufacturing and maintenance. It is desirable to sectionalize or modularize subassemblies into a series of replaceable units, with the parts and wiring arranged to provide for maximum accessibility (Figure 8-1). Each unit should contain all of the components required to perform a specific function, such as amplifying, rectifying, frequency controlling, etc. (Figures 8-2 and 8-3).

The increase in the manufacturing cost introduced into the design due to the unitization of subassemblies, when compared to the cost of the added servicing facilities, will be justified since maintenance costs usually exceed the original cost of the equipment. Furthermore, the size of the overall equipment can be greatly reduced since the increase in permissible density with the ease of maintenance afforded by the unitization techniques offers the designer greater freedom.

The folded sheet-metal chassis is the most common type used today. It offers excellent rigidity with a minimum of metal and is easily fabricated. The larger parts are normally mounted on top of the chassis, and the wiring and smaller components are installed underneath. Many standard sheet-metal chassis are readily available commercially.

A large number of electronics equipments are designed for installation into a standard 19.00 EIA rack or chassis. These racks are commercially available and are in wide use throughout the electronics industry. In many instances, the equipment being designed will be intended as an addition to some existing unit installed into a standard rack. Rack-and-panel equipment offers a proven method of modularizing subassemblies and is installed directly into an existing system (Figures 8-4 and 8-5).

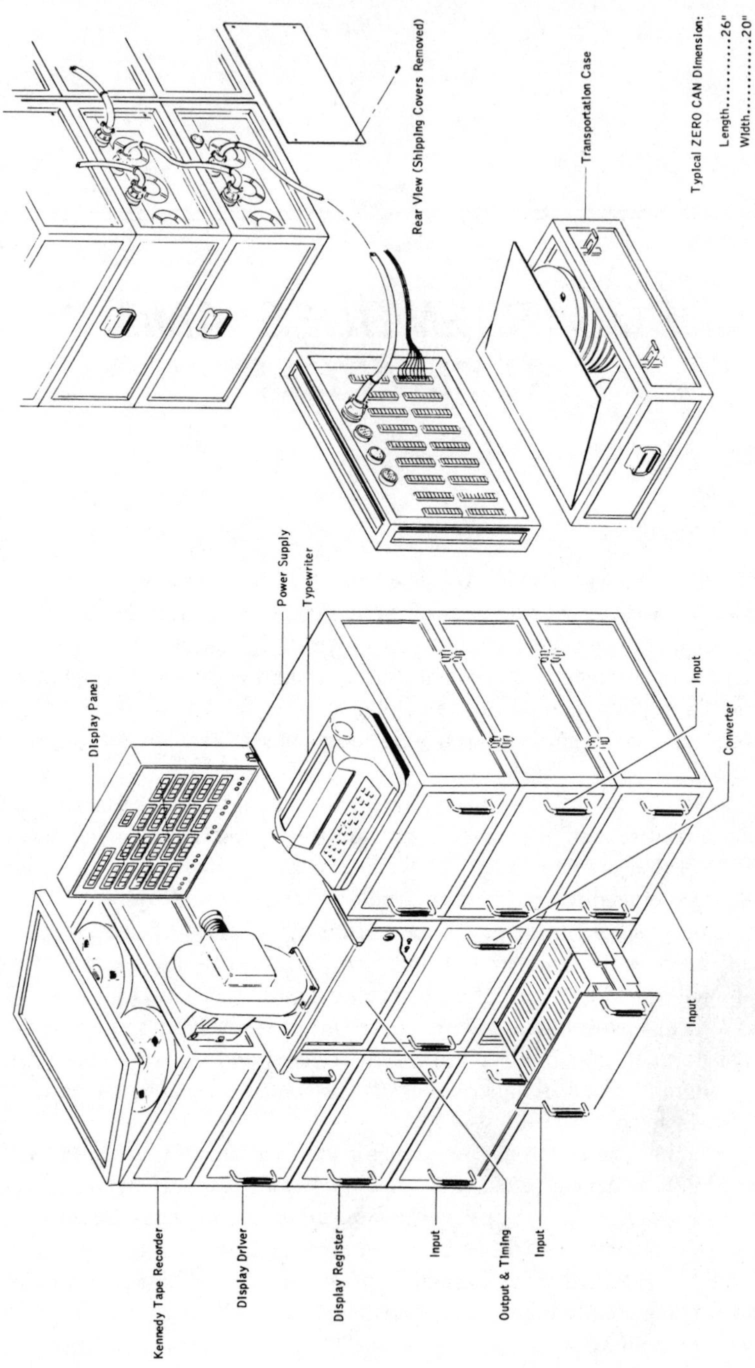

Figure 8-1. Rack and panel field assembly.

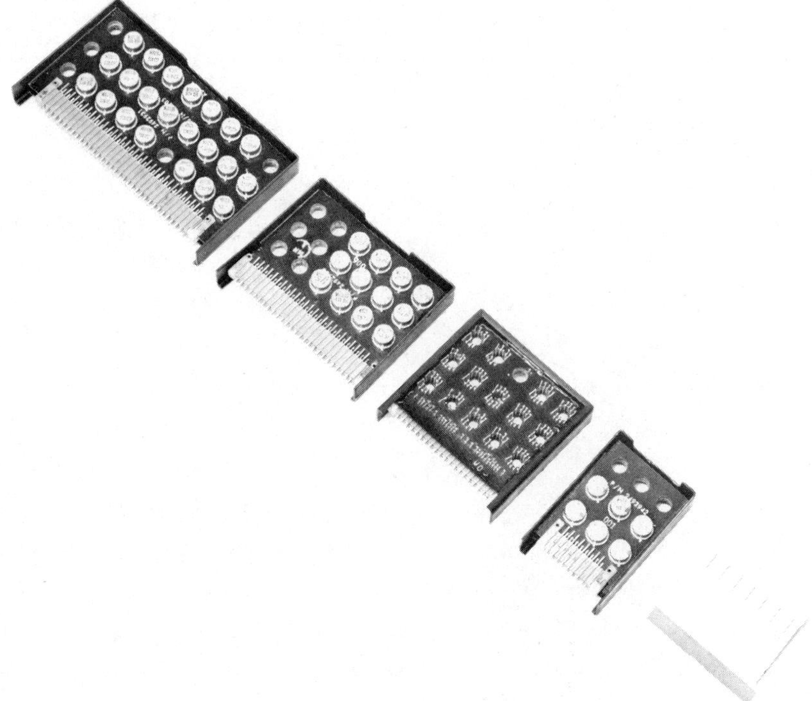

Figure 8-2. Plug-in module assembly.

Mounting boards and strips are often used in mounting small subassemblies. Copper-laminated plastics are of particular value for mounting small components in high-frequency units since they provide a means of signal isolation by facilitating grounding directly to the chassis. This method of circuit isolation is also commonly used in the design of computer equipment to reduce signal interference and magnetic coupling. (More detailed information is presented in Chapter 11.) A small modularized unit may consist of a subassembly encased in a shell or housing containing a plug-in connector (Figures 8-3 and 8-6 through 8-10). Plug-in assemblies offer the advantage of being easily replaced; also, the defective module can be returned to the factory for service with a minimum of machine downtime.

Enclosures may be enclosed or ventilated depending on the heat-removal requirements. When heat dissipation is critical, the enclosure may be perforated to allow free-moving air to pass through (Figures 8-11 and 8-12). Sealed units may be encapsulated with metal-filled plastics to improve the heat-transfer capabilities. These plastic-filled units need not be installed into a metal shell or housing because the unit as cast with the components, held and protected by the

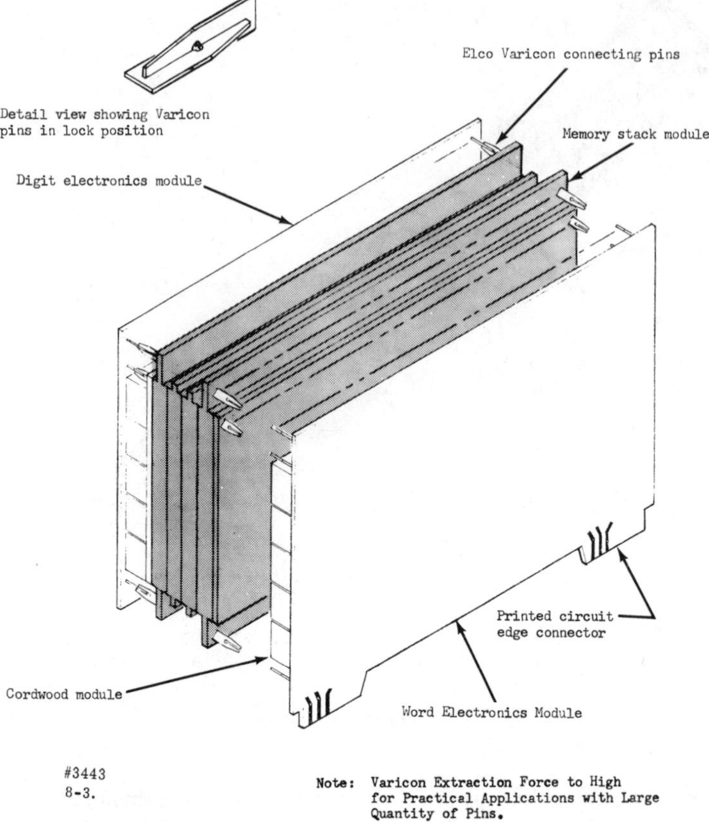

Figure 8-3. Printed circuit plug-in module assembly.

plastic, provides a rugged unit (Figures 8-9 and 8-10). The plastic also provides protection from the elements such as moisture, saltwater, and harsh atmospheric conditions. The one disadvantage of the plastic-embedded unit is that it is not easily repaired.

CONNECTORS

The following merit special consideration in plug-in design:

- A single multiterminal connector simplifies alignment and mating problems (Figure 8-13).
- Exposed pins located in the extractable unit should be recessed to prevent physical or electrical damage that might otherwise result from misalignment.

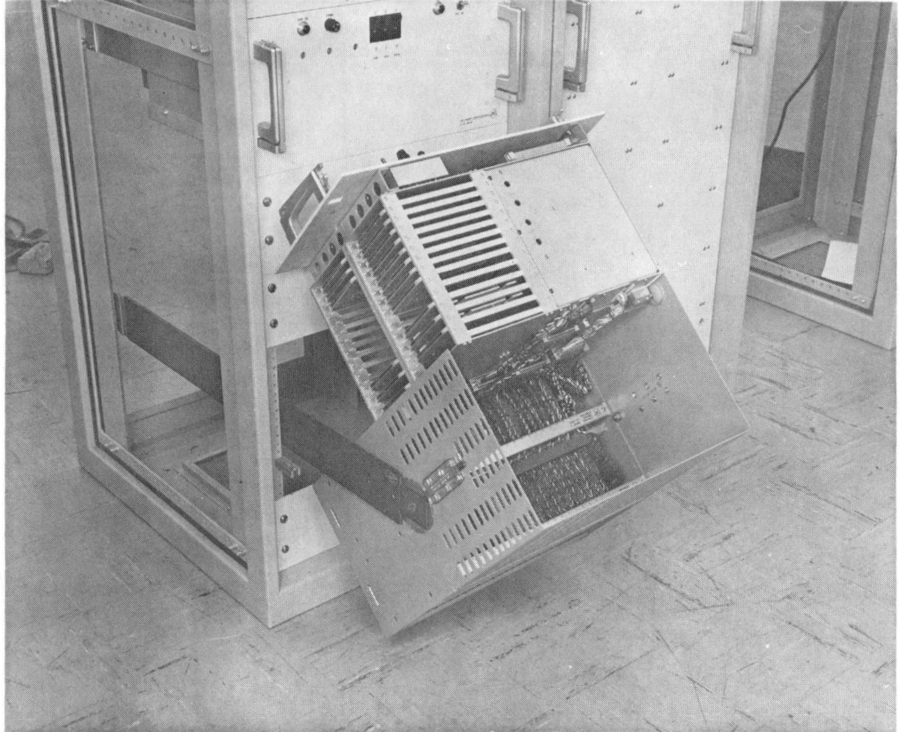

Figure 8-4. Rack and panel tilt assembly.

- Piloting devices and hold-down clamps should be designed to prevent undue stressing of the connector body. Floating contacts are desirable.
- A minimum of 10%, or two excess connector terminals, whichever is greater, should be provided.

Rack-and-panel construction is usually characterized by the location of connectors and terminals at the rear of each assembly. Where accessibility is limited, it is desirable to employ quick-release connectors and doors that are large enough to facilitate easy accessibility (Figure 8-1 and 8-15).

Terminal strips of the barrier type can be employed, if adequate room is provided for making the connections and if some additional length of cable is provided to facilitate withdrawal (Figures 8-16 and 8-17).

CLAMPING AND SUPPORTING

Techniques for clamping and supporting units on flat, sheet-metal chassis vary according to particular circumstances. Mounting projections, such as lugs,

Figure 8-5. Standard 19-inch rack and panel assembly.

brackets, and folds, are used where space permits. Material for projections must not be brittle or of low impact strength. If threaded fasteners cannot be easily or quickly disassembled, bolted tie-rods, slide fasteners (Figure 8-18), knurled nuts, and latches may be employed. Captive types of fasteners are recommended.

CAPTIVE FASTENER

Through-holes in the body can be used if the subassembly is comparatively light and close grouping is desired. Members strong enough to withstand full tension of the mounting bolt should be used (Figure 8-19).

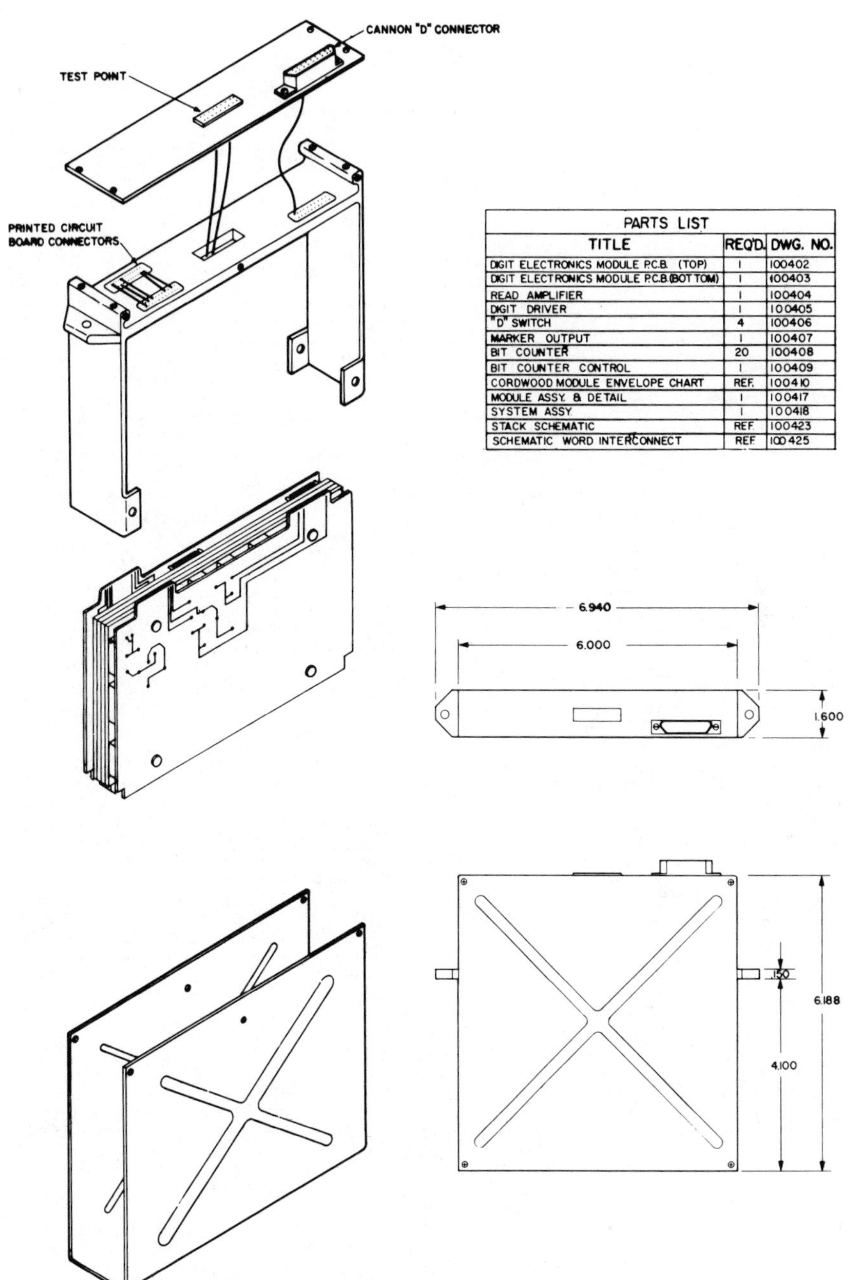

PARTS LIST		
TITLE	REQ'D.	DWG. NO.
DIGIT ELECTRONICS MODULE P.C.B. (TOP)	1	100402
DIGIT ELECTRONICS MODULE P.C.B. (BOTTOM)	1	100403
READ AMPLIFIER	1	100404
DIGIT DRIVER	1	100405
"D" SWITCH	4	100406
MARKER OUTPUT	1	100407
BIT COUNTER	20	100408
BIT COUNTER CONTROL	1	100409
CORDWOOD MODULE ENVELOPE CHART	REF.	100410
MODULE ASSY. & DETAIL	1	100417
SYSTEM ASSY.	1	100418
STACK SCHEMATIC	REF.	100423
SCHEMATIC WORD INTERCONNECT	REF.	100425

Figure 8-6. Plug-in module for severe environments.

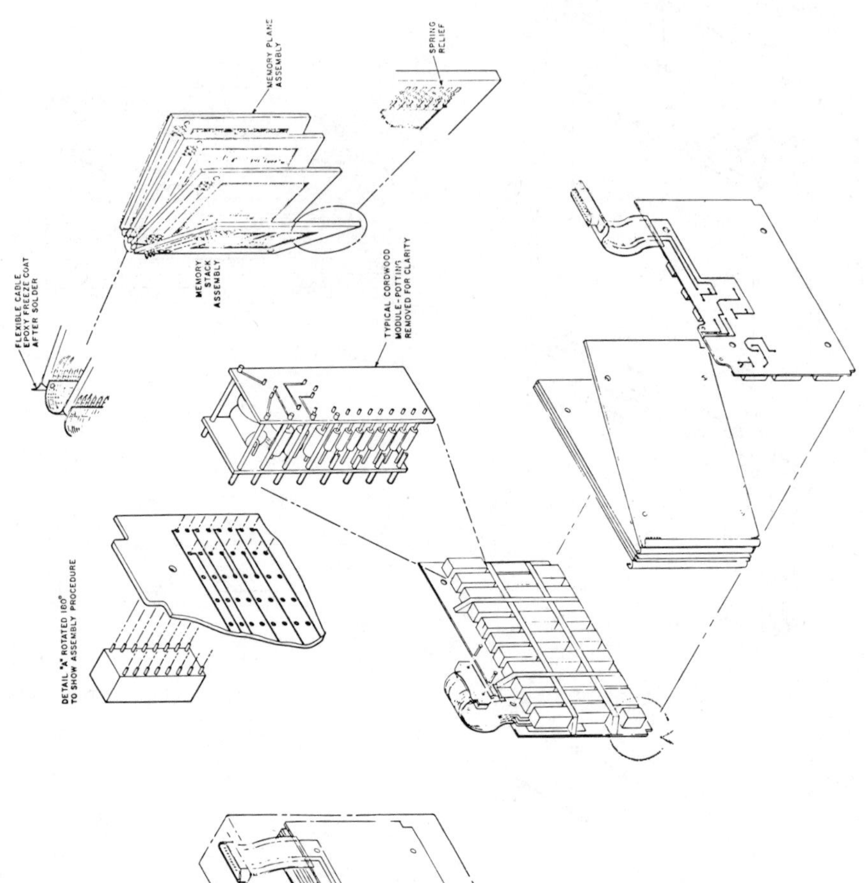

Figure 8-7. Modular ruggedized assembly.

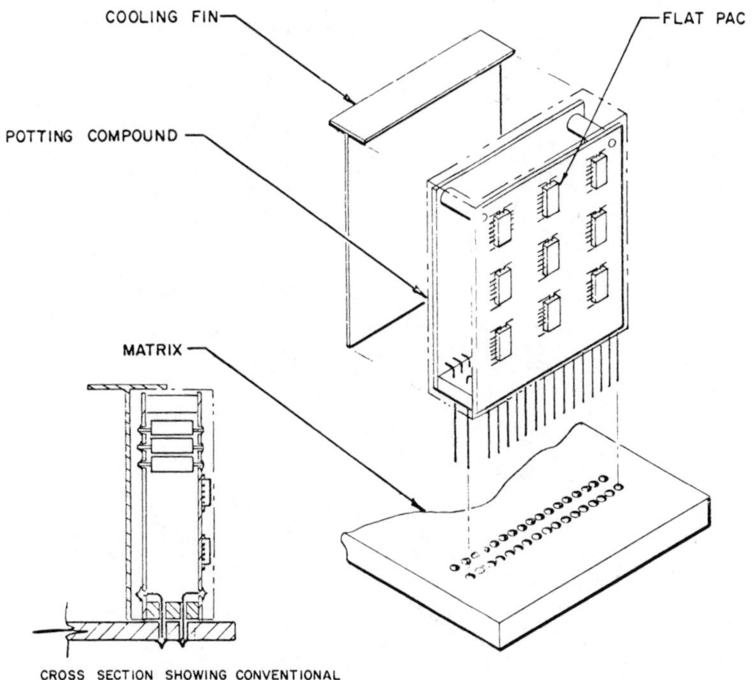

Figure 8-8. Flat pack module.

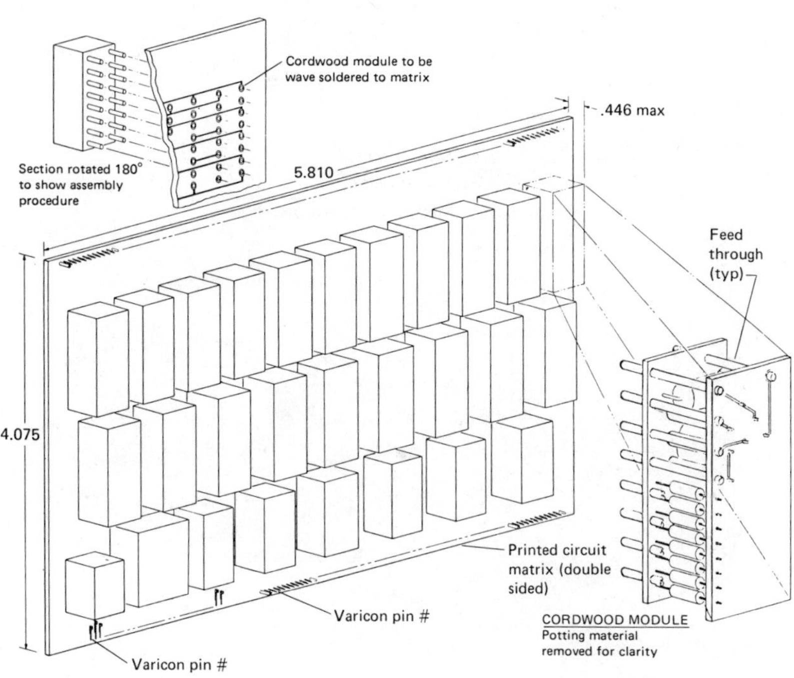

Figure 8-9. Cordwood module assembly.

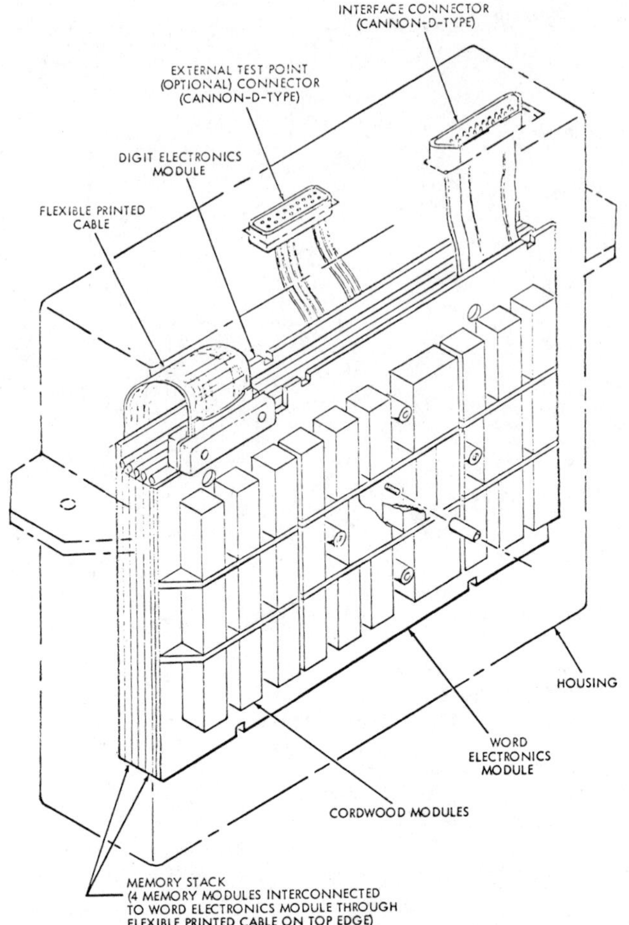

Figure 8-10. Memory system package.

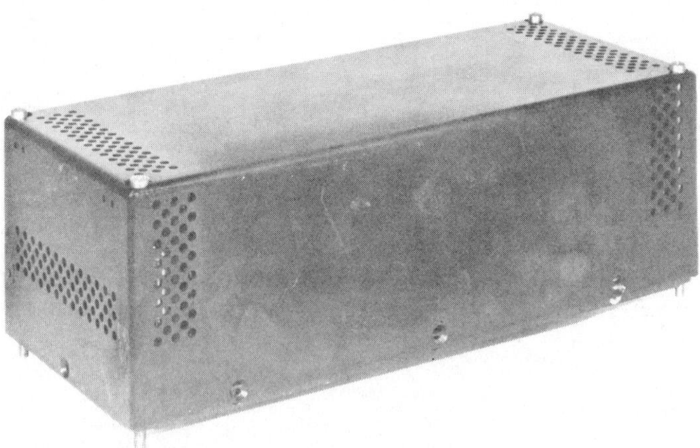

Figure 8-11. High temperature cover.

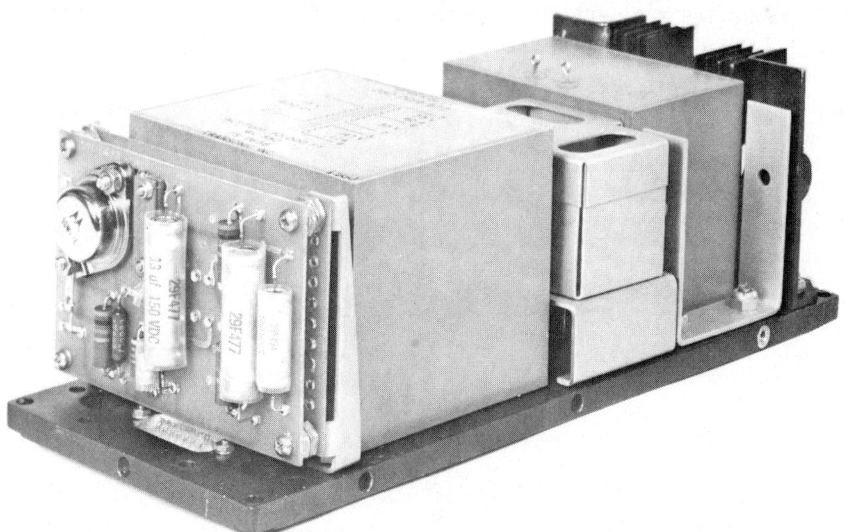

Figure 8-12. High temperature assembly.

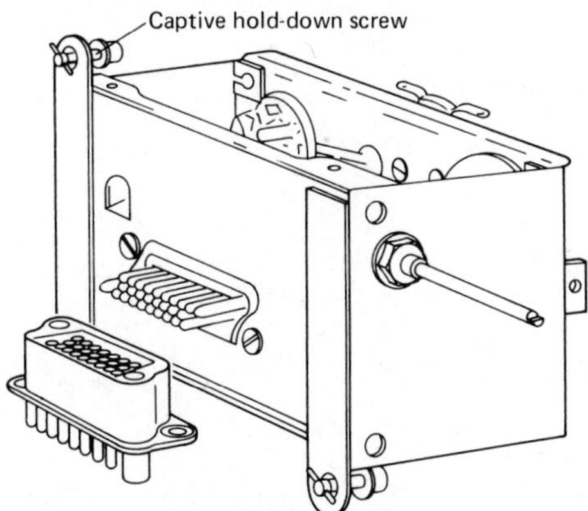

Figure 8-13. Multicontact connector used in plug-in arrangement.

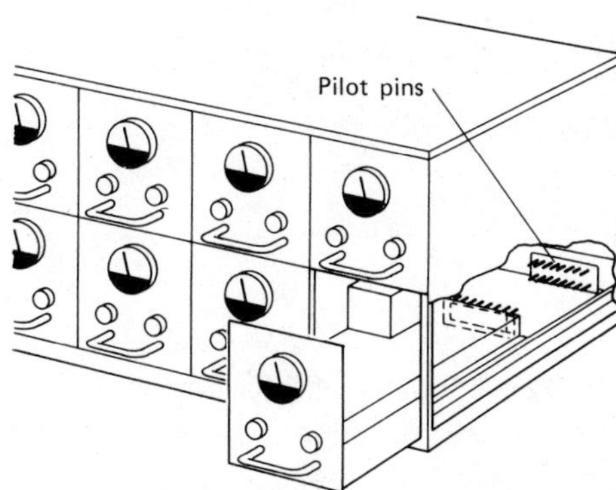

Figure 8-14. Drawer construction utilizing plug-in arrangement.

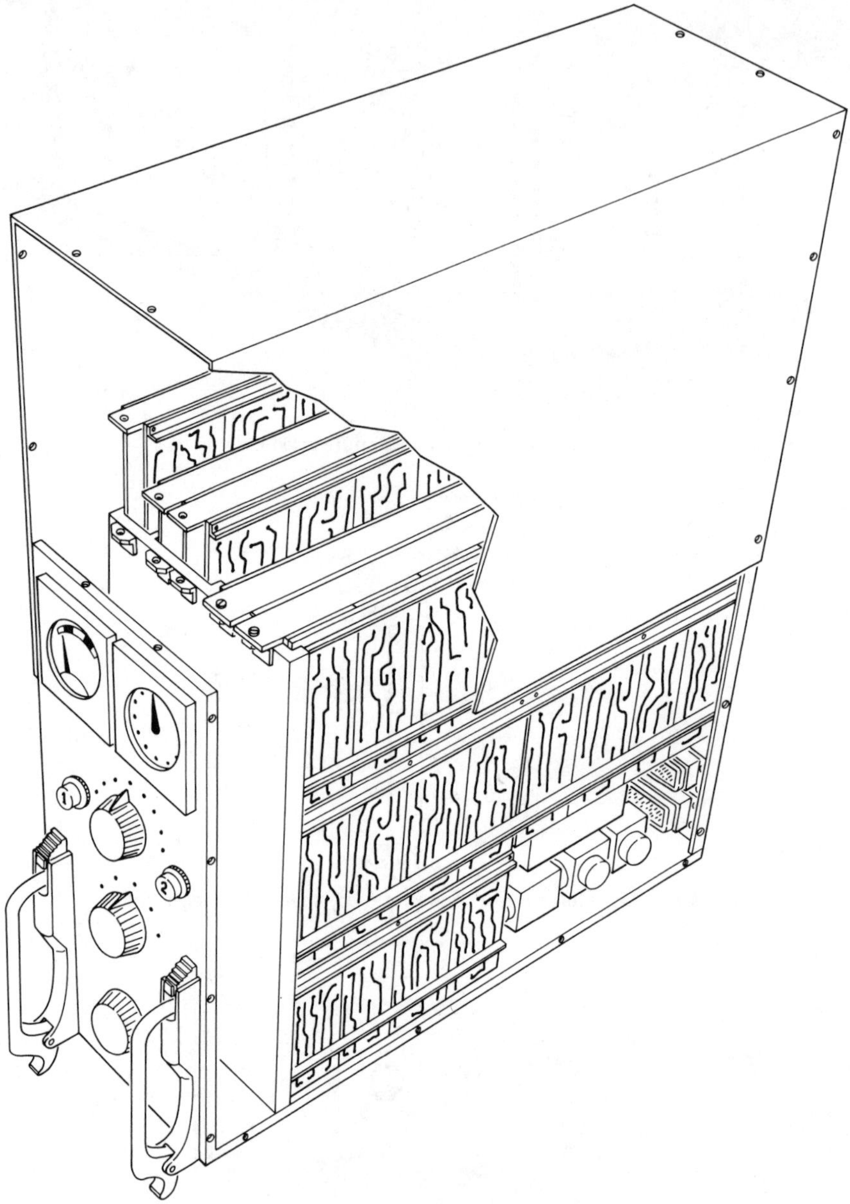

Figure 8-15. Airborne modular assembly.

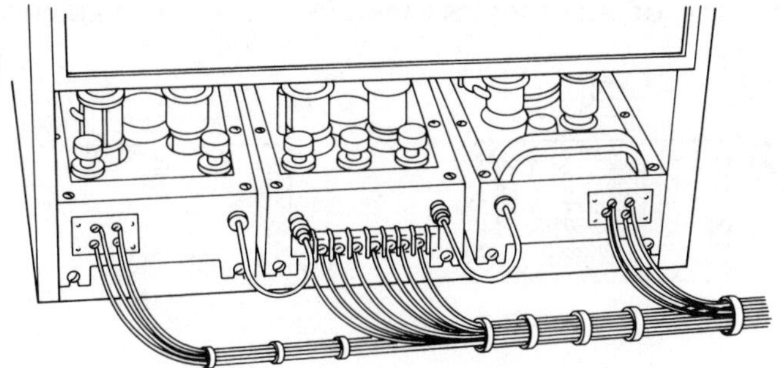

Figure 8-16. Terminal block cabel wiring.

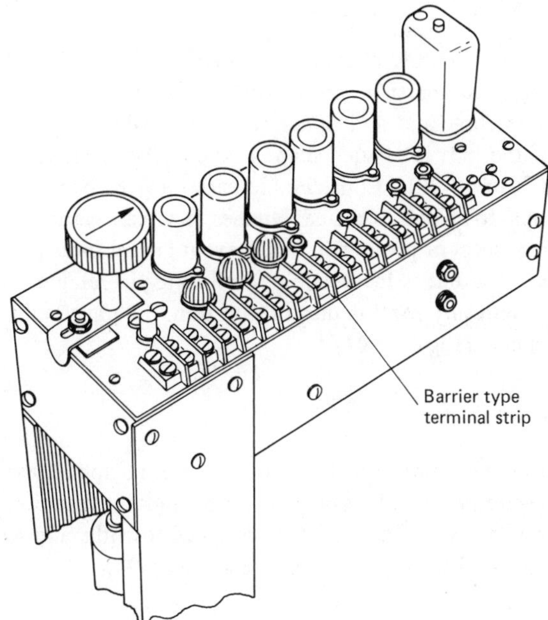

Figure 8-17. Barrier type terminal strip.

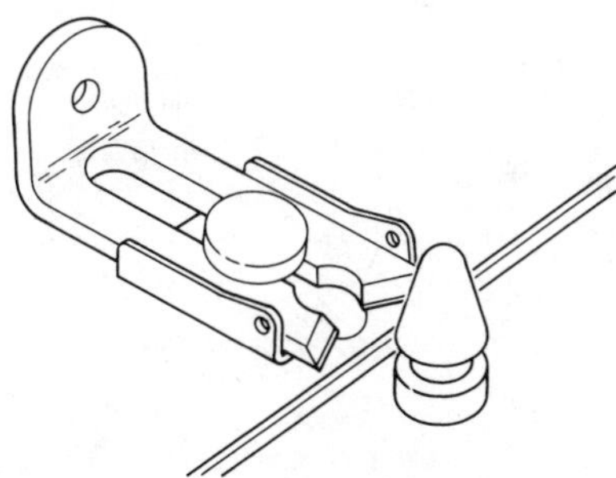

Figure 8-18. Clamp installation.

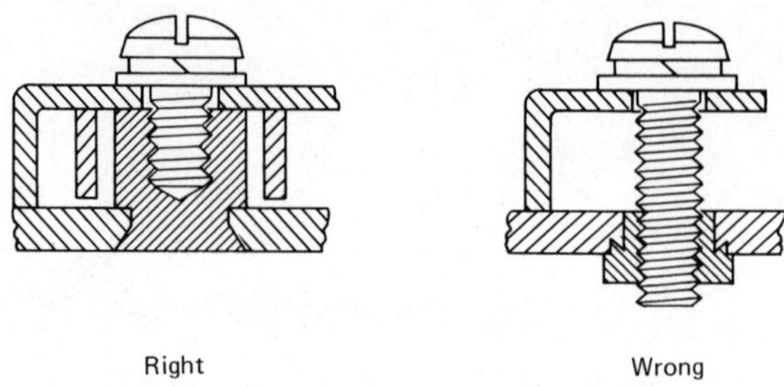

Figure 8-19. Mounting screw installation.

For heavy loads or accurate alignment, through-studs attached to the unit and projecting through the chassis form a satisfactory mounting. Straps, clamps, retaining rings, framebrackets, and similar items are used where structural strength or the profile of the unit is of such shape that customary fastening methods are not applicable (Figure 8-20). Considerable care must be exercised in designing straps and clamps to avoid localized stresses that can cause loosening under shock. A locating recess or projection will prevent lateral slippage.

Where one end of a unit is inaccessible, a piloting-pin retainer may be used at that end. To prevent any motion during vibration, the pin should be tapered to fit snugly in its socket (Figure 8-21).

ENCLOSURES

The primary purpose of enclosures is to protect internal apparatus from external elements. Enclosures vary widely in their shape and construction according to the type of service desired. They must be designed to withstand severe temperature and corrosion conditions (Figures 8-22 through 8-24).

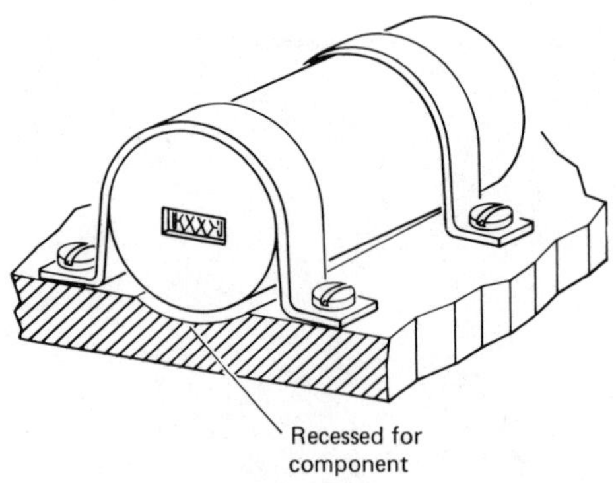

Figure 8-20. Strap mounting.

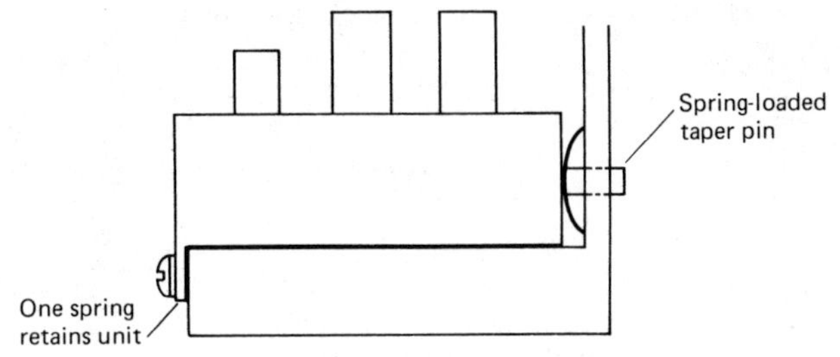

Figure 8-21. Pilot pin installation.

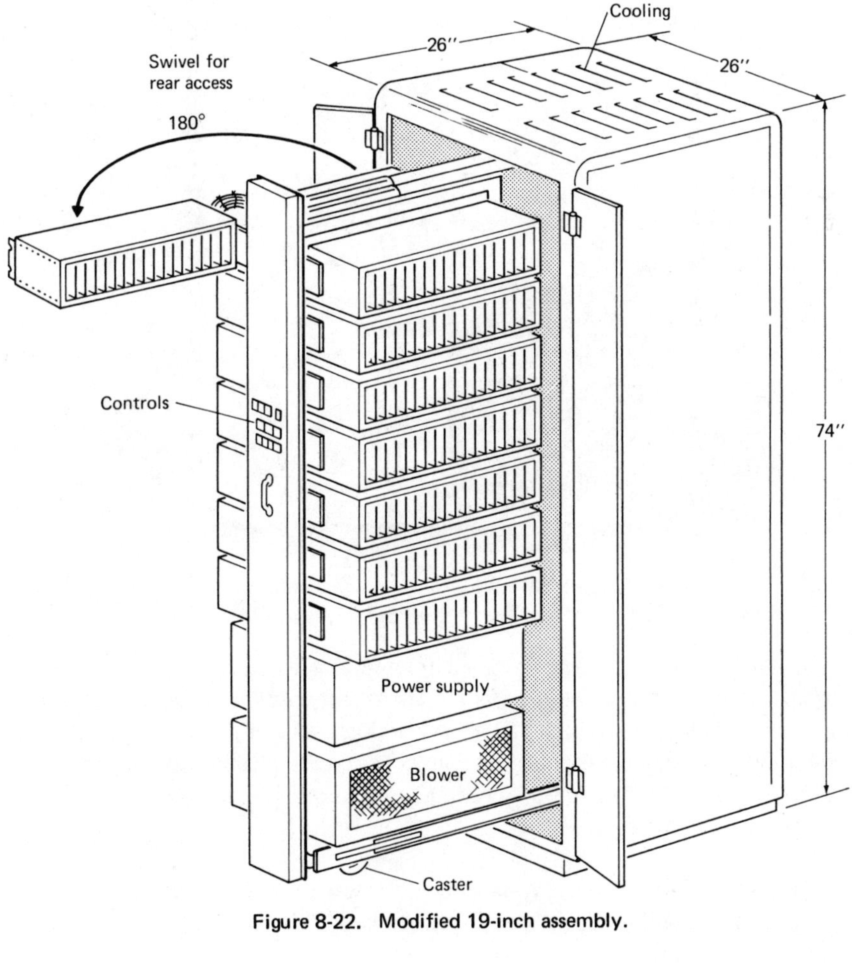

Figure 8-22. Modified 19-inch assembly.

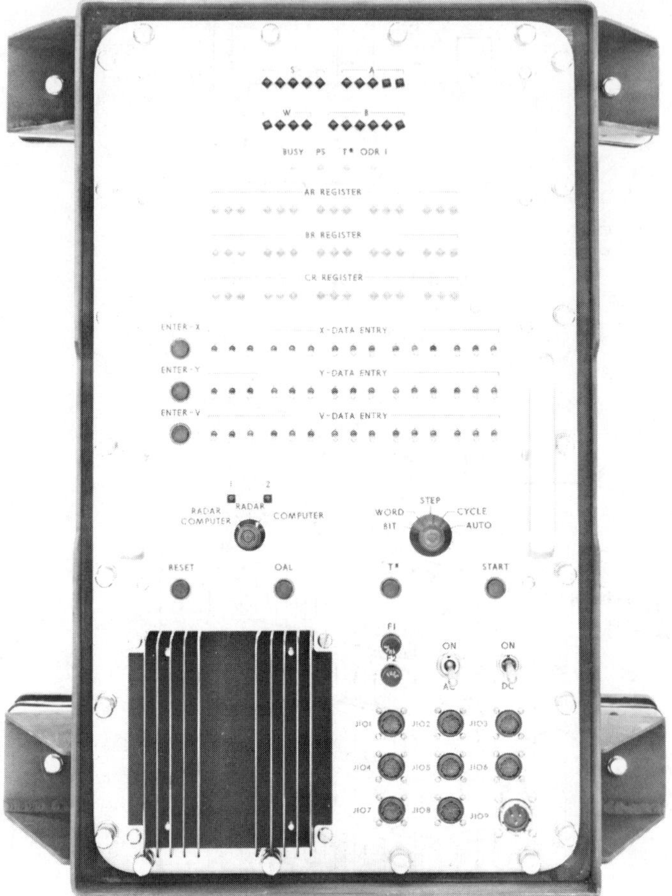

Figure 8-23. Severe environment assembly.

For electronics equipment designed for military environments, the type, or degree, of enclosure may be specified as *protected*, *splashproof*, *fogproof*, *watertight*, *submersible*, or *gunblast-proof.* Accordingly, one or more of the following tests to verify conformance must be conducted on the final design. Military standards are available that specify testing procedures for enclosures and define degrees of enclosure.

1. *Hose test:* water or spray is directed toward the equipment from a hose adjusted at various angles for prescribed periods of time.

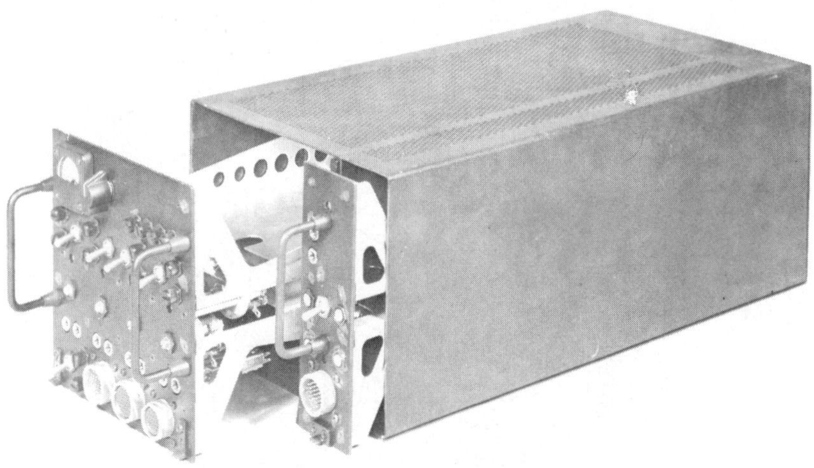

Figure 8-24. Airborne assembly.

2. *Submergence test:* water pressure and duration of test are specified according to the required degree of enclosure.
3. *Gunblast test:* the enclosure is located near a test gun when fired.
4. *Shock, vibration, and inclination tests.*

In addition, in shipboard electronics equipment, for below-deck use, enclosures are usually lighter and more compact, but must be fogproof. More open, or protected, construction may be used for fire-control equipment, and carefully designed to be watertight, and structurally rugged. Submersible equipment, in particular, should be corrosion-proof and obviously pressure-tight.

Dust filters should be used in equipment using forced-air cooling. They should be readily replaceable, easily cleaned, and flush-mounted.

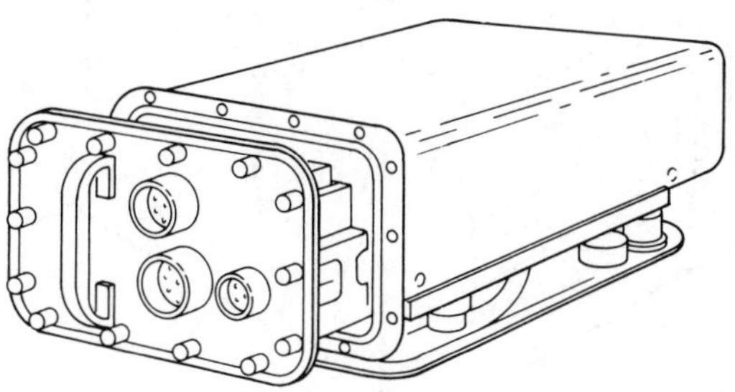

Figure 8-25. Sheet-metal enclosure.

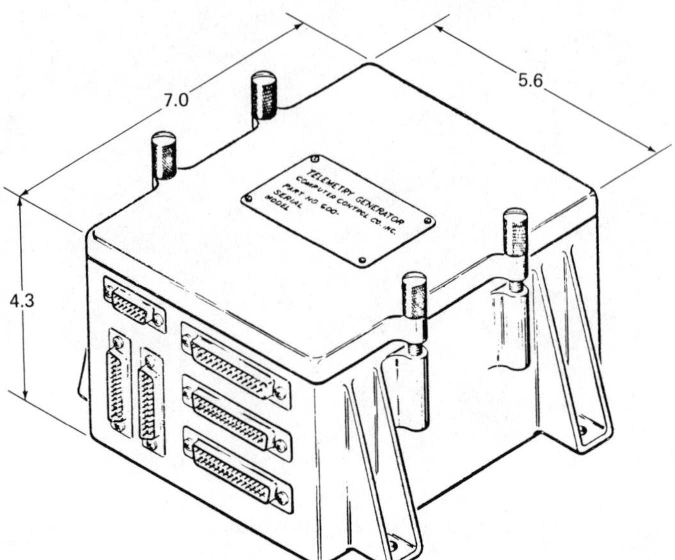

Figure 8-26. Cast housing.

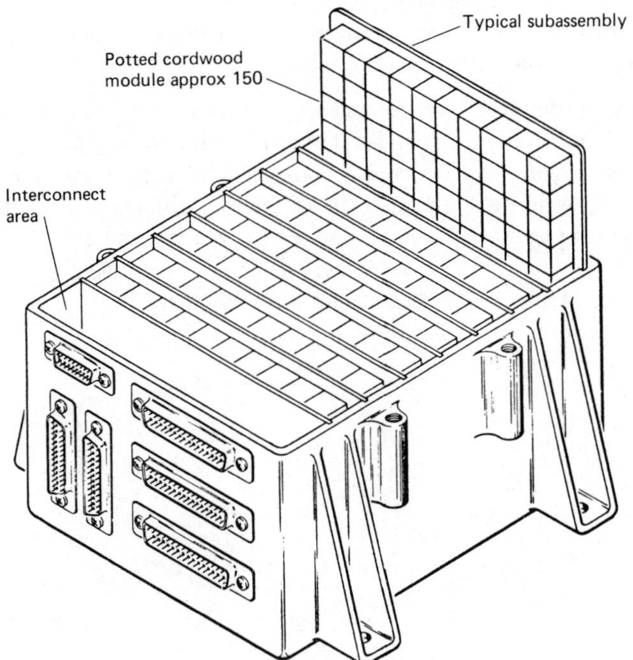

Figure 8-27. Cast housing with cover removed.

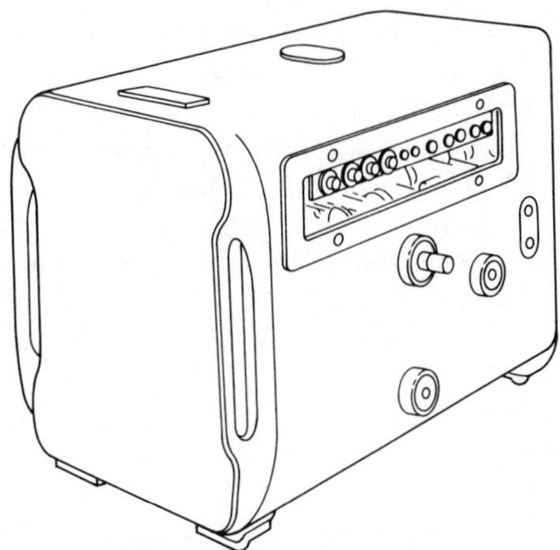

Figure 8-28. Cast enclosure.

STANDARD ENCLOSURES FOR ELECTRONICS EQUIPMENT

Sheet-metal enclosures are preferable to cast enclosures because of advantages in weight and size. The height rather than the depth or width should be extended because of limited floor space. Good practice for many applications is to mount an apparatus on a framework-panel structure that can be inserted into the enclosure. If gasketing is required, the front of the enclosure should be flanged to provide strength and a seating area for the gaskets (Figure 8-25). Chassis runners should be securely fastened to enclosure inner walls. Corners and edges must be rounded to protect personnel.

Internal framing is omitted in small welded consoles where loads are well distributed (Figure 8-27). Sheet material of reasonable thickness can usually provide sufficient rigidity for mounting doors, slides, and covers. Slides and other retractable devices assure accessibility.

Cast enclosures are readily adaptable to sealing. Materials that are weak and brittle, such as cast iron and ordinary die castings, should not be used. Cast steel and certain aluminum alloys are recommended (Figures 8-26 to 8-28).

Large enclosures, designed to carry heavy loads, should be all-welded, heavy sheet construction. Sufficient framing and bracing are recommended to support internal apparatus and to improve overall structural rigidity.

Sectionalizing is necessary where the size of a single-unit enclosure precludes passage through access doors and openings as a single unit.

TEMPERATURE

Care must be taken in locating temperature-critical materials. Many plastics, varnishes, oils, adhesives, and other materials deteriorate at high temperatures.

Loss of the plasticizer is accompanied by brittleness and shrinkage, and oils tend to oxidize and form varnishes. Consequently, these materials must have life-temperature ratings at least equal to those anticipated for the equipment.

Sunlight causes organic materials, sensitive to heat and ultraviolet light, to deteriorate. Vinyl-jacketed cables having polyethylene dielectrics are unsatisfactory for continuous outdoor exposure because their temperature-plasticity effects cause the inner conductor to move off center. Corrosion of the outer braid may also occur because of humidity.

Manually operated controls should be fitted with nonmetallic knobs that have a low heat-transfer coefficient.

Large aluminum or magnesium sections are particularly susceptible to dimensional changes, since their thermal coefficient of expansion is relatively high; it is twice that of steel.

Many plastic materials become brittle at extremely low temperatures and break easily when subjected to undue stresses. This occurs also to an appreciable extent in various types of metal. Polyethylene dielectrics, widely used in coaxial cable, crack readily at $-60^{\circ}C$ if appreciable flexure is encountered.

9

DESIGN CONSIDERATIONS FOR SPACE ELECTRONICS

Not many years ago, electronics packaging engineers were faced only with problems of mechanical integrity and circuit-cooling. They were designing circuits primarily for the earthbound electronics industry; if their packages were destined for use in aircraft, the greatest worry was vibration. For a majority of the products, all that was needed to package was an aluminum chassis, some protection against shock and vibration, an occasional tube shield, and a box, painted black with many holes drilled for ventilation, which would protect the circuits from the wind, rain, and other environmental hazards.

Today's electronics–in space satellite systems for the most part–can be expected to encounter more vigorously adverse environments. Reliable service is one of the basic goals expected of control, guidance, communications, and telemetry equipment in today's space vehicles. A communications satellite, for example, may be expected to provide trouble-free performance for 30,000 hours or more. Electronics comprise a large portion of a satellite system or space vehicle, and extra precautions must be taken to assure successful operation, since the heterogeneous nature of deep space becomes the enemy of humanity's handiwork (Figure 9-1).

The natural environments discussed in this chapter–space vacuum, various types of radiation, thermal shock, and physical erosion from micrometeorites and other high-speed spaceborne matter–were compiled from a literature survey and are presented here as an aid to the design of space-vehicle electronic equipment.

An understanding of the hazards of space environments forms the basis of highly reliable equipment design; and proper material selection determines the life expectancy of the satellite or space-vehicle system. Thermal control devices,

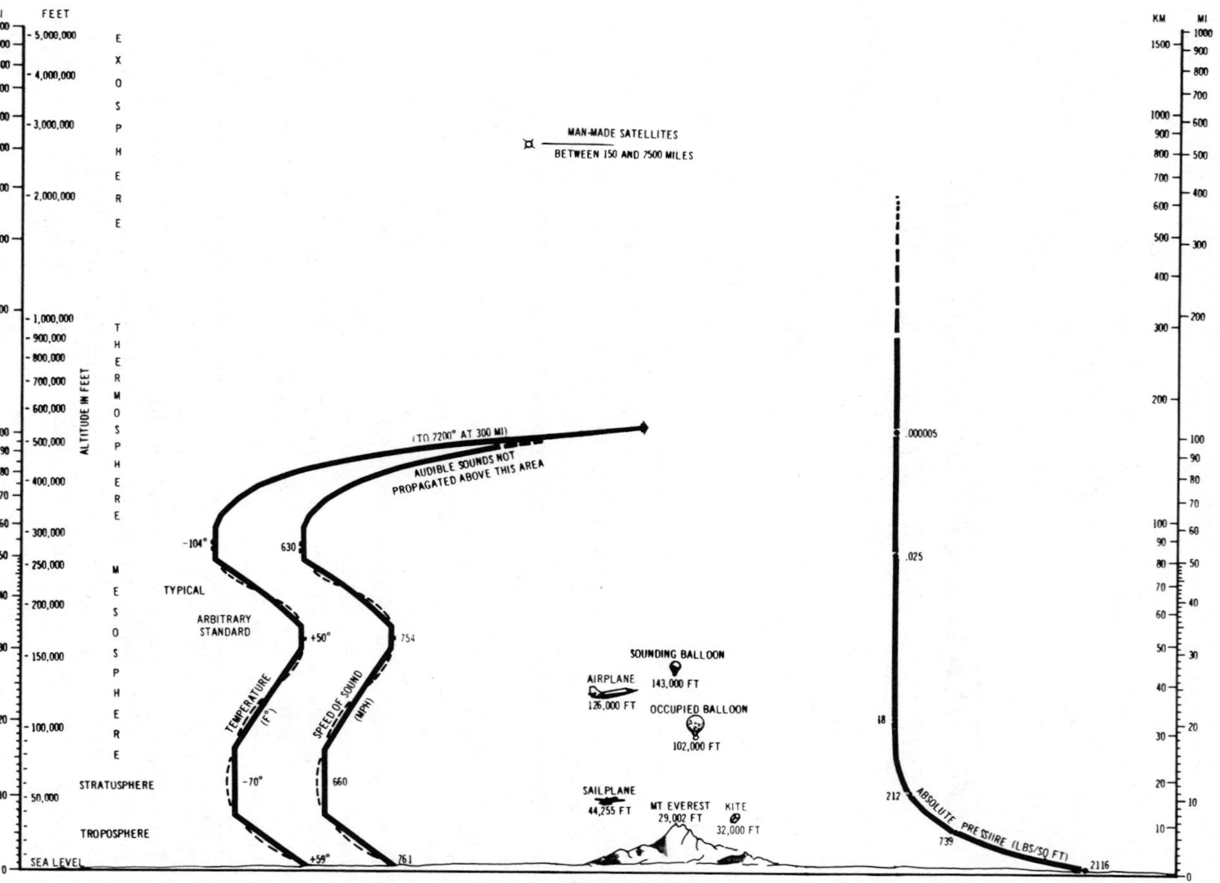

Figure 9-1. Characteristics of the Earth's atmosphere.

radiation shields, and structural members must be efficiently "engineered" into the design of the package for defenses against the hostile environments.

THE VACUUM OF SPACE

Pressures of about 10^{-6} mm of mercury (mm Hg) will be encountered at the altitude of 125 miles (200 km) from earth and will approach 10^{-12} mm Hg at and beyond the altitude of 400 miles (6500 km). The absorbed gas layers, normally present on metal surfaces in the earth's atmosphere, are lost in the vacuum of space. This loss will result in increased friction and possibly "cold welding." This environment also affects the mechanisms of crack propagation, seriously reducing usable strength and fatigue resistance in metals.

Many of the common engineering materials will suffer the loss of surface layers by the means of evaporation and sublimation at high temperatures. Metals and alloys are generally quite stable in the high vacuum of space at normal operating temperatures. Vacuum effects on the organic engineering materials are varied, and extreme care must be exercised in selecting them. These materials are more prone to deterioration from evaporation or sublimation than are metals. Careful selection of materials with low sublimation and evaporation rates, and allowance of sufficient material thickness for the sacrificial deterioration of equipment housing during the expected operating life of the package, will lessen this problem. Arc-over and corona discharge can also be prevented with adequate insulation materials.

Table 9-1 gives various sublimation rates for the more common engineering metals and semiconductors in high vacuum. Note that these rates can be reduced considerably by nonporous coatings, such as oxides of phosphates. These materials and their respective rates are listed in the order of increasing stability under high vacuum conditions. Table 9-2 lists 50 of the more common polymers of organic engineering materials showing the temperature at which a given amount of the polymer decomposes in high vacuum.

THE PROBLEM OF THERMAL SHOCK

For a vehicle or satellite orbiting in the solar system, the sun will be the source of most of the thermal, electromagnetic, and particle radiation. To a less degree—and with much less effect—earthsine (albedo) and direct earth radiations will be encountered. Further down the list of sources appear the stars, other planets, the moon, and interstellar space.

The total resultant energy within the range of several thousand miles near the earth is 7.38 Btu/ft^2/min., or it can be read as 1400 W/m^2. This solar constant is the total radiation energy falling perpendicularly on a unit area of earth at the mean distance of the sun to the earth.

Table 9-1. Sublimation Rates of Metals and Semiconductors in High Vacuum (Jaffe and Rittenhouse 1962).

ELEMENT	SYMBOL	MELTING POINT	10^{-5} CM/YR	10^{-3} CM/YR	10^{-1} CM/YR
CADMIUM	Cd	320°	40°	80°	120°
SELENIUM	Se	220	50	80	120
ZINC	Zn	420	70	130	180
MAGNESIUM	Mg	650	110	170	240
TELLURIUM	Te	450	130	180	220
LITHIUM	Li	180	150		210 280
ANTIMONY	Sb	630	210	270	300
BISMUTH	Bi	270	240	320	400
LEAD	Pb	330	270	330	430
INDIUM	In	160	400	500	610
MANGANESE	Mn	1240	450	540	650
SILVER	Ag	960	480	590	700
TIN	Sn	230	550	660	800
ALUMINUM	Al	660	550	680	810
BERYLLIUM	Be	1280	620	700	840
COPPER	Cu	1080	630	760	900
GOLD	Au	1060	660	800	950
GERMANIUM	Ge	940	660	800	950
CHROMIUM	Cr	1880	750	870	1000
IRON	Fe	1540	770	900	1050
SILICON	Si	1410	790	920	1080
NICKEL	Ni	1450	800	940	1090
PALLADIUM	Pd	1550	810	940	1100
COBALT	Co	1500	820	960	1100
TITANIUM	Ti	1670	920	1070	1250
VANADIUM	V	1900	1020	1180	1350
RHODIUM	Rh	1970	1140	1330	1540
PLATINUM	Pt	1770	1160	1340	1560
BORON	B	2030	1230	1420	1640
ZIRCONIUM	Zr	1850	1280	1500	1740
IRIDIUM	Ir	2450	1300	1500	1740
MOLYBDENUM	Mo	2610	1380	1630	1900
CARBON	C	3700	1530	1680	1880
TANTALUM	Ta	3000	1780	2050	2300
RHENIUM	Re	3200	1820	2050	2300
TUNGSTEN	W	3400	1880	2150	2500

Note: All temperatures are given in °C.

Table 9-2. Temperatures at Which Polymers Decompose in High Vacuum at the Rate of a 10% Weight Loss Per Year (Jaffe and Rittenhouse 1962, Jaffe 1955).

POLYMERS	°CENTIGRADE
POLYAMIDES ("NYLON")	30-210°
SULFIDE	40
CELLULOSE NITRATE ("NITRON")	40
CELLULOSE, OXIDIZED	40
METHYL ACRYLATE	40-150
ESTER	40-240
EPOXY	40-240
URETHANE	70-150
VINYL BUTYRAL ("BUTACITE")	80
VINYL CHLORIDE	90
LINSEED OIL	90
CHLOROPRENE ("NEOPRENE")	90
ALKYD	90-150
METHYL METHACRYLATE ("LUCITE")	100-200
ACRYLONITRILE	120
ISOBUTYLENE-ISOPRENE (BUTYL RUBBER)	120
STYRENE-BUTADIENE ("PLIOLITE 3-6")	130
STYRENE	130-220
PHENOLIC	130-270
BUTADIENE-ACRYLONITRILE	150-230
VINYL ALCOHOL ("ELVANOL")	150
VINYL ACETATE	160
CELLULOSE ACETATE BUTYRATE ("TENITE II")	170
CELLULOSE	180
CARBONATE	180
METHY STYRENT	180-220
CELLULOSE ACETATE ("FIBESTOS")	190
PROPYLENE	190-240
RUBBER, NATURAL	190
ISOPRENE	190
MELAMINE	190
SILICONE ELASTOMER ("SILASTIC")	200
ETHYLENE TEREPHTHALATE ("MYLAR, DACRON")	200
ISOBUTYLENE	200
VINYL TOLUENE	200
STYRENE, CROSS-LINKED	230-250
BUTADIENE-STYRENE (GR-S=SBR)	240
VINYL FLUORIDE	240
ETHYLENE, LOW DENSITY	240-280
BUTADIENE	250
VINYLIDENE FLUORIDE-HEXAFLUOROPROPENE	250
CHLOROTRIFLUOROETHYLENE	250
CHLOROTRIFLUOROETHYLENE-VINYLIDENE FLUORIDE	260
VINYLIDENE FLUORIDE	270
BENZYL	280
XYLYLENE	280
ETHYLENE, HIGH DENSITY	290
TRIVINYL BENZENE	290
TETRAFLUORETHYLENE ("TEFLON")	380
METHYL PHENYL SILICONE RESIN	GREATER THAN 380

*REFERENCES 1 AND 2

For calculations, the sun appears as a 6000°K point source blackbody radiating parallel rays. Similarly, the earth appears as a 250°K: space itself without star heat would read 0°K. For the purpose of electronic packaging for space environments, the latter source will be considered negligible, i.e., with or without stars.

The total resultant energy (7.38 Btu) consists of the following types of electromagnetic radiation by percentage (Lehl et al. 1957):

Type of Radiation	Wavelength (Å)	Percentage
Infrared	7000 and upward	51.0%
Visible	3800 to 7000	41.0%
Near ultraviolet*	2000 to 3800	7.5%
Far ultraviolet*	1 to 2000	0.2%

Near ultraviolet and far ultraviolet radiation must be considered when designing externally-mounted equipment, but their effect on equipment within the payload is negligible and hence may be disregarded. Thermal energy from the X-ray region–1 to 100 Å–can be considered almost nonexistent. Light alone will produce little effect, except in the form of heat, and it exerts as little effect on structures in space as on those situated on the earth's surface. Consequently, any material incapable of withstanding the pressure or other deleterious effects of light at sea level is not likely to be considered for use in space.

Ambient temperatures of space will affect spaceborne electronic systems in any or all of the following ways: loss of mechanical integrity, expansion, creep, fatigue, sublimation, and evaporation. Table 9-3 shows these temperatures and the composition of the gases at given altitudes. Figure 9-1 provides some additional characteristics of the earth's atmosphere.

For exposed equipment, passive temperature control is accomplished by the selection of materials that provide the proper ratio of solar heat absorptivity to infrared emissivity. However, the deterioration of radiating surfaces must be considered. These surfaces may be eroded by the action of micrometeorites and ultraviolet and particle radiation.

For equipment mounted inside the vehicle, active temperature control is necessary. Clearly defined conduction paths with a minimum of thermal joints should be mapped to transfer the excessive heat efficiently to the vehicle's structure or "airframe." The most serious problem in space operation of electronic equipment, or any other heat-generating device, comes from the assembly and installation interfaces. These joints usually provide a high thermal-contact resistance; this break in the heat-flow path can be serious and as dangerous as a broken wire or loose connection in an electronic circuit.

Of the three types of heat movements, i.e., convection, radiation, and conduc-

Table 9-3. Some Properties of the Gases in the Upper Atmosphere.

ALTITUDE	PRESSURE	TEMPERATURE	COMPOSITION
SEA LEVEL	762 1000 mm Hg	−40° TO + 40°C	78% N_2/21% O_2/1% A
100,000 FEET	90 100 mm Hg	−40°C	N_2/O_2/A/CH_3/O_3
125 MILES	10^{-6} mm Hg	1000°C	N_2/O/O_2/O^+
500 MILES	10^{-9} mm Hg	1000°C	O/O^+/H
4000 MILES	10^{-13} mm Hg	1000°C	H^+/H

tion, the method of conduction is the most effective and practical for moving heat from one place to another within the vehicle. Radiation of heat is the single method possible for removing heat from the vehicle itself. Convection is highly improbable by the nature of the airless space.

Organic materials for packaging may be best selected on the basis of results from direct experimental studies of weight losses in space vacuum environments and from information compiled from past vehicle performance. With proper thermal protection, most organic materials can be used without fear of the consequences. However, a word of warning is appropriate if they are to be used for extended periods of time in the orbiting spacecraft. The combined effects of vacuum and radiation—primarily ultraviolet radiation—tend to hasten decomposition of the organic substance, and a radiation shield should be used for protection.

HAZARDS FROM PARTICLE RADIATION

High-energy particle radiation in the region of near space, from cosmic rays and the Van Allen belts, does not constitute a particularly hazardous environment for structural materials. However, secondary radiation (bremsstrahlung) must be considered, and sufficient shielding provided. The high energy of the inner Van Allen radiation belt will produce property changes in certain materials and parts. The extent of these changes depends on the duration of the exposure. The low-energy electrons within the outer Van Allen belt will cause little concern since thin metallic shields provide sufficient protection and will have been already incorporated as part of the chassis or the design.

The Van Allen belts contain charged particles, mainly protons and electrons, that have been trapped by the earth's magnetic field. The inner belt—approximately 250 to 2200 miles above the geomagnetic equator depending on the longitude—consists mostly of protons yielding energies of 40 million electron volts (MeV). The outer belt, which is about 10,000 to 14,000 miles out, consists primarily of electrons in the range of 20,000 electron volts (KeV) to 2.5 MeV, depending on solar activity.

Prolonged exposure is the one condition that causes significant damage to electronic devices—that is, magnets, semiconductors, etc.—that are mainly dependent upon crystal or molecular orientation.

Cosmic flux is the primary source of interplanetary space radiation. During periods of low activity—i.e., when there are no solar flares—this flux has a maximum intensity of 2 particles/cm^2-sec. However, solar flares will increase the intensity by injecting pulses of gamma radiation and additional protons into space, raising the count to about 1000 particles/cm^2-sec with energies of 110 to 220 MeV. These periods of maximum intensity occur only several times a year, and the proton activity lasts approximately one day.

There have been no recorded failures of electronic equipment used on past space probes that were attributed to interplanetary radiation, but during period of solar-flare activity, communications noise is prevalent, and personnel in spacecraft would be subject to serious hazards.

Cosmic radiation will contribute only negligible effects on engineering materials. This radiation is comprised of better than 90% protons, 7% alpha particles, and 1% electrons. However, some wear will appear on external surfaces when these particles bombard the material, displacing some of the surface atoms.

Past experience has shown little indication of an appreciable change in the properties of metals caused by radiation. These properties are thermal conductivity, electrical resistivity, and density. The exposure of metals to ionizing radiation appears as heat and results in a temperature rise. Fast neutrons have a much greater effect on the physical and mechanical properties of structural metals over an extended period of time. High-energy protons and the heavier charged particles damage the mechanisms mainly in the form of atomic displacement. Ceramics, such as beryllium oxide and aluminum oxide; glasses, such as silicates and borates; and minerals, such as mica, suffer atomic displacements that change the density, thermal conductivity, and electrical conductivity of the material. Carbon, silicon, and germanium, and compounds such as silicon carbide, indium antimonide, and zinc oxide experience a decrease of both thermal and electrical conductivity from the effects of irradiation. Elastomers as a class are the most susceptible to radiation damage. Natural rubber is the most radiation-resistant of the elastomers; styrene-butadiene is the most resistant of the synthetics. However, these materials are very susceptible to damage from ultraviolet radiation. Silicones and fluorine-based polymers are well below average in

Table 9-4. Radiation Damage to Common Insulating and Packaging Materials.

MATERIALS	DOSE IN MEGARADS	COMMENTS
	0 10 10^1 10^2 10^3 10^4 10^5	
PTF		LOSES 40% OF ITS TENSILE STRENGTH; CRUMBLES AT HIGHER DOSES
NYLON		LOSES ELONGATION PROPERTIES RAPIDLY; TENSILE STRENGTH REMAINS
METHYL METHACRYLATE		SERIOUS DETERIORATION
BUTYL & FLUOROCARBON PLASTICS		DETERIORATION BEGINNING
SILICONE RUBBER		LOSES ITS ELASTICITY
POLYESTER FILM		THRESHOLD MECHANICAL DAMAGE; BECOMES WEAK AND BRITTLE
PHENOLIC UREA		LITTLE EFFECT
CELLULOSE ACETATE		LOSES 25% MECHANICAL PROPERTIES; ELECTRICAL PROPERTIES UNAFFECTED
VINYLS		LOSES UP TO HALF OF ITS TENSILE STRENGTH AND ELONGATION
POLYETHYLENE		LOSES 75% ELONGATION, 33% TENSILE STRENGTH
NATURAL RUBBER		LOSES 25% OF ITS PROPERTIES
MELAMINE (GENERAL PURPOSE)		LOSES 25% OR MORE OF ITS STRENGTH
POLYSTYRENE		RESISTIVITY DECREASES, MECHANICAL PROPERTIES REMAIN
UNFILLED SEMI-FLEXIBLE EPOXY		LITTLE EVIDENCE OF CHANGE
FILLED SEMI-FLEXIBLE EPOXY		PRODUCES STIFFENING AND SOME BLISTERS
RIGID EPOXY (GENERAL PURPOSE)		MAY CRACK, BECOME WEAK AND BRITTLE
SILICONE GLASS LAMINATE		LITTLE OR NO EFFECT
PHENOLIC MINERAL		LITTLE EFFECT
HEAT-RESISTANT, FILLED EPOXY		NO DETERIORATION
MICA GLASS LAMINATE		VIRTUALLY UNAFFECTED
INORGANICS: MICA, CERAMICS, GLASS		RESISTIVITY TEMPORARILY CHANGED, BUT LITTLE OTHER EFFECTS NOTED

radiation resistance. In general, the effects of radiation will be of minor concern except for the resultant heat. Avoiding the use of tetrafluoroethylene (Teflon) plastic is advisable; oil-filled paper and tantalum capacitors are also especially susceptible to damage from radiation.

Wire insulation is another major problem; polyethylene is recommended as the most favorable and resistant material.

Table 9-4 lists some of the more common insulating and packaging materials with the expected damage from a given dose in megarads. (A *rad* is the unit dose absorbed equal to 100 ergs per gram of energy absorbed.)

MICROMETEORITES AND METEORITES

Space contains high-velocity bits of solid matter. The largest portion of this matter consists of micrometeorites the size of dust and sand particles. These particles present the greatest comparative hazard. Satellites or vehicles traveling through these high-velocity "storms" or "clouds" will suffer from gradual surface alteration, that is, pitting or roughening. The most adversely affected will be the solar reflectors, absorbers, and radiators. Also to be considered is the protection of pressurized containers because of the possibility of puncturing by the larger meteors. Extended duration equipment should include greater outer-

Table 9-5. Probability of Micrometeorite Collision (Whipple 1952).

MAGNITUDE	KINETIC ENERGY (ERGS)	MASS (GRAMS)	RADIUS (CM)	PROBABILITY OF ENCOUNTER IN 24 HOURS	PENETRATION IN ALUMINUM (CM)
0	10^{13}	1.25	0.46	1.2×10^{-8}	10.9
1	4×10^{12}	0.5	0.34	3.1	8
2	1.6×10^{12}	0.198	0.15	7.7	5.9
3	6.3×10^{11}	0.08	0.18	2.0×10^{-7}	4.3
4	2.5	0.03	0.14	4.9	3.2
5	1.0	0.012	0.10	1.2×10^{-6}	2.3
6	4×10^{10}	0.005	0.074	3.1	1.7
7	1.6	0.002	0.054	7.7	1.3
8	6.3×10^{9}	0.0008	0.04	2.0×10^{-5}	0.93
9	2.5	0.0003	0.029	4.9	0.69
10	1	0.0001	0.022	1.2×10^{-4}	0.51
15	1	0.0000012	0.0046	1.2×10^{-2}	0.11

shell thicknesses, or a secondary outer shell to guard against that probability. Table 9-5 provides data on the possibility of a hit from any of several sizes of micrometeorites or meteorites (Whipple 1952). Whipple (1957) stated that "the probable number of micrometeorites near the 13th magnitude capable of penetrating a $\frac{1}{8}$-inch aluminum skin will be .047 per day. This calculation is based on the use of a 118-inch diameter sphere covered with the aforementioned aluminum skin thickness. It would be punctured an average of once every three weeks."

The most important factor affecting the design of electronic packaging is the erosion and general roughening caused by these particles and the resulting effects on the usefulness of thermal radiation devices.

SUMMARIZING THE PRECAUTION

Protection of electronic equipment designed for operation in the upper atmosphere will elicit the largest amount of work for the mechanical designer/packaging engineer. Thermal control will present the major problem in space electronic packaging design. Vacuum and radiation are of secondary importance. Common structural metals and alloys will remain quite stable in the space environment. However, cadmium and zinc plating will present some problems as a result of these materials vaporizing and then "plating out" on cooler insulation surfaces. Most organic materials will be affected at higher altitudes and temperatures with resultant decomposition. If elastomers are required, natural rubber is the most stable. Slight roughening of thermal control surfaces will occur from selective sublimation. The probability of increased friction and cold welding is detrimental to electromechanical devices.

The general effect of radiation will be negligible on inorganic materials. Organics must be carefully selected and shielded to reduce the rate of decomposition. Most damage will be inflicted on semiconductors and magnets, etc. Short-duration exposure in the Van Allen radiation belts will cause little damage. However, currently available materials and requirements for space-vehicle applications make it impractical to operate in the Van Allen belts for extended periods.

"Sand blasting" by micrometeorites is the most probable threat from high-speed solid particles. Meteorites pose a slight but fatal threat. The proper and efficient use of thermal control devices and methods, radiation shields, and suitable structural design and materials will increase the reliability of the electronic circuits and ensure the success of a spacecraft's mission.

10

ENVIRONMENTAL PROTECTION

Electronics equipment must be protected from severe environmental operating conditions. Extremes in temperature, humidity, water damage, ice, wind, and dust are directly attributable to a large number of failures in electronics systems. In Chapters 8 and 12, equipment enclosures are discussed. The material presented in this chapter covers both organic and inorganic environmental coatings commonly used to protect materials and assemblies from harsh environmental exposure.

The finishes used on any electronics assembly or system will be defined by the intended application of the system. For example, will the final units be installed in an office in the Midwest, or on board a naval ship operating in the tropics? The various finishes available vary considerably in cost, and the final selection will depend on the product intent considering the reliability and product life expectancy. Any electronics system using only an enamel finish will have a very good life expectancy if it is intended to be used in a bank or insurance office; however, that same equipment will have an extremely short life if it is installed in a naval facility located close to a saltwater environment.

CORROSION

A part normally must withstand more abuse at its surface than internally. Thus, the surface has more than a decorative purpose. Most functional finishes protect against corrosion. The available finishes provide designers with several functions they may need to produce a reliable product, including:

1. corrosion resistance
2. wear resistance

3. heat and oxidation resistance
4. electrical and magnetic properties
5. soldering aid
6. lubrication
7. reflectivity

Surface treatments of various types play an indispensable role in the reliability and life expectancy of electronics equipment, and the designer should understand the effects of harsh environments and their control.

Corrosion can be simply defined as material deterioration caused by a chemical or electrochemical attack. Although a direct chemical attack can occur in all materials, electrochemical attack normally occurs in metals because, unlike other materials, their electrons are free to move. The main reason that metals are more susceptible to corrosion attack is that they naturally occur as oxides, hydroxides, carbonates, and other compounds of low free energy. In the refined state, metals are in a much higher energy state than their corresponding ores and show a natural tendency to return to their lower energy or combined state. The corrosion of metals therefore can be regarded as the reverse process of reducing metals from their ores.

Atmospheric corrosion is an example of direct chemical attack. Present in the atmosphere are oxygen, carbon dioxide, water vapor, sulfur, and chlorine compounds. The severity of the corrosion is directly related to the amount of water vapor, sulfur, and chlorine compounds present. The three basic elements of atmospheric corrosion are:

1. Positive particles of metal dissolve in the moisture, absorbing oxygen and hydrogen, becoming ferrous ions.
2. Negative-charged electrons flow through steel to copper into the moisture where they combine with oxygen and water, becoming hydroxyl ions.
3. Hydroxyl ions combine with ferrous ions, producing iron oxide (rust), the corrosion product.

The presence of sulfur-containing gases and moisture appreciably increases the oxidation rates of copper, iron and many other metals and alloys in air.

Because metals are such good conductors of electricity, the total resistance of a galvanic circuit is usually controlled by the resistance of the solution or electrolyte. Thus, no appreciable galvanic corrosion is produced in distilled water since it is a relatively poor conductor. Galvanic corrosion is more likely to occur in ordinary tapwater, which contains sufficient ionizable salts to make it moderately conductive.

Galvanic corrosion can even occur with thin films of condensed moisture in the presence of dissolved salts or ionizable gases that cause the corrosion circuit to be completed. However, in this case, corrosion tends to be localized near the points of contact. In solutions of relatively low conductivity, galvanic influences are usually localized so that the less noble metal (the anode) suffers most of its accelerated corrosion in a region in the immediate vicinity of the more noble metal (the cathode). In highly conductive solutions, such as strong salt solutions (brines and seawater) and strong solutions of chemicals (acids and alkalis), corrosion is likely to be distributed over the entire anode surface.

Galvanic corrosion cannot take place unless there is a flow of current. One of the most important factors that determine the magnitude and ease of current flow is the conductivity of the circuit. There must be a complete circuit through the conducting metals and the electrolyte. The rate of this batterylike action is proportional to the electrical potential between the two metals. A strong electrolyte (low resistance to current flow) will increase the corrosion rate.

Two general approaches may be used to minimize or eliminate galvanic corrosion: (1) elimination of the electrolyte and (2) minimization of the degree of dissimilarity of the joined metals (Table 10-1). The corrosion rate increases as the ratio of the cathode area to the anode area increases. Therefore, it is important to keep the cathodic material small and the anodic area large. The anodic material is the one that corrodes (electrons flow from anode to cathode, and ions transfer from cathode to anode). Minimizing the degree of dissimilarity of the joined metals is achieved by selecting metals having the least potential difference in the galvanic series, or by decreasing the potential difference between two metals, for example, by passivating. The galvanic series is a list of metals arranged in descending order of tendency to be cathodic (most noble) or anodic (least noble) (Table 10-1). This table is based on laboratory or actual corrosion tests and should not be confused with the theoretical "electromotive series." The greater the separation of two materials in the galvanic series, the higher is the electrical potential between them.

Where it is necessary that any combination of dissimilar metals be assembled, the following methods or combinations of methods should be employed for the alleviation of electrolytic corrosion, unless electrical considerations prevent it:

1. Interposition of a material compatible with each of the dissimilar metals to decrease electrolytic potential differences (such as cadmium or zinc plating on steel in contact with aluminum).
2. Interposition of an inert material between the dissimilar metals to act as a mechanical and insulating barrier.
3. Application of organic coatings to the contact faces of each of the dissimilar metals (paint coats on steel and aluminum surfaces in contact).
4. Application of corrosion inhibitors to the faces of the dissimilar metals (such as zinc chromate paste on nickel-plated brass screws in contact with aluminum).

Table 10-1. Galvanic Series.

Less noble (anodic)

Current flow	Metal
+	Magnesium magnesium alloys
	Zinc
	Aluminum 1100
	Cadmium
	Aluminum 2024-T4
	Steel or iron cast iron
	Chromium iron (active)
electric current flows from + to −	Ni-resist
no serious corrosion will result if fasteners are selected from alloys in the same group as parts to be fastened or below.	Type 304 stainless (active) Type 316 stainless (active)
	Lead tin solders lead tin
	Nickel (active) inconel
	Brasses copper bronzes copper-nickel alloys monel
	Silver solder
	Nickel (passive) Inconel (passive)
	Chromium-iron (passive)
	Type 304 stainless (passive) Type 316 stainless (passive)
	Silver
	Titanium
−	Graphite gold platinum

More noble (cathodic)

Whenever possible, the engineer should employ the following three recommendations to control the effects of galvanic corrosion.

1. Use the same or similar metals in an assembly, especially where an electrolyte may be present.

2. When dissimilar metals are used together in the presence of an electrolyte, separate them with a dialectric material such as insulation, paint, or coating.
3. Avoid combinations where the area of the less noble material is relatively small. The current density is greater when the current flows from the small area to the large than in the reverse situation.

The galvanic series chart in Table 10-1 indicates relative potential for galvanic corrosion. Coupling of metals widely separated on the chart is most likely to cause corrosion. Under ordinary circumstances, no serious galvanic action will result from the coupling of metals within the same group (such as brass and copper).

ENVIRONMENTAL COATINGS FOR ALUMINUM AND ALUMINUM ALLOYS

There are three types of coatings, other than paint, for aluminum and aluminum alloys: anodize, chromate conversion coating, and plating with another metal. Anodic coatings are good general-purpose finishes with excellent corrosion and abrasion resistance; however, they are electrically nonconductive. If a conductive coating is required, either a conversion coat or a metal plate should be used (Table 10-2).

Anodize

Regular anodize provides a hard, electrically nonconductive coating that is quite corrosion resistant. If extreme wear and abrasion resistance or reliable electrical insulation is required, hard anodize should be used. Anodize coating can be colored with dye. The treatment produces an aluminum oxide coating from 0.00002 to 0.002 in. thick. Anodizing should be specified for parts only after all fabrication operations (heat treatment, machining, forming, drilling, etc.) have been completed.

Sulfuric Acid Anodize

Sulfuric acid anodize is produced by an electrochemical process that provides a hard film on aluminum alloys. This anodize, which is very corrosion resistant, is normally extremely hard and abrasion resistant. It is preferred over chromic acid anodize for dyed applications. Sulfuric acid anodize cannot be used on parts that have recessed or faying surfaces that may entrap some of the coating solution. The thickness of the coating is formed partially at the expense of the base metal. Dimensional change depends on alloy and processing conditions, with buildups of 0.0006 in. common.

Table 10-2. Chemically Deposited Coatings for Aluminum and Aluminum Alloys.

Treatment	Purpose	For Use On	Operation	Finish and Thickness
Zinc phosphate coating	Paint base	Wrought alloys	Power spray or dip. For light to medium coats. 1 to 3 min. at 130 to 135°F	Crystalline, 100 to 200 mg/ft^2
Sulfuric acid anodizing	Corrosion and abrasion resistance, paint base	All alloys; uses limited on assemblies with other metals	15 to 60 min. 12 to 14 amp/ft^2, 18 to 20 V, 68 to 74°F. Tank lining of plastic, rubber, lead, or brick	Very hard, dense, clear. 0.0002 to 0.0008 in. thick. Withstands 250- to 1000-hr salt spray
Chromic acid anodizing	Corrosion resistance, paint base; also as inspection technique with dyed coatings	All alloys except those with more than 5% copper	30 to 40 min. 1 to 3 amp/ft^2, 40 V dc 95°F, steel tanks and cathode, aluminum racks	0.00002 to 0.00006 in. thick, 250-hr min salt spray
Chromate conversion coating	Corrosion resistance, paint adhesion, and decorative effect	All alloys	10 sec to 6 min depending on thickness, by immersion, spray, or brush. 70°F, in tanks of stainless, plastic, acid-resistant brick or chemical stoneware	Electrically conductive, clear to yellow and brown in color. 0.00002 in. or less thick. 150- to 2.000-hr salt spray depending on alloy composition and coating thickness
Electropolishing	Increases smoothness brilliance of paint or plating base	Most wrought alloys, some sandcast and diecast alloys	15 min, 30 to 50 amp/ft^2, 50 to 100 V, less than 120°F, aluminum cathode	35 to 85 rms depending on treatment
Zinc immersion	Preplate for subsequent deposition of most plating metals, improves solderability	Many alloys, modifications for others particularly regarding silicon copper and magnesium content	30 to sec. 60 to 80°F, agitated steel or rubber-lined tank	Thin film
Electroplating	Decorative appeal and/or function	Most alloys after proper preplating	–	Same as on steel

Table 10-2. *(Continued)*

Treatment	Purpose	For Use On	Operation	Finish and Thickness
Copper	–	–	Directly over zinc, or follow with copper strike, then plate in conventional copper bath	–
Brass	–	–	Directly over zinc, 80 to 90°F, 2 to 3 V, 3 to 5 amp/ft^2	–
Nickel	–	–	Directly over zinc, or follow with copper strike, then plate in conventional nickel bath	–
Cadmium	–	–	Directly over zinc, or follow with copper or nickel strike, or preferably cadmium strike, then plate in conventional cadmium bath	–
Silver	–	–	Copper strike over zinc using copper cyanide bath, low pH, low temperature, 24 amp/ft^2 for 2 min., drop to 12 amp/ft^2 for 3 to 5 min., plate in silver cyanide bath. 75 to 80$^\circ_0$F, 1 V, 5 to 15 amp/ft^2	–
Zinc	–	–	Directly over zinc immersion coating	–
Tin	–	–	Directly over zinc immersion coating	–
Gold	–	–	Copper strike over zinc as for silver, then plate in conventional bath	–

Chromic Acid Anodize

Chromic acid anodize is produced in a manner similar to sulfuric acid anodize except that a chromic acid electrolyte and a higher voltage is used. Chromic acid anodize is preferred and should be used when small dimensional change (less than 0.0001 in.) is necessary for highly stressed parts, or on assemblies that could entrap some processing solution. Chromic acid anodize can be dyed, but the resulting colors are inferior to those obtainable with dyed sulfuric acid anodize.

Hard Anodize

Hard anodize provides an extremely hard finish with excellent abrasion resistance and poor electrical insulation characteristics. Its principal disadvantages are the brittleness of the coating and the reduced strength of the coated part. The loss of strength is due to a loss in the material of the coated part during processing. As an approximation, for each 1 mil of coating thickness, 0.5 mil of material is lost. When the finish is used in friction applications under load, the anodize has a tendency to spall from the aluminum alloy. Hard anodize treatments produce oxide coatings from 0.0005 to 0.0045 in. in thickness, with 0.002 in. most common. The coating is formed at the expense of the substrate metal, with dimensional growth equal to about one-half the coating thickness. Buildup is almost perpendicular to the surface being coated; therefore, at sharp edges, corners, or fillets, the coating is either porous or incompletely formed and easily chipped. Routine rounding of all edges to a radius of 0.010 in (minimum) is generally adequate to maintain coating integrity.

Table 10-3 is the accepted military specification and the typical callout used on engineering drawings in both commercial and military electronics systems.

ALUMINUM PLATING

Metal plating is the deposition of one metal upon another. It can be accomplished by electroplating, catalytic chemical reduction vacuum deposition, or hot dipping. Electroplating, by far the most common process, is applied by

Table 10-3. Anodized Aluminum Specifications.

Finish	Specification MIL-A-8625	Typical Drawing Callout
Anodized Aluminum	Type I (chromic acid) Type II (sulfuric acid) Type III (hard coating) Class 1 (nondyed) Class 2 (dyed)	Anodize per MIL-A-8625 Type II (specify dye color, if desired, in accordance with FED-STD-595)

making the base metal cathodic in an aqueous solution of the salt of the plating metal and then passing direct current through the solution to apply the coating metal by electrolysis. These processes change the reaction of the base material to its environment.

Hardness, conductivity, solderability, and corrosion resistance are generally the most important plating properties since coating life usually depends upon these factors. Although hardness gives good wear resistance, it may also cause brittleness, high internal stresses, low ductility, and poor buffability. Bright deposits are usually characteristically hard, brittle, and stressed. Many metals can be deposited with a bright, reflective surface eliminating the need for buffing to obtain these properties. It is common practice to buff a softer substrate metal and then to plate a bright, hard deposit.

It is recommended that the entire part be plated, if possible, because of the high added cost and time required to mask areas not to be plated. If plating must be omitted in some area of the part, the engineering drawing must indicate these areas and contain the note: *Area to be free of plating.*

Where applicable, the engineering drawing should specify whether the dimensions apply before or after plating. (For example, sulfuric acid anodizing adds considerably to the dimensions of a part.)

MAGNESIUM PLATING

Like aluminum, there are three types of coating for magnesium other than paint: anodize, conversion coating, and plating with other metals. Magnesium is an extremely active metal that is difficult to protect in a corrosive environment. Plated surfaces, which couple a more noble metal with the magnesium, can, by galvanic action, significantly accelerate corrosion. Therefore, it is recommended that plated magnesium be avoided in corrosive applications. Conversion coatings will provide a minimal amount of corrosion protection in relatively mild environments and are useful as pretreatments for surfaces to be painted.

Anodizing is a suitable process for all magnesium alloys, but shall not be used for parts with steel or copper alloy inserts, with deep blind holes, or with other features that might interfere with circulation or removal of the coating solution. Table 10-4 is the accepted military specification and the typical callout on the engineering drawings in both military and commercial electronics systems.

The light coating (Type I, Class C) is preferred for a paint base under corrosive conditions. The heavy coatings (Type II) provide maximum wear and abrasion resistance and are more corrosion resistant than Type I coatings, but they reduce the fatigue life of the part. They are quite brittle, powdery, and easily cracked or chipped, and should not be used on parts that are subject to flexing. Since all anodic coatings are good insulators, do not specify these coatings for applications requiring electrical conductivity. These treatments alone will inhibit, but

Table 10-4. Anodized Magnesium Specifications.

Finish	Specification MIL-M-45202	Typical Drawing Callout
Anodized Magnesium	Tvpe I (light coating) Class C (light green) Type II (heavy coating) Class A (hard brown) Grade 1 (without posttreatment) Grade 2 (with chromate posttreatment) Grade 3 (bifluoride dichromate posttreatment) Class D (dark green) Class E (high-density dark green)	Anodize per MIL-M-45202 Type I, Class C

Type I

Class A: 0.0001 to 0.0003
Class C: 0.0001 to 0.0005

Type II

Class A: 0.0013 to 0.0017
Class D: 0.0009 to 0.0016

not prevent, galvanic corrosion when contacted with most dissimilar metals or platings. Since these coatings are brittle and hard, do not use these treatments for any applications that might spall or chip the coating. Avoid line or point contacts. Specify treatment only after all fabrication operations have been completed. Parts containing inserts should be treated prior to the installtion of these inserts.

CHEMICAL CONVERSION COATINGS

Chemical conversion coatings are extremely thin. Their color usually ranges from irridescent yellow to golden brown. Dyed finishes are not recommended because they provide inferior corrosion resistance and unattractive colors. No dimensional change is caused by the film. Table 10-5 presents a selection guide for conversion coatings. Chemical conversion coatings provide an easily and cheaply applied corrosion-resistant finish for aluminum. Certain coatings are excellent pretreatment for surfaces to be painted. These films have poor abrasion resistance and are easily scratched or marred.

Chemical-film surfaces will conduct RF and power frequencies. The average electrical resistance is from 0.0001 to 0.001 ohm/in.^{2} at 100 psi, depending on thickness of coating. Chemical-film-coated aluminum will often have a surface

Table 10-5. Selection Guide for Conversion Coatings.

Conversion Coating	Color	Structure	Typical Applications	Base Metals
Zinc phosphate	Medium gray	Microcrystalline to crystalline	Automobiles, home appliances, ordnance components, coil stock	Steel, aluminum
				Zinc, galvanized surfaces, cadmium
Chromate	Colorless through gold-tan	Amorphous	Aircraft components, architectural hardware, coil stock, fasteners	Aluminum
				Zinc, galvanized surfaces, cadmium
				Magnesium, copper, brass, bronze
Amorphous phosphate	Colorless through medium green	Amorphous	Aircraft components, architectural hardware, general OEM parts	Aluminum

[a]These additional pretreatments may be needed, depending on alloy and surface condition of substrate.
[b]Special application, costs depend on individual application.

Service	Coating Weight (mg/ft^2)	Cost (c/100 ft^2)	Pretreatments	Post Treatments	Government Specifications
Underpaint corrosion and paint–adhesion aid	100–300	5–10	Alkaline clean, activate	Passivating rinse	TT-C-490 MIL-S-5002
Unpainted corrosion improvement, appearance	1,000	20–30	Alkaline clean, acid or alkaline pickle	Rust–preventive oil	MIL-P-16232 MIL-P-50002
Drawing aid	1,000	20–30	–	Drawing lubricant	–
Underpaint corrosion and paint–adhesion aid	100–300	5–10	Alkaline clean	Passivating rinse	MIL-T-12879 QQ-P-416 QQ-Z-325
Underpaint corrosion and paint–adhesion aid, appearance	10–50	3–10	Alkaline clean, deoxidize[a] alkaline etch, desmut[a]	Passivating rinse	MIL-C-5541 MIL-S-5002 MIL-STD-171 MIL-STD-186 MIL-STD-193 MIL-STD-808 AMS-2473 AMS-2474
	5–10	SA[b]	Alkaline clean	–	QQ-P-416 QQ-Z-325 MIL-T-12879 MIL-C-17711
	Varies	SA[b]	Alkaline clean, acid pickle[a]	–	–
Underpaint corrosion and paint–adhesion aid	5–50	3–5	Alkaline clean	Passivating rinse	–
Unpainted corrosion improvement, appearance	200–600	30–50	Alkaline clean, deoxidize[a]	Passivating rinse	–

Table 10-6. Conversion Coating of Aluminum.

Finish	Specification MIL-C-5541	Typical Drawing Callout
Conversion coating, aluminum (chemical film)	Class 1A: For maximum protection against corrosion, painted or unpainted. Class 3: For protection against corrosion where low electrical resistance is required.	Chromate conversion coating per MIL-C-5541, Class 1A

conductivity greater than bare aluminum. All fabrication of parts should be specified before chemical-film coating, although mild forming, binding, and arc welding may be done after coating. Spot welding may be performed on chemical-film-coated assemblies, but is not recommended. Chemical coatings are usually defined by the use of the specifications and drawing callout shown in Table 10-6. Chromate conversion coating should not be specified for parts that are to be exposed to temperatures in excess of 250° F—e.g., if a rubber seal is to be vulcanized to an aluminum part at 300° F, specify mask seals, and perform chromate conversion coating after vulcanizing.

Conversion Coating of Magnesium

Types I and III. Mil-M-3171 is the preferred specification for conversion coatings for magnesium. The four most commonly used conversion coatings (Table 10-7) should be used to provide some corrosion protection and as a paint base. Type I may be used under very mild corrosive conditions only and results in a stock loss of about 0.0003 to 0.0008 in. Type III cannot be applied to certain magnesium

Table 10-7. Conversion Coating of Magnesium.

Finish	Specification MIL-M-3171	Typical Drawing Callout
Conversion coating, magnesium	Type I (Dow 1, chrome pickle) Type III (Dow 7, dichromate treatment) Type IV (Dow 9, galvanic anodizing) Type VI (Dow 19, chromic acid brush-on)	Dichromate treat per MIL-M-3171, Type III.

alloys, notably HK31A, HM21A, and HM31A. For most other alloys, Type I provides long-term protection and causes no dimensional change. Parts processed by this treatment are not to be subjected to temperatures above 550° F.

Type IV (Dow 9) is for general long-term protection of all alloys when close dimensional tolerances are required. Parts processed by this treatment are not to be subjected to temperatures above 550°F. A black finish may be obtained on some alloys.

Type VI. Chromic acid brush-on treatment is used for temporary storage, protective touchup of previously treated work, and brush application where parts and assemblies are too large to be immersed. This method is applicable to all alloys when close dimensional tolerances are not required. Parts processed by this treatment are not to be subjected to temperatures above 450° F. Chromic acid should be used for spot touchup of all anodized magnesium.

CADMIUM-PLATING

Cadmium-plating is used extensively to protect steel against corrosion and as a coating on copper and ferrous alloys to provide dissimilar metal protection when these metals are in contact with aluminum (Table 10-8). When applied to copper and ferrous alloys, no underplate is required. Cadmium-plating is seldom used as an undercoating with other metals, and its resistance to chemical corrosion is low. It should be applied after all welding, soldering, brazing, and, if possible, machining operations have been completed because the temperatures generated by these processes will cause release of toxic cadmium vapors.

Use of cadmium-plating in high-temperature applications is quite limited. It may be used as a finish up to about 450° F. It cannot be used near its melting point (approximately 610° F) on certain steels under stress because it will attack the substrate and cause intergranular corrosion, which may result in part failure. The maximum continuous operating temperature for cadmium-plating with a supplementary chromate coating (Type II) is 150° F. Along with these temper-

Table 10-8. Cadmium-Plating.

Finish	Specification QQ-P-416	Typical Drawing Callout
Cadmium-plating	Type I (as deposited) Type II (chromated) Type III (phosphated) Class 1 (0.0005 in. thick) Class 2 (0.0003 in. thick) Class 3 (0.0002 in. thick)	Cadmium plate per QQ-P-416, Type II, Class 2 (Specify embrittlement relief if applicable.)

ature considerations, there are a number of additional limitations on the use of cadmium plating:

1. It should never be used in high vacuum (space) applications because of its high vapor pressure.
2. It should not be used in friction applications because of its softness.
3. To provide embrittlement relief, a postplate bake of 375° ±25° F for 3 hr immediately after cadmium-plating and before chromate treatment is required for steel parts with tensile strengths from 180,000 to 220,000 psi or with Rockwell hardness from C39 to C47. All cadmium-plated steel springs must be baked.
4. Cadmium-plated parts should not be used in direct contact with any synthetic base hydraulic fluid, such as silicone or silicone ester-based fluid.

Plating Thickness

Cadmium plating should normally be used in thicknesses less than 1 mil; 0.0003-in. (Class 2) is suitable for general use. Where low dimensional change is important (such as threaded parts), 0.0002-in. (Class 3) may be specified. A thickness of 0.0005-in. (Class 1) should be specified when maximum corrosion protection is needed, such as when equipment is exposed to outdoor weathering.

Supplementary Treatments

Chromate (Type II) should be applied when the cadmium is to be unpainted. Handling of cadmium without a conversion coating will cause the plating to tarnish and fingerprint. The chromate film is normally yellow to brown.

COPPER-PLATING

Copper-plating is used primarily as an underplating for other electrodeposits. Some of the more common uses of copper-plating are:

1. As a base for silver or nickel plating on ferrous alloys.
2. As a base plating, when very rough surfaces (such as sand castings) are to be plated.
3. For electrical conductivity purposes.
4. As a base plating for electrodeposited tin to prevent poisoning of solderability.

Copper-plating can be used as an in-process protective coating during heat-treatment and case-hardening operation. Copper-plating must be protected from tarnishing with an overplate of a more inert metal or an organic finish.

Table 10-9. Copper-Plating.

Finish	Specification MIL-C-14550	Typical Drawing Callout
Copper-plating	Class 1 (0.001 in. thick) Class 2 (0.0005 in thick) Class 3 (0.0002 in. thick) Class 4 (0.0001 in. thick)	Copperplate per MIL-C-14550 (Specify overplate)

Plating Thickness

To improve adhesion of subsequent coatings, a 0.000010-in. copper strike should be used. A thickness of 0.0005 in. (Class 2) should be specified as an underplate on ferrous alloys. For stopoff during carburizing and nitriding, a thickness of 0.001 in. (Class 1) should be specified. The copper-plating classes (see Table 10-9) are intended for such applications as:

- Class 1 (0.00100-in. thickness) for carburizing shield or brazing operations
- Class 2 (0.00050-in. thickness) as an undercoating for nickel and other metals
- Class 3 (0.00020-in. thickness) and Class 4 (0.00010-in. thickness) to prevent base metal migration into the tin layer to impair solderability

A copper strike is a common method for preparing nickel and high-nickel alloys for subsequent plating. Ferrous alloys (not including stainless steel) are ordinarily given a copperplate prior to plating with nickel, gold, or tin. Copper-plating is also used where electrical conductivity is required.

Surfaces requiring a bright polished surface, such as decorative chromium, are often produced by underplating the part with copper and buffing the copper smooth before the final finish is added. These applications require fairly thick copper; the exact amount depends upon the degree of buffing required, but 0.001 in. (Class 1) is the minimum that should be considered.

GOLD-PLATING

Electrodeposited gold-plating provides excellent tarnish resistance and is especially useful in high-frequency electronic applications where good electrical properties that do not change with time are required. Gold-plating also provides high infrared wavelength reflection and is an effective radiant heat shield. The recommended electrodeposited gold-plating specification is MIL-G-45204 (Table 10-10).

Gold-plating is easily scratched and should not be used on rough, burred, or scratched parts. Because of its cost, gold-plating should be limited to small

Table 10-10. Gold-Plating.

	Specification MIL-G-45204				
Finish	Type	% Gold (min)	Grade	Knoop Hardness	Typical Drawing Callout
Gold-plating, electrodeposited	I	99.7	A	90 max	(Specify underplate if required) Goldplate per MIL-G-45204
	II	95.5	B	91–129	
	III	99.9	C	130–200	
			D	201 up	

parts or to thin coatings over copper or nickel underplates. Gold-plating does not oxidize or tarnish when exposed to elevated temperatures. However, it tends to diffuse with copper or silver underplates even at moderate temperatures. This diffusion is a time-temperature phenomenon; rigid limits cannot be defined, but general guidelines can be offered. Temperatures at or above 400° F will cause serious diffusion within a short time. Therefore, these underplates should not be used when these temperatures are expected. With temperatures up to 400° F, copper or silver underplates may be acceptable depending upon the length of exposure. However, the copper and silver will react with chemicals in the air, thus propagating corrosion products through a porous goldplate. This type of reaction has, in many instances, been mistakenly interpreted to be metal-to-metal diffusion.

Underplate Selection

Copper, electrodeposited nickel, and electroless nickel are the normal underplates for gold. Certain substrate materials (notably aluminum, ferrous alloys, and magnesium) cannot be goldplated directly because of a formation of a poorly adherent immersion plate; therefore, an underplate must be used.

Gold-plating may be used for solderability. However, thickness must be carefully controlled in these applications, or an excessive amount of gold will alloy with the solder, causing a dull, grainy solder joint that is often brittle. Gold-plating is also used as an underplating for rhodium. Specification MIL-G-45204 contains three types of electrodeposited gold-plating. Type I, an unalloyed soft plate, is essentially 100% gold and is quite ductile. Type II, an alloyed deposit (usually silver, nickel, or cobalt) is bright, hard, and much more wear-resistant than Type I. Type I should be specified for general use and is preferred. Hard plate (Type II) should be used where a nonporous coating is desired, or where abrasion and wear resistance are required. Type III is extremely soft and can be applied as Grade A only.

Table 10-11. Type of Nickel-Plating.

Type	Used On
I	Steel–heavy duty
II	Steel–limited wear
III	Steel–standard
IV	Steel–threaded parts
V	Copper alloy–standard
VI	Copper alloy–threaded parts
VII	Copper alloy–threaded parts where copper alloy contains 40% or more zinc
VIII	Zinc alloy–limited wear
IX	Zinc alloy–standard
X	Zinc alloy–threaded parts

Plating Thickness

Since gold-plating is quite expensive, the minimum thickness needed should be specified. For general use, including solderability applications, 0.000050 to 0.00010 in. (Class 1) is suitable. Heavier deposits may be specified as shown. Copper cannot be used under gold at temperatures above 400° F because of diffusion. Nickel should be used when the gold-plating will be exposed to potential diffusion-causing temperatures. It may also be used when a bright, decorative gold finish is needed. Either electrodeposited or electroless nickel is acceptable; however, electrolytic is preferred for decorative purposes.

NICKEL-PLATING

Electrodeposited nickel-plating (Table 10-11) provides a hard, dense coating that will usually seal off the base metal from corrosive media. It is recommended (1) for corrosion protection in such applications as high-temperature or vacuum environments where cadmium is not suitable, (2) for wear and abrasion resistance, and (3) in applications for which chromium-plating is not suitable because of corrosion problems. It may also be used for buildup of worn or overmachined parts prior to rework and is the recommended underplating for chromium and rhodium. Electrodeposited nickel plating has two major advantages over electroless: it is much cheaper, and it can be applied in greater thicknesses more satisfactorily. These advantages should be balanced against those of electroless plating before a final selection is made.

Class 1 nickel-plating is intended as a bright decorative coating and may be used either as an underplate for chromium or by itself. Class 1 has type designa-

Table 10-12. Nickel-Plating, Electrodeposited.

Finish	Specification QQ-N-290	Typical Drawing Callout
Nickel-plating, electrodeposited	Class 1 (decorative) Type I (0.001 Nickel, 0.002 Nickel + Copper) Type II (0.0006 Nickel, 0.00125 Nickel + Copper) Type III (0.0004 Nickel, 0.00075 Nickel + Copper) Type IV (0.0002 Nickel, 0.0004 Nickel + Copper) Type V (0.0005 Nickel) Type VI (0.0003 Nickel) Type VII (0.0001 Nickel) Type VIII (0.005 Nickel, 0.00125 Nickel + Copper) Type IX (0.0003 Nickel, 0.00075 Nickel + Copper) Type X (0.0003 Nickel, 0.0005 Nickel + Copper) Class 2 (thickness as specified)	(Specify underplate if required.) Nickel-plate per QQ-N-290, Class 2 (Specify thickness)

tions for use in specifying plate thicknesses and, for some types, a copper underplate. Class 2, to be used where decoration is not the primary function, has no applicable type designations; plating thicknesses must be specified. The engineering plating (Class 2) may be plated from 0.0005 to 0.080 in. as required. For the larger thicknesses, a 0.0002-in. underplating of copper may be used for greater adhesion. The thicker platings tend to build up at the corners. The Class 2 coating protects the base metal by providing an impervious, nonporous barrier. Because of its hardness and density, it may be used on sliding and wearing surfaces that require electrical conductivity and that cannot be lubricated. Table 10-12 lists the types of electrodeposited nickel-plating and their uses.

Electroless nickel-plating is a chemical reduction process that deposits a coating of nickel and varying amounts of phosphorus, (6% to 10%) to a base metal (Table 10-13). Electroless nickel-plating has three major advantages over electrodeposited nickel-plating:

1. *Uniform coverage.* It plates by chemical reduction without the need for electrical current. Therefore, the entire surface in contact with the solution is plated uniformly and does not have the edge buildup characteristics of electrolytic coatings.
2. *Deposition directly on aluminum is possible.* Thus, electroless nickel can be used as an underplate for other coatings on aluminum.

Table 10-13. Electroless Nickel-Plating.

Finish	Specification MIL-C-26074	Typical Drawing Callout
Nickel-plating, electroless[a]	Class 1: As coated, no subsequent heat treatment Class 2: Steel, copper, nickel, and cobalt-base alloys heat treated for hardness of nickel deposit[b] Class 3: Aluminum alloys other than 7075, heat-treated to improve adhesion of nickel deposit Grade A: 0.001-in. minimum thickness Grade B: 0.0005-in. minimum thickness	Electroless nickel-plate per MIL-C-26074, Class 1, Grade A

[a]The melting point of electroless nickel (1635° F) is considerably below that of electrodeposited nickel (2651° F). This limits service temperature and precludes its use where high temperatures are required for fabrication processes.
[b]Because of the heat treatment required for the hardened coating. Specification MIL-C-26074, Class 2, plating is magnetic and cannot be used in applications requiring a nonmagnetic film.

3. *Greater hardenability.* Electroless nickel is an alloy responding to heat treatment, which can be conditioned to a Vickers hardness number of 1000.

Specification and Class Selection

Class 1 is an as-deposited coating, whereas Class 2 is heat-treated to improve hardness (a maximum of 1000 Vickers). Class 2 should not be used on aluminum alloys because optimum metallurgical properties of the aluminum will be affected by the heat-treatment temperatures required for the plating.

Plating Thickness

A plating thickness of 0.005 in. is considered a practical maximum. In some applications, however, electrodeposited nickel may be plated over the electroless nickel to build up to required dimensions. A coating thickness of 0.0005 in. should ordinarily be used for aluminum; a thickness of 0.001 in. is recommended for ferrous parts. Specify 0.0003-in. thickness for threaded parts. Electroless nickel should not be plated to greater than 0.0003-in. thickness on steels where fatigue is a consideration at operational temperatures in excess of 550° F because of detrimental effects on fatigue life.

Electroless nickel has a high and variable electrical resistivity. Figure 10-1 shows typical curves that relate resistivity, phosphorus content, and heat-treatment temperatures. Since the precise resistivity will vary depending upon process conditions, the values shown can be used as guidelines only.

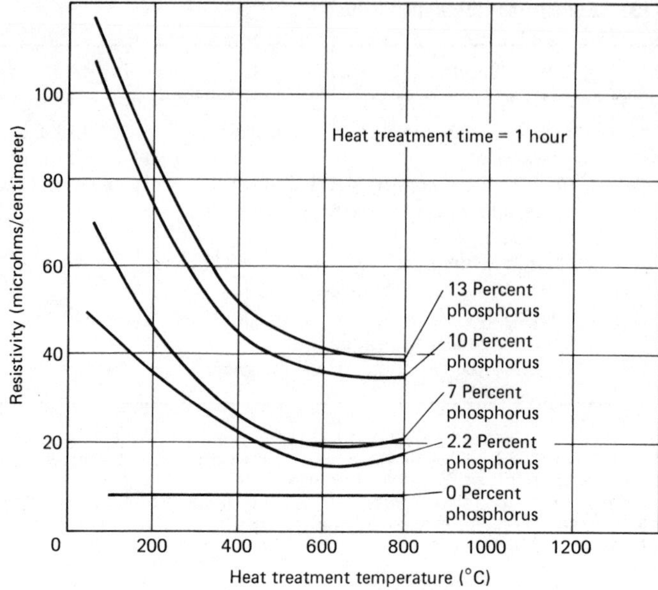

Figure 10-1. Typical variations of electroless nickel plating electrical resistivity.

BLACK OXIDE COATING

Black oxide coatings are particularly suitable for steel parts that cannot tolerate the dimensional buildup of more corrosion-resistant finishes. The oxide does not change the dimensions of the part or affect the result of previous heat treatment. Most iron and steel alloys can be black oxide coated. However, certain high-nickel steels do not respond satisfactorily to the blackening solution. When long-term storage is anticipated, a corrosion-preventive oil should be specified after blackening. Paint and solid film lubricants may be applied over the black oxide coating. See Table 10-14.

Table 10-14. Black Oxide Coating.

Finish	Specification MIL-C-13924	Typical Drawing Callout
Oxide, black, for ferrous alloys	Class 1: For wrought iron and plain carbon and low alloy steels Class 2: For corrosion-resistant steel Class 3: For all grades of corrosion-resistant steel alloy and carbon steel with draw temperature above 900° F Class 4: For corrosion-resistant steel	Black oxide coat per MIL-C-13924

Table 10-15. Phosphate Coatings for Ferrous Alloys.

Finish	Specification TT-C-490	Typical Drawing Callout
Phosphate coatings for ferrous alloys	Method VI	Phosphate coat per TT-C-490

PHOSPHATE COATING

Phosphate coating is recommended for ferrous alloys and zinc-coated steels. It is a chemical treatment that produces a crystalline phosphate-base coating on steel surfaces, giving an excellent base for organic coatings. For steel parts that are not plated and that will subsequently be painted, a phosphate coat prior to pretreat and painting should be specified.

Zinc phosphate coating should be specified after all welding and forming operations and before any organic coatings. A phosphate coating should not be specified for iron or steel that has been plated with any metal other than zinc. See Table 10-15.

PASSIVATION OF STAINLESS STEELS

Corrosion-resistant steel is passivated for three reasons: (1) to remove any particles of steel embedded from machining operations, (2) to clean any scale from heat treatment, and (3) to form the uniform protective oxide film that gives these alloys their corrosion-resistant qualities. This inexpensive, relatively simple process should be specified for all corrosion-resistant parts after machining and other forming operations.

Corrosion-resistant steel assemblies brazed with silver or copper cannot be passivated since the solution will attack the brazing metal. Certain of the heat-treatable steels will be embrittled if passivated after heat treatment. This presents a problem in removing the light heat tint that may form during heat treatment. Mechanical cleaning techniques (such as liquid honing, sandblasting, or rubbing with emery cloth) must then be used. See Table 10-16.

SILVER-PLATING

Silver-plating may be used to reduce friction, to prevent galling, and to provide a solderable surface. It must be given a supplementary chromate finish to reduce tarnishing. However, this added protection is only nominal, and, in all but the most protected conditions, the silver should be overplated with a more inert metal or coated with an organic resin.

Silver-plating to increase electrical conductivity is generally not recommended

Table 10-16. Passivation of Corrosion-Resistant Steel.

Finish	Specification MIL-S-5002 and QQ-P-35	Typical Drawing Callout
Passivation for corrosion-resistant steel	Type I: All AISI 300 series CRES Type II: Precipitation hardening steels including: PH15-7 Mo, 17-4 PH, A-286, 17-7 PH, AM 350, and AM 355 and 400 series CRES	Passivate per MIL-S-5002

because other platings perform the same function at lower cost. Where high electrical conductivity is necessary, copper can normally be used with satisfactory results. For pure wrought metals, silver conductivity is only 5% better than that of copper. For electrodeposited metals, this difference can be changed by variations in the plating process; e.g., silver may be more resistive than copper under certain conditions of electrodeposition. In addition, silver is apt to migrate under certain conditions of temperature, humidity, and electrical stress, degrading the circuit.

A bright finish (Type III) should be specified for all applications requiring a polished appearance (Table 10-17). Where desired, a matte finish (Type I) or a semibright finish (Type II) is also available. Where the silver-plating is to be the final coating or will be subjected to extended periods of storage before application of an overplating, the silver plating should be specified with supplementary chromate treatment (Grade A). Above 300° F, silver forms a nonadherent diffused alloy with copper. For elevated temperature applications, an underplate of nickel is recommended.

For threaded parts, 0.0002-in.-thick silver-plating should be specified. For general applications, a 0.0005-in.-thick silver-plating should be used. On ferrous parts, 0.0005-in.-thick copper-plating should be applied prior to silver-plating.

Table 10-17. Silver-Plating.

Finish	Specification QQ-S-365	Typical Drawing Callout
Silver-plating	Type I: matte Type II: semibright Type III: bright Grade A: with chromate Grade B: without chromate	(Specify underplate if required) Silver plate per QQ-S-365. Type II, Grade B (specify minimum thickness)

Table 10-18. Tin Plating.

Finish	Specification MIL-T-10727	Typical Drawing Callout
Tin-plating	Type I (electrodeposited) Type II (hot-dipped)	(Specify underplate if required) Tin plate per MIL-T-10727, Type I, Fused. (Specify minimum thickness)

On aluminum parts, a 0.0005-in. electroless nickel-plating is used prior to silver-plating.

TIN-PLATING

Tin-plating provides excellent solderability because it fuses and alloys with the solder during soldering. It also provides a limited amount of corrosion protection. The tin finish is soft, provides poor abrasion resistance, has excellent ductility, and is a good paint base. It is recommended that tin-plating be specified for all solder applications that do not require the special properties of gold.

Copper should be specified as an underplate for the tin since it will prevent dewetting of the tinplate and alloy with the tin to produce a smooth, bright finish.

Type I plating (Table 10-18) is electrodeposited and has a matte finish in the as-plated state. Normally, this finish is fused, producing a bright, reflective surface. Because tin is normally fused after electroplating, the thickness that can be applied is limited to a maximum of 0.0003 in. If any thicker, the tin flows to the edges of the part. If thinner than about 0.0001 in., it may dewet.

Because of its low melting point (450° F), tin-plating cannot be used in high-temperature applications. Also, at low temperatures, under certain conditions, it tends to undergo a phase change to powdery gray tin. The use of this coating on threaded parts should be avoided.

Wherever possible, electrodeposited tin should be fused (by melting and flowing at about 500° F) to minimize porosity, improve corrosion resistance, and enhance solderability (particularly after relatively short periods of storage).

ORGANIC COATINGS

Organic finishes are classified as paints, enamels, lacquers, varnishes, primers, and zinc chromate. A brief description of these and their application follows:

- *Paint.* A pigmented liquid composition that is converted to an opaque solid film after application as thin layer.
- *Lacquer.* A liquid coating composition containing, as the basic film-forming

ingredients, cellulose esters or ethers and plasticizers. Apparatus such as networks and components that cannot be heated or that must be dried quickly are usually finished with a pigmented fast air-drying lacquer.

- *Varnish.* A liquid composition that is converted to a transparent or translucent solid film after application as a thin layer. Impregnating varnishes, of either the air-drying or baking type, may be dried by oxidation or polymerization.
- *Enamel.* A paint that is characterized by the ability to form an especially smooth film. An enamel may be dull, semiglossy, or glossy and may be air-drying or baking type. Equipment is usually finished with enamel of the baking type. Air-dry enamel is used for touch-up or for equipment too large for baking ovens, or where components may be damaged by baking temperatures.
- *Primers.* Primers are extremely important and should not be omitted. They provide for good adhesion and are part of the entire paint system.
- *Zinc chromate.* Zinc chromate primer is used as a corrosion-inhibiting coating on metal surfaces. May be used either as an undercoat or as a finish for metallic surfaces.

The proper callout of paint or lacquer finish involves the consideration of several factors that influence the final finish system to be used. These factors are:

1. The kind of base metal passivation or treatment required.
2. The type of primer or surfaces required.
3. The types of final paint and color required by the customer.

When satisfactory protection and appearance standards can be met by the application of one color to a part or place of equipment instead of two or more colors, only one color should be specified. This procedure will reduce masking costs. Paint should be omitted from threads and grounding surfaces. A semigloss paint may be specified for equipment components such as cabinets, frames, covers, etc., that require an organic finish and on which a gloss finish is not specifically required. Lower gloss paints usually hide surface imperfections in the base metal better than glossy paints and will minimize the use of fillers to obtain a smooth surface appearance.

Organic finishes provide surface finishes ranging from flat or dull to high gloss and may be baked or air-dried. Baked finishes are preferred since they are more durable than air-dried finishes, but baked finishes cannot always be used because:

1. Limitations exist in the dimensions of available ovens.
2. Components of the item being finished may deteriorate when subjected to baking temperature.

11

RADIO FREQUENCY AND ELECTROMAGNETIC SHIELDING

An ideal electronic device would neither radiate unwanted energy, nor would it be susceptible to unwanted radiation produced elsewhere. Even if such a device were possible, it would be impractical. However, there are many techniques that can be used to prevent unwanted radiation from degrading performance or causing a malfunction. This chapter provides concise information on techniques for controlling and suppressing electromagnetic radiation.

ELECTROMAGNETIC INTERFERENCE

When electromagnetic energy from sources external or internal to electrical or electronic equipment adversely affects that equipment, causing it to have undesirable responses such as malfunctions or degraded performance, the electromagnetic interference is known as *electromagnetic interference* (EMI). EMI may leave a source or enter a susceptible equipment by radiation, conduction, or coupling. EMI may occur between various equipments in close proximity, or between one part of the equipment and another, such as between the power supply and nearby circuitry.

EMI is radiated through openings of any type in the equipment enclosure: ventilation, access, meter holes, the edges around doors, hatches, drawers, and panels. And EMI will travel through imperfect joints in the enclosure. Another source of EMI is through leads and cables.

The problems of EMI and radio frequency interference (RFI) often fall upon the packaging engineer whose electrical or electronics background is limited. Even though the principles of RFI/EMI shielding involve concepts with which the packaging engineer is familiar, the terminology is basically electrical, and communications is difficult.

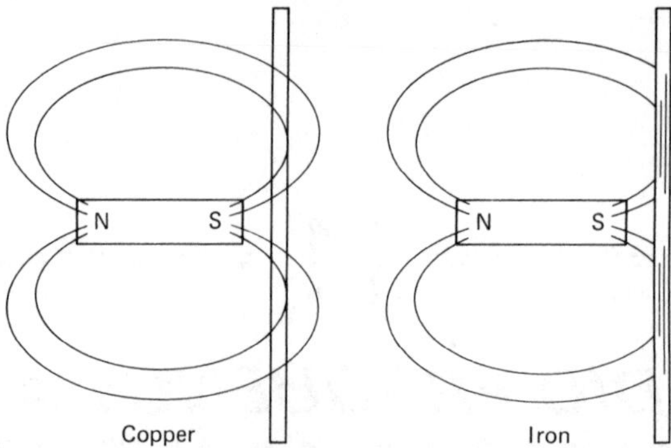

Figure 11-1. Magnetic shields.

RFI is defined as any radio frequency that causes an undesirable response. Some well-known examples are the television interference caused by "ham" radio operators, radar-scanning noise in communications receivers, and even the accidental detonation of explosives. The serious consequences of uncontrolled RFI are evident when one considers the abundance of electronics systems, communications, computers, control systems, radar, etc., in use today and their close proximity. Because of simultaneous operation of many types of equipment, at one location, radiation or electrical noise conduction of each unit must be held at an absolute minimum to prevent interference.

RFI is electromagnetic; therefore, the shielding must consider both the magnetic and electrostatic problems. An example is the permanent magnet in free air that has a magnetic field around it. A copper sheet that is nonmagnetic will not alter the field when placed within its boundaries (Figure 11-1). On the other hand, a sheet of iron or other magnetic material will distort the field and reduce or eliminate it entirely on the far side of the magnetic sheet (Figure 11-2). Therefore, we must use a magnetic material to shield a stationary magnetic field.

A plate charged with a high direct-current voltage has an electrostatic field surrounding it. A nonconducting material will not alter the field; however, any conducting material, magnetic or nonmagnetic, will short-circuit the field and eliminate it on the far side of the barrier. Therefore, an electrically conductive material may be used to shield a stationary electrostatic field.

Three factors are employed by the designer both in controlling electromagnetic energy from emanating from the equipment and in protecting it from external sources:

1. Electromagnetic energy levels can be minimized through proper circuit design and the choice of components.
2. Conducted electromagnetic energy can be suppressed or controlled by using filtering devices.

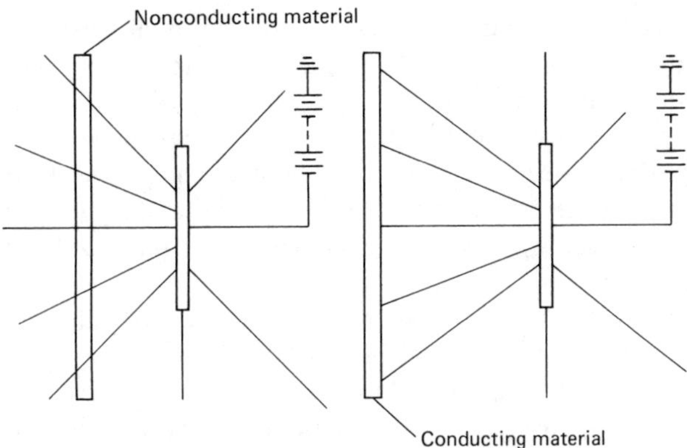

Figure 11-2. Electrostatic shields.

3. *Electromagnetic energy can be confined by the use of shields.* The third factor is the subject of this chapter since this area is the most common to the electronics packaging engineer.

To perform effectively, the selected shielding must;

- Allow only the unwanted internal energy to pass that will not affect the performance of nearby equipment.
- Prevent harmful energy from passing through the shield into susceptible equipment.
- Be compatible with the equipment's functional intent.

TEN RULES OF THUMB FOR RFI PREVENTION

1. Keep the number (and size) of openings in a shielded structure to a minimum.

2. Every opening in a cabinet must be RFI-proofed. This includes every panel, orifice, door, cooling vent, power outlet, etc.

3. RFI stripping must make good electrical contact with the surface or surfaces it adjoins. To assist in achieving this aim, cabinet surfaces are frequently plated with a metal having a low contact resistance. Cadmium is such a metal, and it is fairly economical.

4. Constant physical pressure should be maintained on all RFI stripping to insure good electrical contact. Make certain the applied pressure is great enough to insure adequate electrical contact, even when nonconducting corrosion is present.

5. Continuously soldered or welded seams are preferred; spot-welded or screw-fastened seams should be avoided unless absolutely necessary. In the latter instance, keep the spacing between fasteners or spot welds small.

6. Make certain RFI electrical filters are inserted in all power lines entering the equipment to be shielded. (All signal leads attached to the equipment should be carefully shielded.)

7. Unless complete electrical integrity is assured, panel connectors should have a wire-mesh gasket between the connector mounting flanges and the panels.

8. Where there are several radiating sources in a piece of equipment, it may be desirable to shield each signal source separately rather than to shield the entire enclosure.

9. Inlet holes for the introduction and exhaust of cooling air should be small compared to the shortest wavelength of signal developed in the equipment. RFI leakage through a hole is proportional to the cube of its diameter.

10. In dealing with RFI gaskets, the greater the compressibility of the gasket, the higher is its effectiveness.

A shield reduces equipment's radiation or susceptibility to radiation by both reflecting and absorbing the energy contained in the radiating electromagnetic field. The measure of a shield's capability to reduce electromagnetic field strength is called its *shielding effectiveness*. This can be defined as the reduction in transmitted radio-frequency (RF) power as a result of inserting a shield between the undesirable radiating source and the equipment, and is expressed in decibels (Table 11-1).

CLASSES OF APPLICATION

All shielding problems may be divided into four major categories [see Tables 11-1 and 11-2] by the nature of the result required.

1. *Shielding for Containment*. In these applications, the designer is required to prevent the escape of the interference generated by a particular device, circuit, or system . . . , to nullify its potential behavior as a "disturbing source." This large group of applications is characterized by high energy–the source of interference generally is one that operates at relatively high voltage, current, or power levels. For that reason, most of them dissipate relatively high power per unit of volume or scanfree area; and this fact often imposes critical thermal constraints on the shielding structure–it is necessary to contain the interference without impeding heat flow. This conflict between electromagnetic containment and thermal transfer creates the need for many special materials, components, and structures . . . for one example, the need for cooling air filters that are "opaque" to EMI, but that allow relatively free passage of cooling air.

2. *Shielding for Exclusion*. In these applications, the designer is required to prevent the intrusion of anticipated or known levels of interference on the performance of a specific device, circuit, or system. Starting with a presumed level and spectrum of interference, and information on the tolerable level at each frequency of interest, he must devise effective means of attenuating the ambient interference in which the protected equipment must operate. This group of applications is not as easily char-

Table 11-1. Application Constraint Summary.

Classes and Subclasses of Applications	Thermal Transfer by Conduction	Thermal Transfer by Forced Air	Visual Access	Physical Access by Doors, Covers, etc.	Shield Isolation from Enclosed Circuitry	Hermetic and/or Fluidic Sealing
Shielding for Containment	–	Note 1	–	–	–	Note 2
• Components: relays, motors, etc.	Rarely	Rarely	Sometimes	Often	Rarely	Rarely
• Circuits: power supplies, pulse generators, etc.	Usually	Sometimes	Rarely	Often	Sometimes	Rarely
• Systems: controllers, computers, etc.	Rarely	Often	Sometimes	Always	Rarely	Sometimes
Shielding for Exclusion	–	–	–	–	–	–
• Components: IC's, networks, etc.	Rarely	Never	Never	Rarely	Often	Never
• Circuits: amplifiers, detectors, signal converters, etc.	Rarely	Never	Never	Rarely	Often	Never
• Systems: controllers, computers, etc.	Rarely	Often	Sometimes	Always	Rarely	Sometimes

Table 11-2. Characteristics of Specific Applications.

Classes and Subclasses of Applications	Digital Computers, Instruments, Control Systems	Analog Instruments and Control Systems	Aircraft, Spacecraft, Automotive Electronics	Communications Systems and Signal Instrumentation	Secure Environments and Shielded Rooms	Home Appliances
Shielding for Containment	–	–	–	–	–	–
• Components: relays, motors, etc.	Note 3	Note 6	Note 3	–	–	Note 3
• Circuits: power supplies, pulse generators, etc.	Note 4	Note 4	–	Note 4	–	Note 8
• Systems: controllers, computers, etc.	Note 5	–	Note 5	Note 7	–	–
Shielding for Exclusion	–	–	–	–	–	–
• Components: IC's, networks, etc.	–	Note 11	Note 11	Note 13	–	–
• Circuits: amplifiers, detectors, signal converters, etc.	Note 9	Note 9	Note 9	Note 14	–	–
• Systems: controllers, computers, etc.	Note 10	Note 10	Note 12	Note 12	Note 15	–

acterized as the preceding one, because the protected equipment may be operating at both high sensitivity and high power density. But it may or may not impose thermal-transfer constraints as well.

One aspect of exclusion shielding is worth special mention here: the question of undesired interaction between the shield and the protected equipment. While this is sometimes also a problem in containment shielding, it is a much more severe constraint when one must deal with high-sensitivity (and therefore high-impedance) circuitry. A shield that is effective as a shield may introduce intolerable parasitic capacitance; it may lower the "Q" of reactive and/or resonant circuits; and it may create ground-loop paths that introduce intolerable coupling between two parts of the protected circuit.

3. *Exclusion/Containment Shielding plus Pressure Sealing.* In these applications, the shielding structure must provide "pressure containment" as well as EMI/RFI attenuation. While the most commonly encountered requirement is the maintenance of an air-pressure seal (at seams, doors, etc.) to preserve the integrity of an air-flow pattern used for cooling, there are many other applications that require hermetic and/or fluidic sealing. To cite a few familiar examples, the dust-proofing of enclosures; the sealing of the access door of a microwave oven against the escape of cooking odors and steam; the prevention of gas leakage from high-voltage apparatus; and the exclusion of moisture from electrical and electronic assemblies.

4. *Grounding and Contacting.* This class of applications uses much the same technology as shielding, but for another purpose, . . . to conduct electrical energy. Among the many tasks the designer faces that fall into this category are: discharging electrostatic energy from imperfectly grounded structures to ground, providing a dependable, low-impedance connection between removable (pluggable) circuit assemblies and the host-circuit ground plane and/or power buses; and ensuring good contact between large or massive electrical structures that do not necessarily mate mechanically.

SHIELDING TECHNIQUES

1. *Digital Computers, Peripherals, and Instrumentation*

   Interference sources: switching power supplies; clock-pulse and address-strobe generators; drivers for impact printers, card punches, solenoid controls, and tape-drive controls; brushes on drive motors for printers, discs, card readers, card punches, and drums; high-voltage power supplies for CRT terminals; relays and contactors; teletypewriters.

   Susceptibility: tape and disc reading heads and amplifiers; fast logic circuits; A/D input circuits; sense amplifiers; modem inputs; electrooptical isolators; regulator circuits in power supplies; cabling between enclosures.

2. *Analog Instrumentation and Process-Control Systems*

   Interference sources: relays and contactors; solenoid drives; switching power supplies; brushes on motors; switches and program drums. (Typical interference spectrum: 10 kHz to 50 MHz.)

   Susceptibility: power-supply regulator circuitry; signal conditioning and amplifying circuitry; cabling to transducers; reading heads on analog tape drives.

3. *Digital Process Control Systems*

   Interference sources: essentially the same as item 1.

   Susceptibility: fast logic circuits; multiplexer, signal-conditioning, sampling, and preamplifier circuits on analog data-acquisition "front-end"; electrooptical isolators; regulator circuits in power supplies; input and intercabinet cabling.

4. *Automotive Electronics*

   Interference sources: ignition system; switches; alternator slip rings; motors for windshield wipers, seats, windows, fans, compressors; flashers. (Typical interference spectrum: 20 kHz to 100 MHz.)

   Susceptibility: radio, tape player, automatic regulators on fuel injection, differential traction, brake pressure, carburetion.

5. *Communications Systems and RF Instrumentation (Including Radar)*

   Interference sources: oscillators; pulse modulators; HV power supplies; switching power supplies; corona discharge from HV RF points; arcing at high-current RF junctions. (Spectrum is highly dependent on carrier and modulation parameters.)

   Susceptibility: Receiver "front-end," IF, and detector/discriminator circuits; power-supply regulator circuits; built-in digital logic.

6. *Medical Electronics*

   Interference sources: HV power supplies in CRT and X-ray equipment; switching power supplies; diathermy oscillators; corona discharge from X-ray circuits; flyback and scan-drive circuits in CRT displays; pulse generators; switches; motion. (Typical interference spectrum: 10 kHz to 50 MHz.)

   Susceptibility: electrode cabling and amplifier input circuits for EKGs and EEGs; regulator circuits in power supplies; computing circuits (analog and digital) in automatic analyzers, titrometers, and monitors.

7. *Security Systems (Military and Commercial)*

   Interference sources: switching power supplies; fluorescent lighting; oscillator and pulse-generator circuits; transmitters (see item 5); computers and peripherals (see item 1); CRT-terminal power supplies and scan drivers; appliances, tools, and business machines. (Spectrum not specifically definable, but generally very broad.)

   Susceptibility: receivers; computer circuitry; peripheral recorders and printers; antenna lead-ins; intercabinet cabling.

8. *Home Appliances and Business Machines*

   Interference sources: motors; relays; switches; fluorescent lighting; gas-discharge lamps; driver circuits; HV power supplies; switching power supplies; microwave oscillators. (Typical interference spectrum: 1 kHz to 100 MHz.)

Susceptibility: almost none.

9. *Shielded Rooms*

Interference sources: fluorescent lighting; fan motors; instrumentation containing oscillators and/or pulse generators; switches. (See related items for typical spectrum).

Susceptibility: determined by application. See items 1 to 8 for details.

10. *Aircraft and Spacecraft*

Interference sources: engine electrical systems; aerodynamic control systems; radars; altimeters; transponders; communications systems; lighting; air-conditioning systems; switching; galley equipment; passenger-entertainment electronics; on-board computers; telemetry systems. (See related items for typical spectrum.)

Susceptibility: communications systems; control systems; computers; and flight instrumentation.

APPLICATIONS

Table 11-2 presents an application constraint summary. Table 11-3 presents characteristics of specific applications.

Notes for Tables 11-2 and 11-3

1. Pressure sealing is usually necessary. Use combination gaskets and gasketing materials.
2. Certain of the combination gaskets and gasketing materials referred to in Note 1 are suitable for fluidic and absolute hermetic sealing. Check for material's compatibility with the insulating fluid!
3. Generally, any component using a moving contact (brushes, relay contacts, switches) may require containment in itself, to prevent interference with nearby circuitry.
4. Generally, any switching power supply or power pulse generator with more than 10 W output should be considered a likely candidate for individual containment; also, HV rectifiers and very-high-current capacitor-input rectifier-filters. In computers, printer-drive and card-reader advance drives should be considered for self-containment.
5. Digital logic, particularly clock-pulse generators and data strobe circuits, including core-memory address generators, justify shielding of the complete system enclosure, as well as signal and power cabling between the computer and its peripherals.
6. Process controllers often involve relay panels, actuators, and motor contactors, all of which are logical candidates for self-containment. See Note 3.
7. RF transmitters and signal generators almost always require comprehensive containment, i.e., design problems. Radar and sonar transmitters always require

Table 11-3. Comparative Data on Commonly Used RFI Shielding Materials.

	(Attenuation (db))				
Frequency	Galvanized Steel (22 gauge)	Aluminum Sheet (0.026 in.)	Copper Mesh (No. 22)	Electrosheet Copper (1 oz)	Electrosheet Copper (4 oz)
500 kc	75	71	66	75	65
1000 kc	80	80	71	70	71
0.15 Mc	100	100	100	100	100
0.50 Mc	100	100	100	100	100
1.5 Mc	100	100	100	100	100
5.0 Mc	100	100	100	100	100
10 Mc	100	100	100	100	100
60 Mc	87	76	70	87	77
100 Mc	77	69	48	74	64
400 Mc	42	61	64	59	57
750 Mc	55	54	64	73	61
1000 Mc	42	49	62	81	68
5000 Mc	30	23	32	29	39
10000 Mc	20	9	13	22	28

self-containment. Even receivers, which contain local oscillators or super-regeneration, frequently demand self-containment.

8. Microwave ovens are a good example of the urgent need for self-containment to avoid EMI "pollution," and television receivers are also worthy of consideration in this category, although less powerful.

9. Amplifiers designed to accept low-level, broadband signals (100 mV and less), including the comparators used in A/D converters and limit detectors, are particularly sensitive, and often require exclusion shielding from their host equipment's power supplies and from other sources of interference. Core-memory sense amplifiers are particularly vulnerable.

10. Computer logic demands exclusion shielding, as well as electrostatic-discharge protection. In process instrumentation and control, the low circuit impedances and narrow bandwidths associated with strain-gauge, thermocouple, potentiometric, and conductivity transducers and amplifiers make them relatively insensitive, despite their low signal levels. On the other hand, data loggers, with thin associated broadband buffers, are vulnerable.

11. The scaling networks associated with D/A and A/D converters, as well as certain high-gain IC comparator and amplifier chips, are likely candidates for exclusion shielding. Hybrid RF amplifier ICs and various receiver components are also moderately sensitive.

12. The high density of equipment packaging in modern aircraft and space-

craft (in combination with high ambient interference levels) demands complete exclusion shielding of receivers, flight computers, and instruments. In automotive systems, the sensitivity is lower, but the ambient interference levels are very high, demanding careful shielding and grounding to the vehicle frame.

13. Hybrid microwave and RF amplifier chips, and similar broadband "front-end" and IF components may have to be individually shielded, despite effective enclosure shielding, to prevent interaction effects (images, spurious responses, etc.).

14. It is almost always necessary to provide some degree of exclusion shielding for receiver "front-end" circuits, as well as the IF and even detector/discriminator sections of a receiver. See Note 13.

15. Secure environments, particularly those used for surveillance and spectrum survey, present the most demanding needs for highly effective, very-broadband exclusion shielding. (120 dB from 0 to 18 GHz is not an uncommon requirement!) Shielded rooms for instrument calibration are not quite as demanding, but are still major problems, worthy of every precaution.

GENERAL SHIELDING PRACTICE

Full and completely enveloping shielding is the only sure type of magnetic shielding, and the less this envelope is broken by holes and cutouts, the better the protection will be. Shielding materials in the form of thin tapes and foils are quite soft and can be used to wrap items to be shielded, thus furnishing an all-enveloping shield.

Lapped joints can function well if the main portion of the shield is adequate for the disturbing flux present. Small gaps in seams can also be tolerated (0.005 in., for example). It is for this reason that welds in the shield joints can also be tolerated, provided the recommended heat treatment follows. Placement of magnetic shielding in a single-plane partition is not too satisfactory unless all the factors are considered (Figure 11-3). Single-plane shielding is particularly bad from the standpoint of symmetry: the lines of flux, which would normally pass out beyond the shield, will instead tend to concentrate along it. If an insufficient thickness of shielding metal is used, the shield will appear to be transparent beyond the point of saturation.

The ability of the magnetic field to pass through a shield depends upon the permeability and saturation level of the shield material (Table 11-4). Ideally, it should have high permeability, low retentivity, and the ability to carry a large magnetic-field density before saturating. The saturation level is determined by the thickness as related to the characteristic magnetic property and frequency of the disturbance (other than direct current). Economy in shielding can sometimes be gained by laminating the shield. For example, a thin low-permeability material can be used to attenuate low-level high-frequency fields, and a heavier,

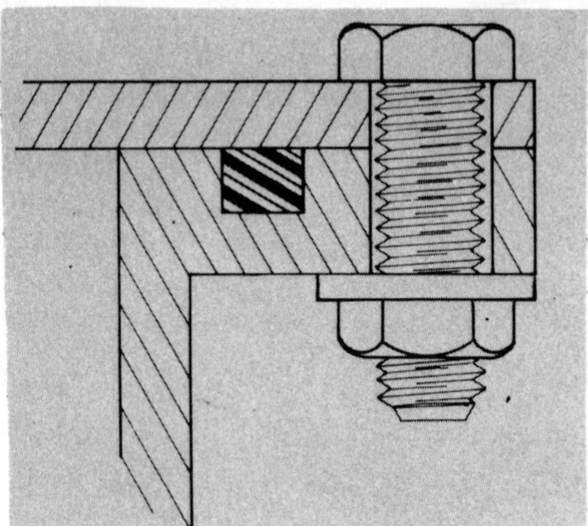

Figure 11-3. Die-cast joint.

medium-permeability material can then stop the heavy d-c fields. Prior to establishing the sequence of the laminations, a choice must be made between excluding or containing the disturbing field.

Although enclosures can be made of any material, practical considerations sharply restrict the selection, limit some materials to certain design categories, and dictate the fabrication techniques for others. Steel and aluminum are found in every enclosure size from tiny cans and boxes to large buildings and missile silos. Most electronic equipment is housed in aluminum enclosures. Metals of high magnetic permeability–mu metal, grain-oriented silicon steel, etc.–are used mostly in smaller-scale enclosures. Exotic metals such as gold, silver, and platinum are encountered chiefly in the smallest enclosures, usually in the form of an electroplated or sprayed coating on a base-metal or nonmetallic housing.

In selecting an enclosure material, the equation

$$A = 3.3t\sqrt{\mu f G}$$

shows the significant factors in the shielding problem. A (the attenuation in decibels) is directly proportional to t (the thickness of the material in mils), to F (the frequency of the EMI/RFI field in megahertz), to G (the conductivity of the material referred to that of copper), and to μ (the relative magnetic permeability of the material referred to that of free space) (Table 11-4).

Usually, the frequency of the field to be dealt with is known, and the desired degree of attenuation has been established. Any material whose properties satisfy the requirements of the equation can be used for the shielding enclosure. If there is a free choice of materials, one with high μ and/or G could be used in a thinner gauge than one with low permeability and/or low conductivity. If the material has been specified, then the required thickness will depend upon the EMI/RFI frequency; if the construction requirements limit the thickness, then

Table 11-4. Characteristics of Various Metals.

Metal	Relative Conductivity	Relative Permeability at 150 kc	Penetration Loss (db/mil) at 150 kc
Silver	1.05	1	1.32
Copper–annealed	1.00	1	1.29
Copper–hard drawn	0.97	1	1.26
Gold	0.70	1	1.08
Aluminum	0.61	1	1.01
Magnesium	0.38	1	0.79
Zinc	0.29	1	0.70
Brass	0.26	1	0.66
Cadmium	0.23	1	0.62
Nickel	0.20	1	0.58
Phosphor-Bronze	0.18	1	0.55
Iron	0.17	1,000	16.9
Tin	0.15	1	0.50
Steel, SAE 1045	0.10	1,000	12.9
Beryllium	0.10	1	0.41
Lead	0.08	1	0.36
Hypernick	0.06	80,000	88.5
Monel	0.04	1	0.26
Mu-metal	0.03	80,000	63.2
Permalloy	0.03	80,000	63.2
Steel, stainless	0.02	1,000	5.7

the designer may have to settle for less attenuation, or shield certain equipment components or subassemblies individually.

At frequencies above 100 MHz, conventional thicknesses used in standard construction will usually provide more than adequate shielding. At lower frequencies, magnetic fields begin to predominate; these are more difficult to suppress than are electric or plane-wave fields—the lower the frequency, and the higher the intensity, the bigger the problem. High-permeability materials provide the greatest magnetic-field attenuation for a given thickness. However, it must be remembered that these materials decrease in effectiveness as the intensity of the field increases. When the material reaches its saturation level, the permeability is no better than that of ordinary steel.

When the enclosure design is completed, consideration can then be given to the design of the openings and access ports, and their covers or doors. No matter how well they are built or how tightly they are closed, the seams or joints at these ports represent anomalies in the continuity and integrity of the enclosure, and are subject to leakage and penetration. They must be designed, and their gaskets selected and installed, with the purpose of restoring the original continuity and integrity.

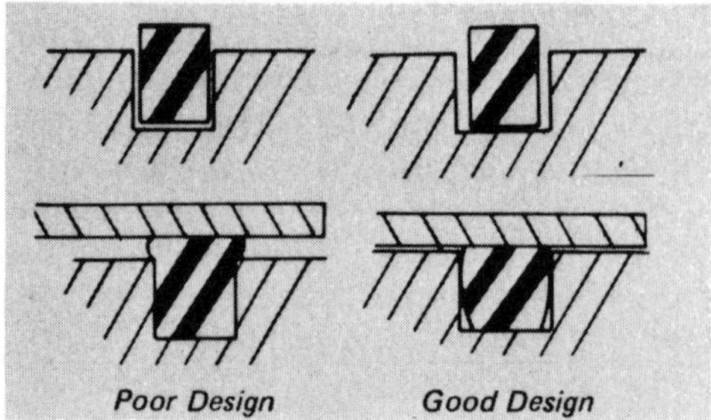

Figure 11-4. Making allowance for solid elastomer flow.

DIE-CAST VS. SHEET-METAL JOINTS

The most reliable and satisfactory joint is achieved with a die-cast flange (preferably cast as an integral part of the enclosure), in which a groove or slot is cast or machined to accommodate the sealing gasket (Figure 11-4). The joint, the gasket, the method of closure, and the degree of closure force are all selected with one fundamental purpose in mind: to achieve maximum shielding effectiveness by maintaining the highest practical conductivity across the joint interface, at every point (Figure 11-5).

In sheet-metal enclosures, where practical, a formed flange may be welded into the enclosure. In the majority of cases, an overlap flange joint (Figure 11-6) will be used. This joint can be reinforced with metal strips, bars, or angles, but obviously this does not offer the same rigidity as a cast flange, and places greater reliance on the compressibility and resiliency of the gasket material in achieving a full and uniform seal.

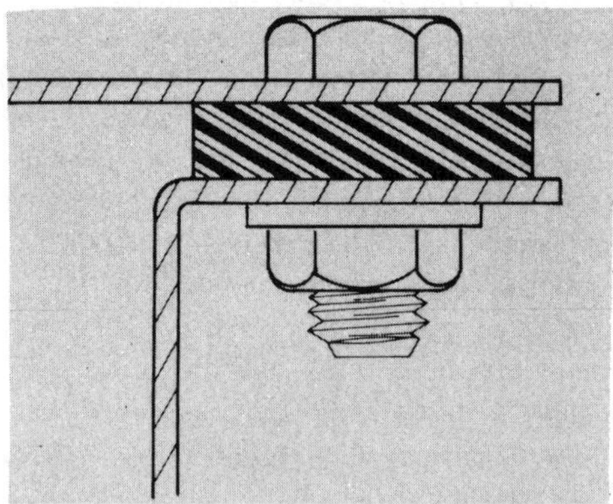

Figure 11-5. Sheet-metal joint.

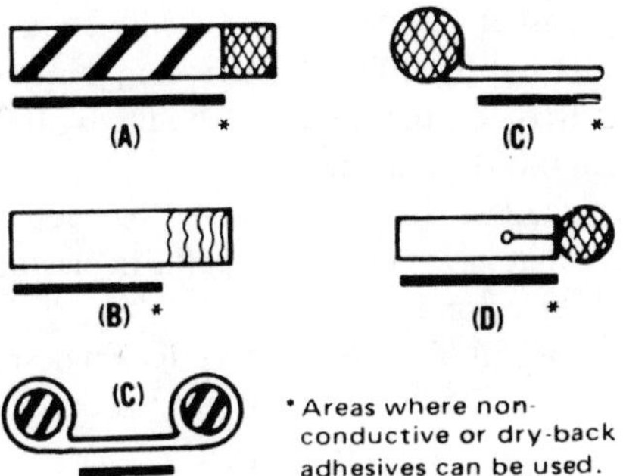

Figure 11-6. Application of nonconductive adhesive.

Pressure-Sensitive Adhesive

Pressure-sensitive adhesive is often the least expensive method for attaching EMI gasket materials; installation costs are often drastically reduced, with only a slight increase in cost over a material without adhesive backing (Figure 11-7).

Caution: In all cases, the designer specifying nonconductive adhesive attachment must include adequate warnings in the applicable drawings and standard procedures for instructing personnel that the adhesive is to be applied only to the portion of the gasket material not involved with the EMI/RFI gasketing function.

In the past, many installation workers have, either through carelessness or a misguided desire to do a better job ("this gasket would hold better if I glued all of it rather than half of it"), applied the nonconductive adhesive to the EMI gasket portion also. This seriously degrades the EMI/RFI performance.

The following effects of adhesives must be carefully considered when bonding of RFI materials is to be used.

- Most conducting adhesives are hard and incompressible. Thus, if too much adhesive is applied, and it is allowed to soak too far into the EMI/RFI gas-

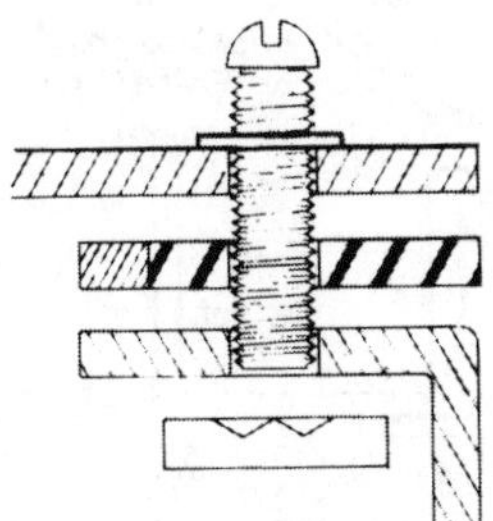

Figure 11-7. Bolt-through gasket mountings.

ket material, the gasket's compressibility will be destroyed. Irregularly applied adhesive also has the effect of increasing joint unevenness.

- The volume resistivity of the adhesive should be 0.010 ohm-cm or less, preferably less than 0.001 ohm-cm.
- Most conductive adhesives do not bond well to neoprene or silicone. This is why all products that have conductive paths in elastomer are rated "poor" for conductive adhesive bonding.
- Applying a $\frac{1}{8}$-to-$\frac{1}{4}$-in. (3.67-to-6.35 mm) diameter spot of conductive adhesive every 1 or 2 in. (25.40 to 50.80 mm) is preferred over a continous bead.
- Conductive epoxies will attach the gasket permanently. Removal of the gasket without destroying it is almost impossible. These adhesives must also be very carefully applied, to avoid postcure hardspots that can destroy gasket compressibility.

Bolt-Through Holes

A bolt-through hole is a common and inexpensive way to hold gaskets in position (Figure 11-8). Providing bolt holes involves only a small initial tooling charge. Bolt holes can be provided in the fin portion of EMI/RFI mesh strips, or in rectangular cross-section EMI/RFI mesh strips if they are wide enough–minimum width $\frac{3}{8}$ in. (9.52 mm).

Shielding

Poor seam construction, air vents, meter faces, fuses, shaft openings, and so on, create equipment discontinuities through which interference energy can pass.

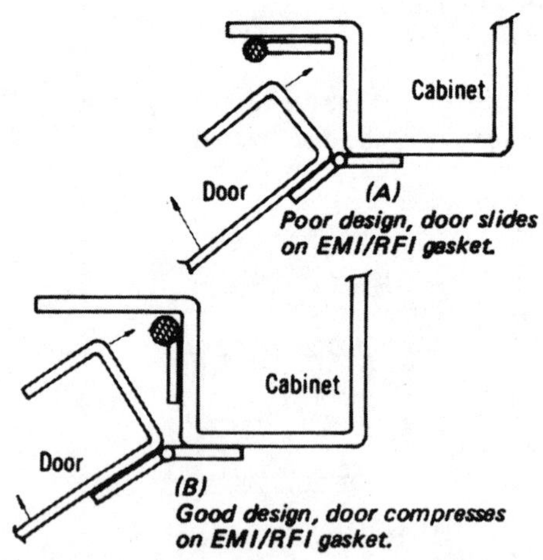

Figure 11-8. Sliding motion vs. straight compression.

Proper use of seam gasketing, conductive glass, and grounding can reduce the problem, but the metal of the enclosure itself is the most important factor.

Metal shielding cuts interference considerably. To make the best use of it, consider what type of metal would be best, the thickness of the metal, the best types of openings, and reflection and absorption losses. RFI can leak through seams, holes, and other openings, and can penetrate thin walls. Copper, aluminum, and magnesium have proved best for high-frequency shields; iron makes the best low-frequency shield.

Some general rules for shielding design are:

- Solder, weld, or rivet all joints.
- Use screen or honeycomb for ventilation openings.
- Use spring fingers or conducting gaskets on lids and panels.
- Keep mating surfaces free of paint and anodized finishes.
- Shield or filter all external leads.
- Install conducting gasket washers under connectors.

In all cases where an overlap of material can be tolerated at a seam, this method of maintaining seam continuity should be used in preference to the insertion of gasket materials. The most important factor is that the materials be fitted as tightly as possible to ensure maximum electrical contact between the two mating surfaces (Figure 11-9).

If bolts are used, it is necessary to distribute the pressure along the joint evenly. The bolt spacing should be consistent with the amount of electromagnetic energy to be contained: the greater the energy, the closer the bolt spacing. When bolts are used to join cabinet parts, electrical contact can be maintained by using resilient conductive gaskets between the two surfaces of the seam. To achieve an optimum insertion loss with a metal gasket, a pressure of 20 psi should be uniformly applied. Pressures in excess of 30 psi tend to destroy the elastic properties of the metal gasket and result in decreased insertion-loss characteristics (Figure 11-10 and 11-11).

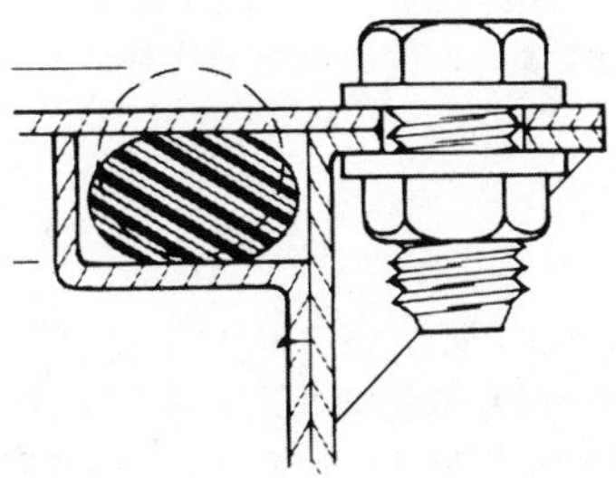

Figure 11-9. Typical sheet-metal slot flange.

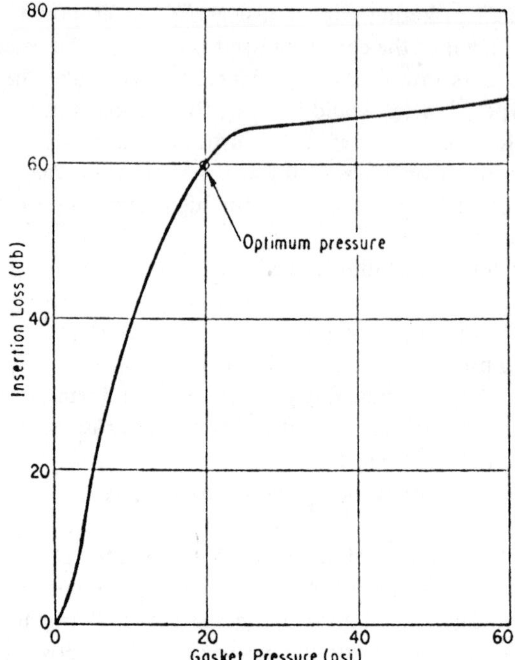

Figure 11-10. Typical plot of insertion loss vs. gasket pressure.

SHIELDING REQUIREMENTS

It is difficult to give a general specification for the amount of attenuation or shielding effectiveness required. In general, it is good design practice to obtain 120 to 150 db attenuation of electrical fields at 1 MHz. In addition, if intense magnetic fields must be attenuated, a minimum of 60 to 80 db at 1 kHz is recommended. It is usually the task of a specialized group to measure radiated interference or susceptibility levels to help the designer estimate the amount of shielding effectiveness that will be required.

In most military applications, radiation specifications must be complied with. Standards of measurement are called out and the designer's problem is well defined.

CABLES

To protect sensitive, low-frequency (DC to 10 mc) data circuits from large, unbalanced AC currents in the ground plane, use a shielded twisted pair, which is available with low loss in this frequency range. Ground the shields at each end of the cable run and ground the signal return at the source only.

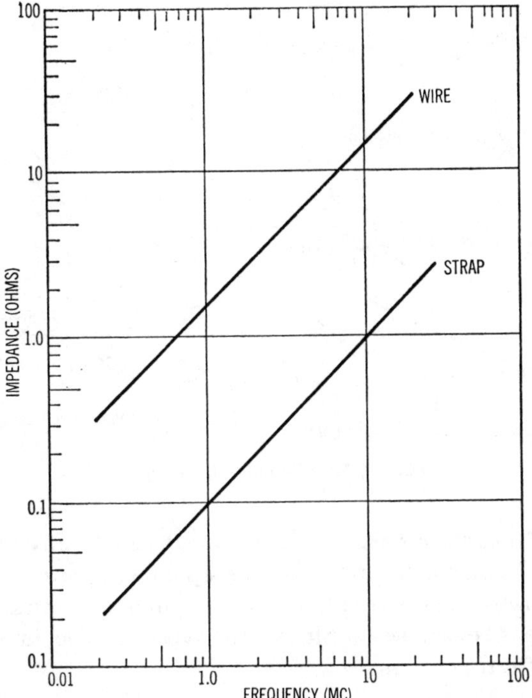

Figure 11-11. Impedance advantage of a $5\frac{1}{2} \times 1 \times \frac{1}{16}$-in. beryllium copper bonding strap over an 8-in. length of No. 12 wire.

Multipoint grounding has clear advantages over a floating, or single-point, system:

- It provides short ground connections.
- It allows use of a low-impedance ground return circuit.
- It provides low-impedance connections between components.
- It improves RF filtering.
- It reduces radiation and cross talk in cables.

Cable shield continuity and shield ground connections directly affect cable radiation and cross talk. Cable shields must be continuous throughout the length of the cable run and grounded at each end. At connectors, they should be carried through on at least two cable pins or on the connector shell itself.

Interconnecting cables can transfer interference from one circuit to another by radiation, induction, or direct connection. To get isolation between cables, you must consider the types and functions of the cable (power, control, signal), the

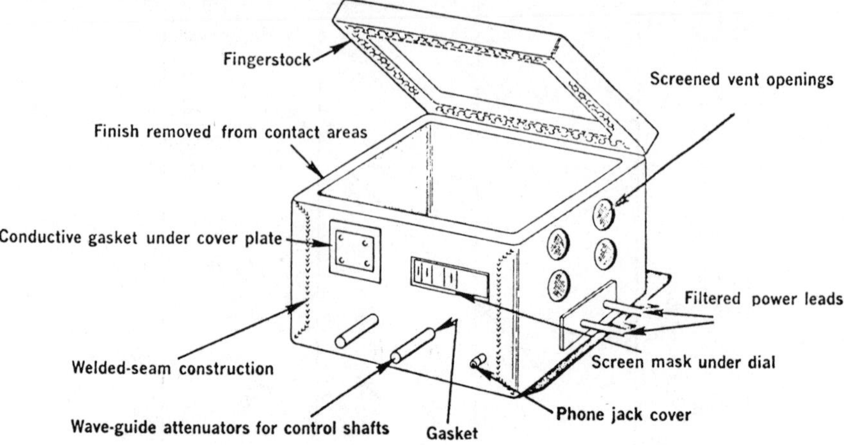

Figure 11-12. Grounding and bonding.

signal sources (transmitters, receivers, computers), and the other electromagnetic parameters that describe the cable circuit's signature. The amount of isolation depends on how the cables are routed, their separation and class, their shielding effectiveness, and sometimes on filters. The types of cable, in order of increasing shielding effectiveness, are:

- unshielded wire (for noncritical power and control applications)
- twisted pair (for audio frequency circuits)
- shielded wire (for critical power and control applications)
- twisted shielded pair (for audio- and video-frequency multipoint grounding)
- coaxial line (for video and radio frequencies)

Grounding and bonding. A ground plane that extends throughout the installation is the most fundamental requirement of system grounding. Good rules for bonding are:

- Bond equipment housings directly to the ground plane.
- Make sure that mating surfaces are clean and free from paint and nonconductive finishes. The bonds should be metal-to-metal.
- Use Iridite or Alodine finishes for corrosion protection and good electric conductivity. Or, use at least two solid, flexible straps with a length-width ratio of 5 or less (Figure 11-12).

CONTROL BOX SOURCES

In control boxes, interference sources are the panel lights, controls, meters, rotary switches, and fuses. To suppress such interference, follow these steps (Figure 11-13):

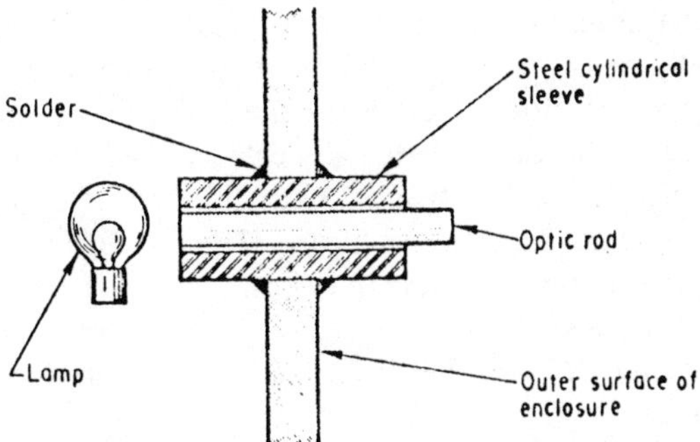

Figure 11-13. Indicator lamp shielded by installing a nonconductive opticrod in a metal waveguide with minimum length-to-diameter ratio of 3 to 1.

- Filter the panel-light leads. The filter mounting should be in shielded-hat form if rear-panel lights protrude.
- Ground the control shafts as they leave the cases.
- Handle meters and fuses in the same way as the panel lights.
- Handle rotary and snap switches as relay contact transient sources.

RFI sources in power supplies are silicon-diode, thyratron, and gas-tube rectifiers. Silicon-diode rectifiers are inherently noisy; the suppression methods for them range from brute-force input and output filtering to diode selection on the basis of operating characteristics.

INDICATOR RFI SOURCES

- Use AC rather than DC blower motors.
- Use conductive, coated Plexiglas with peripherally sputtered silver for the indicator face. Use metal gaskets or spring fingers for mating the face to the case.
- Ground the control shafts as they leave the case.

MODULATOR AND RECEIVER-TRANSMITTER RFI

Interference sources in the modulator and receiver-transmitter are thyratrons, magnetrons, and local oscillators. To suppress such interference, follow these steps:

- Put the modulator, modulator supply, and receiver-transmitter in the same case.

- If the modulator and its supply are separated, use an RF filter at the high-voltage supply input to the modulator. If the modulator is separated from the transmitter, use triaxial cable for the high-voltage pulse to the transmitter. Ground the cable's internal shield at the thyratron cathode, and carry it through to the low side of the transmitter input pulse transformer. Ground the external shield at each case. Use RF filters for all control and power circuits entering or leaving the cases.

Sources of antenna RFI are the drive and servo motors and switching. The suppression steps are:

- Filter DC drive motors at the motor. If the filter is remotely located, use shielded leads between it and the motor.
- Shield the servo leads.
- Use coaxial cable carried through coaxial connectors for all signal wiring.

PANEL-MOUNTED HARDWARE

Control panels are extremely difficult, but not impossible, to shield. Every control panel contains a number of controls, such as indicators, switches, and meters that must be properly shielded. The use of waveguide sleeves in all the control openings of the panel, surrounding nonmetallic control shafts, helps to maintain the shielding integrity of the control panel. Ferrous sleeves are recommended at low frequencies with a minimum length-to-diagonal ratio of 3 to 1.

An example of an opening required for a power on-off indicator is shown in Figure 11-14. An exposed indicator-lamp filament, for example, would serve as a radiating element for RF. The sleeve enclosing the glass rod should have a length-to-inner-diameter ratio of at least 3 to 1.

Indicators should employ magnetic-waveguide shafts having a minimum length-to-diameter ratio of 3 to 1 where low-frequency magnetic fields have to be

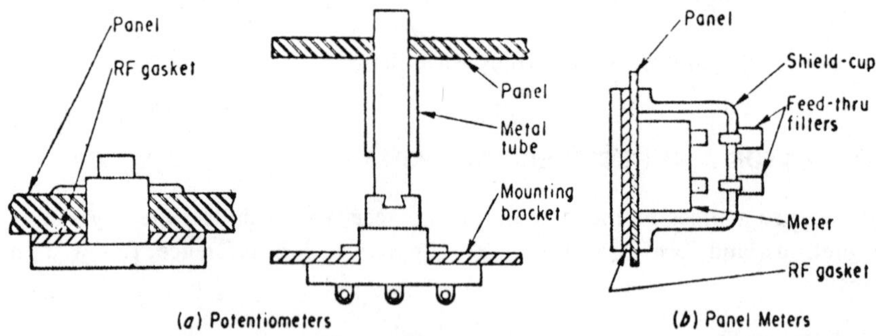

Figure 11-14. Left, proper shielding technique for (*a*) potentiometers, (*b*) panel meters.

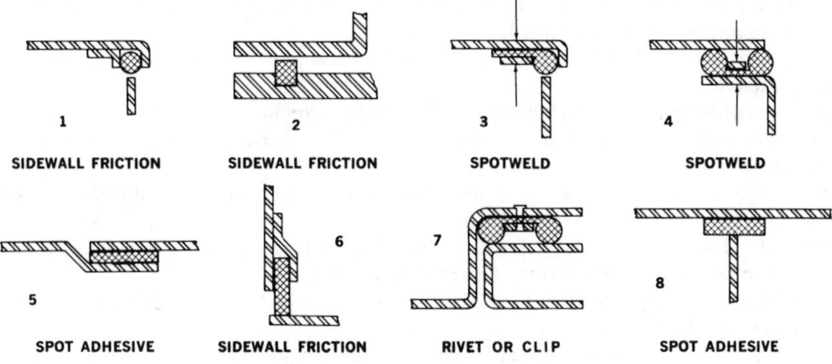

Figure 11-15. Typical mounting methods.

attenuated. In addition, these shafts provide good attenuation for both electrical and magnetic fields over a wide frequency range.

Similarly, push-button switches should employ magnetic-waveguide shafts as previously described. However, where waveguide shafts are used for controlling the leakage of electromagnetic energy, the switch shaft should be nonconductive. Toggle switches present more difficult shielding problems, especially at low frequencies, because they do not lend themselves to the use of waveguide shafts. Push-button switches with light indicators should be used instead of toggle switches whenever possible.

Several techniques can be used to prevent the leakage of electromagnetic energy when installing potentiometers and indicator meters on the front panel of an enclosure (Figure 11-15). It is difficult to properly shield a unit and, at the same time, provide visual access. At high frequency, standard wire-mesh configurations used in conjunction with glass provide adequate attenuation. Shielding against magnetic fields at low frequencies is an extremely difficult problem. Combining a steel grid, having waveguide properties, with conductive glass provides some attenuation of magnetic fields at low frequencies.

GLOSSARY OF EMI TERMS*

Absorption. Transfer of electromagnetic wave energy to a substance being traversed by the wave.

Anechoic enclosure or chamber. A low-reflection enclosure (chamber) whose surfaces are lined with wave-absorbing material.

Attenuation. The ratio, expressed in decibels, of received-to-transmitted power; a measure of the amount of wave energy absorption.

*From *A Short Glossary of EMI Terms*. Huntingdon Valley, Pa.: Ace Engineering and Machine Company, Inc., November 1963.

Built-on-site shielding. Shielding fabricated in the field, as opposed to prefabricated modular components assembled in the field.

Cell-type enclosure. Prefabricated basic shielded enclosure of double-walled, copper-mesh construction.

Conducted interference. Interfering signals carried by a conductor.

Conductive gasket. Special highly resilient gasket used to reduce r-f leakage in shielding which has one or more access openings.

Contact finger. A conductive metal strip used as a compression-wiping door and access-port seal.

Cross talk. Interference on signal wiring caused by electromagnetic or electrostatic coupling to other signal leads in wiring harnesses or at interfaces.

Double-shield enclosure. A type of shielded enclosure in which the inner shield is connected to the outer shield at one point.

Electromagnetic environment. The r-f field or fields existing in an area or desired in an area to be shielded; an essential factor to be determined in designing enclosures.

Electrostatic induction. Capacitive induction of interfering signal over a dielectric gap separating circuits.

Enclosure, shielded. Essentially, a six-sided metallic barrier to radio-frequency interference.

Field strength meter. Measurement device used with antennas in making field strength measurements in RFI investigations.

Field strength. The intensity of RFI energy as determined by measurement or prediction.

Filter. Capacitive-inductive network or electronic or semiconductive elements used to reduce interference on conductors.

Frequency range. That portion of the frequency spectrum over which a particular shielded enclosure is to be effective.

Ground loop. A source of interference when a system is grounded improperly at several points.

Hash. A completely random interfering signal usually caused by arcing and occasionally by natural environmental disturbances.

Insertion loss. The ratio of received powers before and after the insertion of shielding between a source and a receiver of electromagnetic energy. More often applied to loss occurring in a filter passband.

Interference. A general term covering random noise signals emitted unintentionally from a transmitter, and radio-frequency signals accidentally picked up by a receiver tuned to a point on the radio spectrum. Specific designations for interference include the following:

TVI. Specialized designation for interfering signals showing up on television screens.

RFI. Any interfering signal capable of being detected on a receiver tuned to a radio frequency.

QRM. Obsolete term for any type of man-made interference.

PRF. Pulse recurring frequencies; type of interference found at radar installations.

Mutual. Transmitters in close proximity causing interference between themselves.

EMI. Electromagnetic interference; proposed designation for all kinds of interference occurring at any point in the electromagnetic spectrum.

Interference investigation. Usually a formal effort by trained engineers and technicians to determine the source of interference, measure its values, suggest remedial measures, and, if required, design appropriate equipment to carry out the recommendations.

ISM equipment. Federal Communications Commission designation for industrial, scientific, and medical equipment capable of causing interference.

Lighting, interference-free. Special lighting fixtures that do not in themselves constitute a source of high-level interference; normally, not completely interference-free.

Leakage. Failure of a shield to attenuate RFI due to imperfection in design, engineering, construction, or maintenance; most trouble is at access openings such as doors, power-line openings, utility penetrations, etc.

Microvolts/meter/mile. One method of stating field strength of radiated field. Radiation from industrial heating equipment, for example, must be suppressed so that radiated field strength does not exceed 10 microvolts per meter at a distance of 1 mile from the source.

Reflection. The return of electromagnetic waves from a surface which presents an impedance mismatch to the incident wave.

Shielding effectiveness. The reduction, expressed in db, of the intensity of an electromagnetic wave at a point in space after a metallic barrier (shield) is inserted between that point and the source.

Shot noise. Noise caused by thermal agitation in vacuum tubes and transistors.

Sky noise. Noise produced by radio energy from stars.

Sourcing. Generally, the redesign or modification of existing equipment to eliminate a source of RFI; when sourcing isn't feasible, engineers are forced to resort to suppression, filtering, or shielding.

Spectrum signature analysis. Evaluation of EMI from transmitting and receiving equipment in order to determine operational and environmental compatibility.

Suppression. Elimination of unwanted signals or interference by means of shielding, filtering, grounding, component relocation, or, sometimes, redesign.

Transients. Momentary surge on a signal or power line; may produce false signals or triggering impulses.

Transverse interference. Interference occurring across terminals or between signal leads.

Waveguide. A hollow, metallic duct used to confine and guide energy at microwave frequencies; at lower frequencies waveguides attenuate the energy.

12

DESIGN AND DEVELOPMENT OF MINIATURE ELECTRONICS SYSTEMS

To develop a packaging technique for miniaturized electronics using integrated and flat-pak circuit elements in cordwood packaging (Figure 12-1) methods, the designer should strive for reliability, weight reduction, volume reduction, and maintainability. The design should provide enough freedom to adapt to any application that may be specified, such as space vehicle, ground support equipment, vehicle installations such as truck or jeeps, and heavy industrial applications, with compatible costs regardless of the final application. A standard module assembly should be a design goal. This standard module should be capable of installation into any existing standard installation, such as a standard 19-inch EIA (Electronics Industries Association) rack and panel, ATR (air transport racking) cases (Figure 12-2), or commercial instrument cases.

The advantage of the compact system is the ease with which the system may be environmentally ruggedized; the smaller the unit, the easier it is to hermetically seal the package and to shield it against hazards. In addition, damage induced by vibration and shock is considerably reduced because the package mass and weight is decreased.

One of the fundamental problems in miniaturized electronics packaging is interconnection. The relationship of the interconnection volume to the total package volume tends to be very high. The following method outlines a practical and relatively efficient technique of miniaturized electronics design. The principal advantages of this technique are the simplicity in design and manufacturing and the ease of maintainability.

The interconnection of a miniaturized modular system presents a problem in

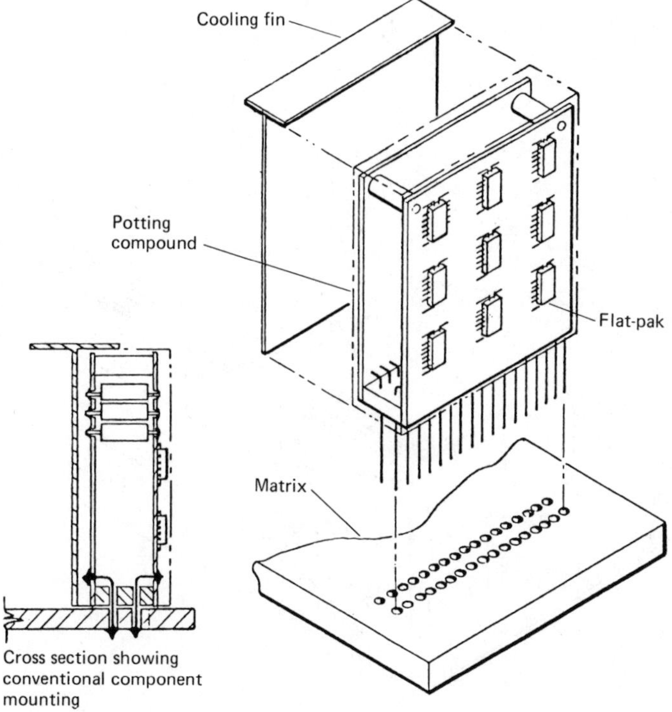

Figure 12-1. Functional module assembly.

the efficient use of available space. Unless the design is carefully planned, the volume occupied by the interconnections will approach that of the modules to be interconnected.

The basic requirements for the design of a miniaturized package using the modular concept are:

1. ease of maintenance and modulale replacement
2. reliability
3. efficient use of available space
4. efficient thermal controls
5. mechanical resistance (shock and vibration)
6. manufacturability

One technique that fulfills most of the basic requirements is the use of a mother board (interconnect matrix) on which the system of modules are mounted (Figure 12-3). In this technique, the interconnect matrix is usually the module

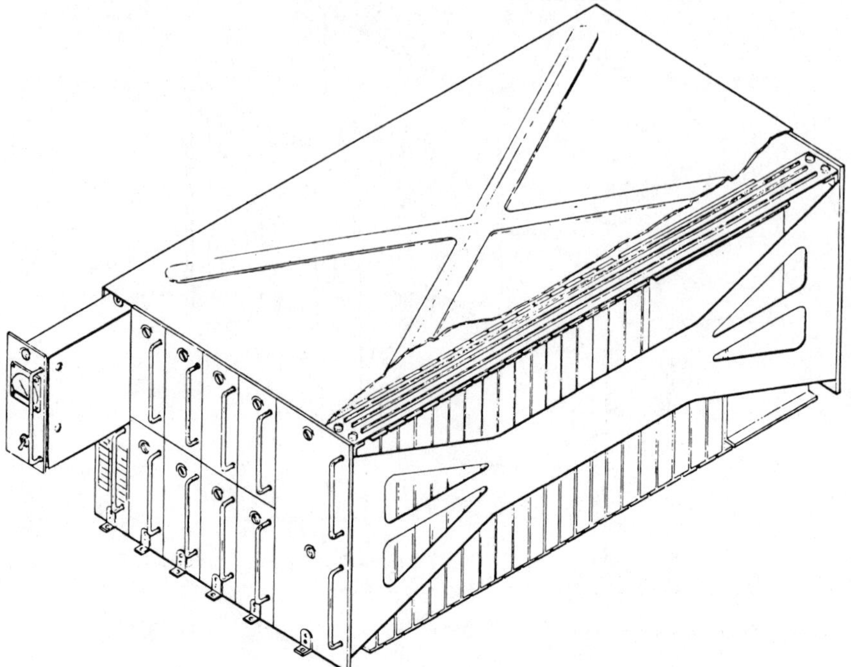

Figure 12-2. General assembly, ATR case.

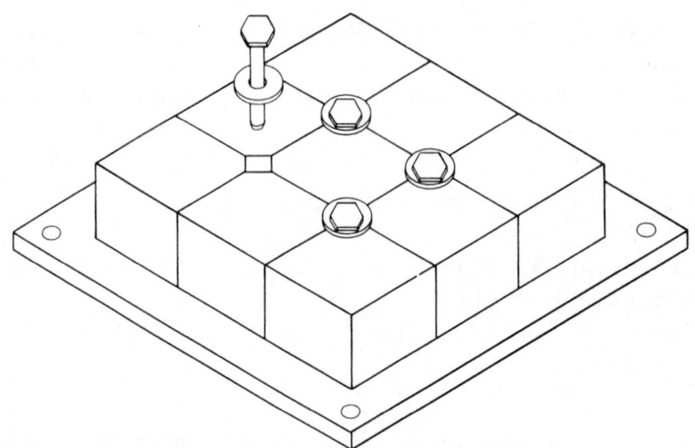

Figure 12-3. Corner clamped modules.

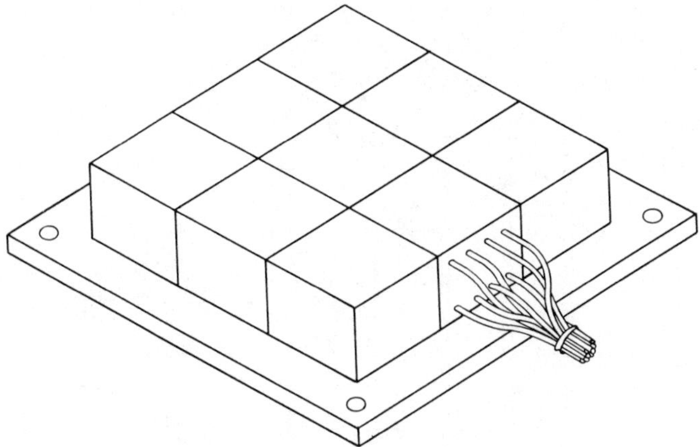

Figure 12-4. Assembly—motherboard mounting.

structural support (Figure 12-4). A very useful rule of thumb is that for any given circuit, the larger the module size, the less the interconnect problem.* However, this rule must also consider the complexity of the intraconnect within the module itself.† One criterion for developing the module size is the permissible throwaway cost. The engineer must consider this cost based on current component pricing since this will vary. The throwaway cost should be as large as possible to allow the maximum in design freedom and simplicity. At the start of the design phase, certain standards should be established. For example, to avoid a disorganized combination of odd-shaped modules, the specific size and shape of the modules must be determined. The modules should maintain a constant height, and should be allowed to vary only in length and width. The module spacing should be established by the use of a grid pattern. The acceptable grid patterns in use throughout the industry are 0.05, 0.10, or multiples thereof.

This method may employ the standard printed circuit matrix or, when the unit is to be used in severe environments such as space applications, the welded wire method can be used.

SYSTEM ANALYSIS

1. The system should be divided into major functions, and circuits should be modular; the power supply should be a separate functional module to facilitate manufacturing and maintenance. The modular concept is used because it offers

*_Interconnect_ refers to external connections between modules.
†_Intraconnect_ refers to internal connections within a module.

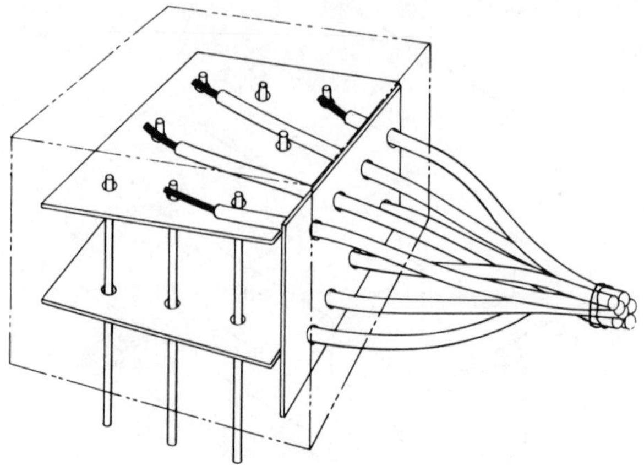

Figure 12-5. Interconnection module.

greater reliability, volume reduction, and weight reduction, higher heat dissipation potential, design flexibility, cost, and performance.

2. The modules should be as large as practical. Since large modules will contain a greater portion of the potential interconnections within themselves, this will simplify the interconnection problems.

3. The modules should be flat assemblies. Three-dimensional modules containing many expensive components are themselves expensive, and are not considered expendable. The modules must be repairable during and subsequent to manufacture. Also, three-dimensional modules tend to bury heat-producing

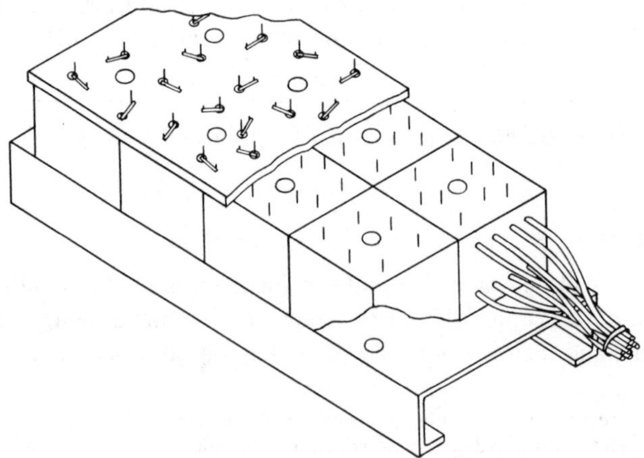

Figure 12-6. Assembly—cold plate mounting.

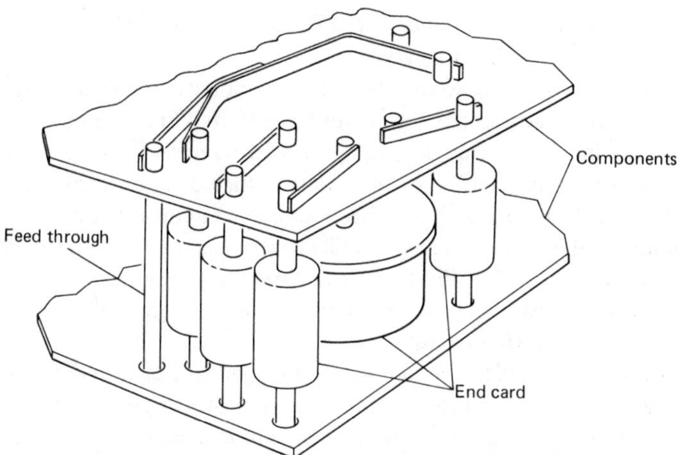

Figure 12-7. Cordwood module—unpotted.

elements and place some components into positions where their removal for repair is impractical.

4. Use of friction connectors should be kept to a minimum. Friction connectors are known to be less reliable than solid connections. This decision is a compromise since it tends to make the system less repairable (Figure 12-5).

5. Prewiring, such as printed circuits, flexible cables, and welded matrix techniques, reduces the possibility of manufacturing errors and improves manufacturing efficiency. A standard pattern will reduce the expense of generating artwork for each new assembly and facilitate handling by providing a single stock item.

6. Conductive cooling is the basic mode to be employed in removing excessive heat from the system, because the miniaturization may preclude the use of fans or blowers due to available space and power requirements (Figure 12-6).

The cordwood/flat-pak concept is a sound and reasonable approach to miniaturization of digital equipment. It offers a useful design tool for miniaturized

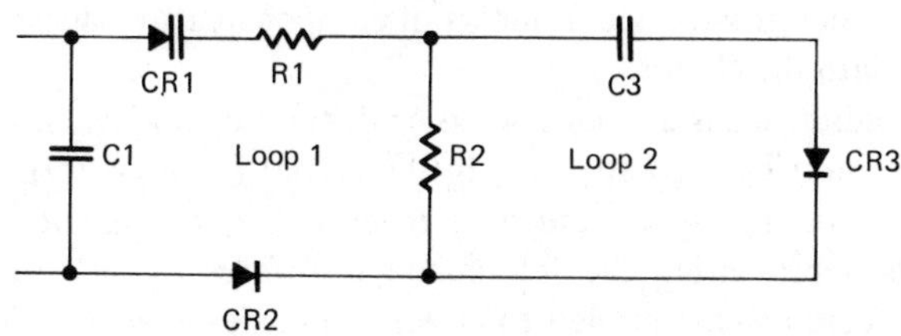

Figure 12-8. Dividing the circuit into loops.

electronic packaging and contains state-of-the-art components that are expected to be extremely valuable for many years.

"Cordwood" assemblies, as an approach to miniaturization and higher reliability, have become a standard throughout the United States armed services and the electronics industrv. This technique provides increased maintainability based on the "throwaway module" concept. Rather than component servicing, maintenance is carried to the next level where complete circuit functions are replaced with minimum downtime and semiskilled technicians, both of which greatly reduce maintenance costs. Production techniques have improved considerably over the past few years, and as reliability becomes more stringent, the use of cordwood techniques will increase.

Each cordwood module is a completely independent electrical function, and the module circuits are repeated segments of the overall circuit. In digital systems, they are flip-flops, gates, inverters, etc., and in analog circuits, they are any recognizable repeated portion of the overall circuit. Each module should be designed within certain throwaway cost limits, which are established by the current manufacturing costs. After the general appearance of the equipment is determined and the mounting arrangement and heat-conducting paths are planned, the modules are then designed to the available space envelope. Where possible, all components are chosen for minimum size and uniformity in body length (extra-long components may be laid crosswise to cut down volume, if necessary). The heat-generating components must be separated from heat-sensitive components and placed near one of the module surfaces to allow proper heat-sinking.

The relative component layouts may be done without consideration of the size of the components. Before the component layout is started, the number of "feed-throughs," (leads from one intraconnect surface to the other should be determined. The feed-throughs are shown in Figure 12-7 as distinct components. The number of feed-throughs is determined by dividing the circuit into "loops" (Figure 12-8). For example: Loop 1 is a five-element loop; loop 2 is a three-element loop. Both require a feed-through. (Note that R_2 is common to both loops and is counted as an element of both.) Additional feed-throughs may be required to bring the input and the output leads to the same side of the module. Loop circuits with even numbers of elements normally do not require feed-throughs. Loops with odd numbers of components are balanced by placing feed-throughs into the circuit.

Since the feed-throughs are considered as distinct components, their position in the module must be accounted for in the layout (Figure 12-9). An alternate method of determining feed-throughs is to draw the components between two phantom lines representing the interconnect surfaces. Connections are then made and the components juggled to determine the least required feed-throughs (Figure 12-10).

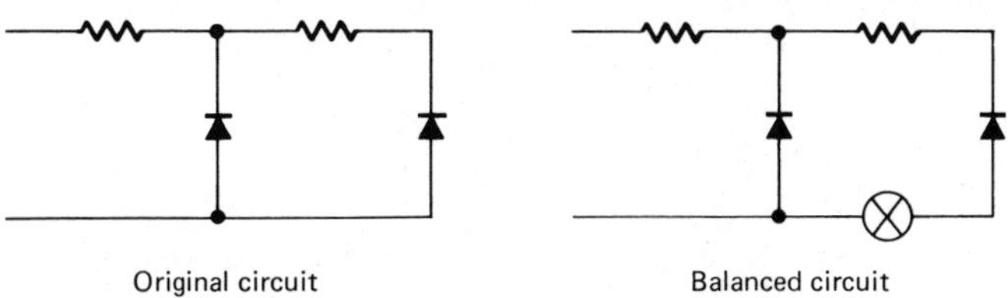

Figure 12-9. Balancing the loop.

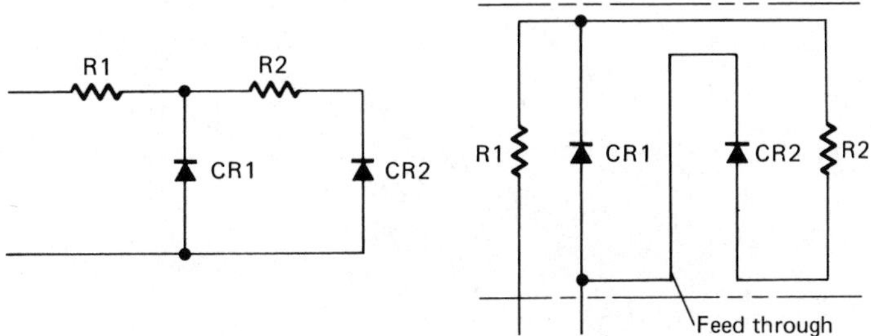

Figure 12-10. The feed-through as component.

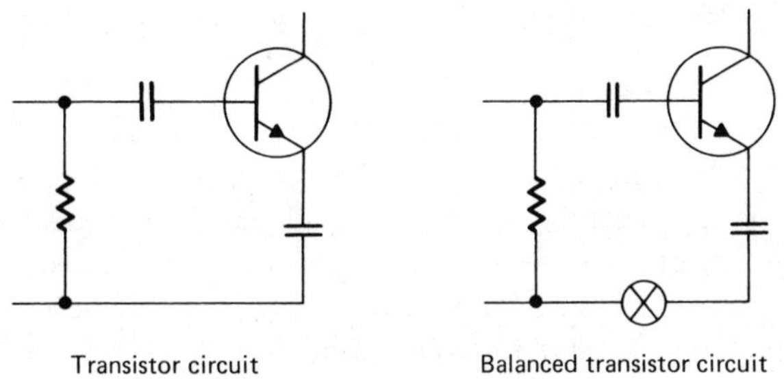

Figure 12-11. Transistor loop.

Components, such as transistors having three leads on the same end of the part, will have all three leads on the same side of the module and will require one feed-through for balancing the circuit (Figure 12-11). Accomplishment of the optimum relative layout is a trial-and-error process of shifting components until the best arrangement is found.

The final layout locating all the components must take into account component size and alignment of component leads, and feed-through wires with specific connectors or other interconnect devices to be used. The final layout or general assembly is approached by using a standard grid. The outline of the envelope is lightly sketched over a preselected grid, and following the original

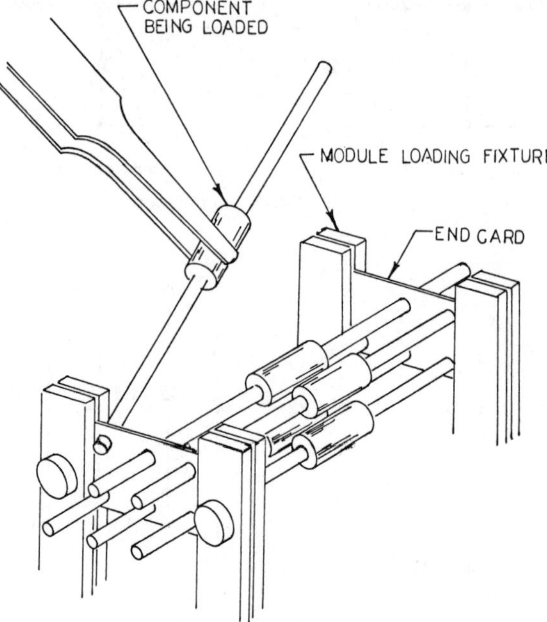

Figure 12-12. Cordwood module assembly fixture.

relative layout, the components are located to provide the most efficient assembly. The general assembly is drawn on a dimensionally stable material, such as Mylar, with tape or ink to facilitate clear photo reproductions.

In welded modules, the end cards (plastic films) are used to locate, hold components in place, and provide for connection information (Figure 12-12). They are photo-reduced from the final general assembly. Component lead holes are punched in the end cards and connections are made against the end cards to the component leads. After assembly, joining, and embedding, the end cards will remain as a part of the module, for soldered modules printed circuit boards are used as the end cards.

Holding jigs similar to those shown in Figure 12-12 are used to retain the end cards during assembly, and all components are weaved into place by following the information photo produced on the end card. After all circuit elements are properly positioned, the end cards are moved into the final assembly position and the interconnections are made.

The method of assembly must be determined based on the specific requirements of each piece of equipment. The simplest technique is to use printed circuit cards as the end cards. Printed circuits in ground-based and aircraft applications will in most cases be sufficient.

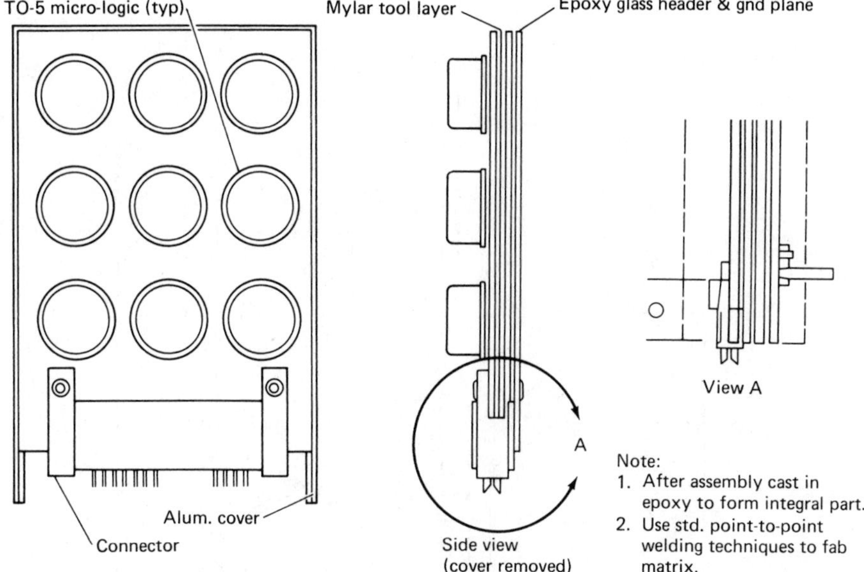

Figure 12-13. Micrologic plug-in functional block.

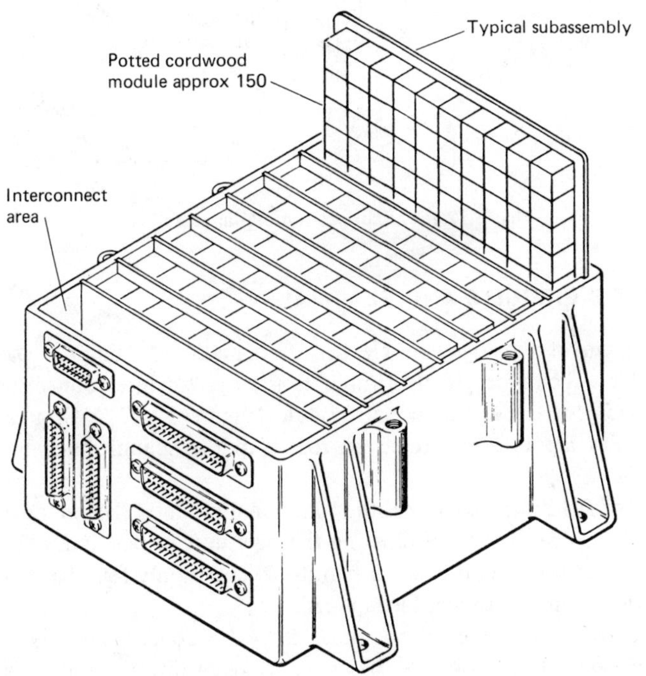

Figure 12-14. Ruggedized cordwood module assembly—(cover removed).

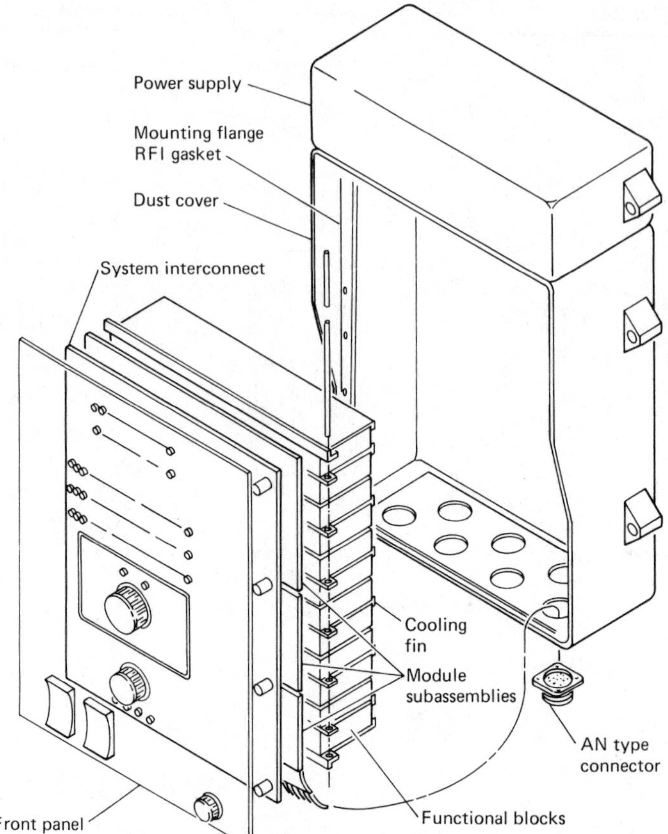

Figure 12-15. Fabricated modular assembly.

INTEGRATED CIRCUITS AND FLAT-PAKS

The breakdown of the logic and electrical schematics to minimize the number of repeatable subcircuits greatly simplifies the system interconnection design. The basic element is the functional module (Figure 12-13). Groups of functional modules are interconnected to make up the subunit assembly (Figure 12-14).

The level of field maintenance is the functional module. The modules may be replaced by unsoldering and resoldering the new unit in position. This may be accomplished with standard existing handtools commonly found at any installation where electronic equipment is repaired.

There are two basic types of components in this concept—the flat-pak solid-circuit component to perform the basic functions of flip-flop and gates, and the

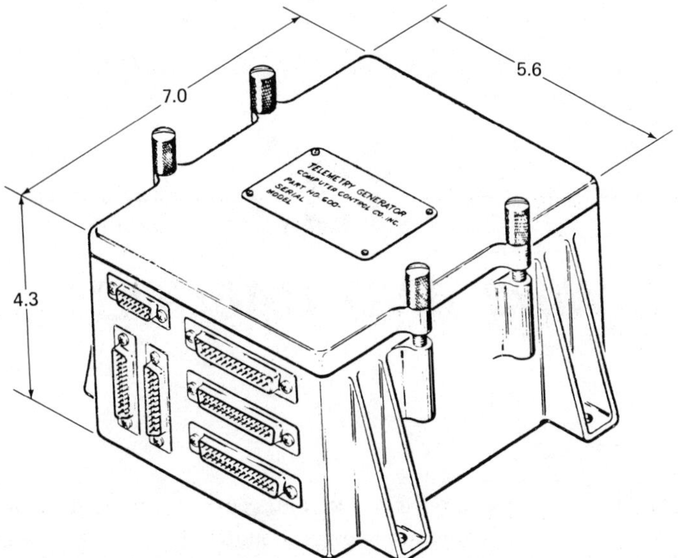

Figure 12-16. Ruggedized cordwood module assembly—(cover in place).

use of cordwood modules to facilitate circuit-design flexibility for peripheral circuits.

Heat transfer through an aluminum fin provides a metallic conductive path directly to the chassis structure. Encapsulation of the functional modules reduces the probability of hot spots and improves the heat transfer.

The planar construction of the substrates offers two relatively large sides for heat conduction. One-half of the substrate thickness is the maximum distance that heat must travel to a heat-conducting fin (Figure 12-15). Environmental protection for the system is provided by an aluminum cover (Figures 12-2 and 12-15 to 12-16). All of the exposed parts are hermetically sealed. RF control is accomplished by designing in a minimum of mechanical joints. The front panel, connector panel, and switch indicator assemblies are the only probable areas of concern. They are sealed by RF gasket materials (Figure 12-15). Shock and vibration damping is provided by the use of potting compounds (Figure 12-1).

13

WIRE AND CABLING

Conductors play a vital role in reliable electronics equipment. With the advances over the past few years in equipment design, conductor variations have become as complex as the electronics components they serve, and requirements have become correspondingly exacting. Conductors are selected for their current-carrying capacity, mechanical strength, and the properties of their insulation. However, unique situations may require the use of wire with special characteristics, and several types of special wires may be needed in a single chassis. The electrical problems may include current-carrying capacity, impedance, voltage, RFI, operator protection, and special placement of components. These problems are best approached by both the electronics engineer and the mechanical engineer working as a team.

Mechanical design problems include resistance to shock and vibration, placement of the wires, types of connections, servicing, marking, replacement of components and parts, environmental protection, separation, and insulation. The environmental conditions to be considered include temperature, humidity, abrasion, fungus, shock, and vibration.

TYPES AND USES

Electrical conductors are available as solid or stranded, bare or insulated, individual or cabled wire. Soft, annealed copper is most commonly used in making wire because of its high conductivity and ductility, resistance to corrosion and mechanical fatigue, and ease of soldering. Aluminum is sometimes used where weight is a primary concern, and various other types of materials are used for specific applications.

Stranded wire is preferable to solid wire due to its greater flexibility. It can be easily bent and formed into wire assemblies. Also, stranded wire is less apt to break when unsoldered and unwrapped during servicing. The most frequently used wire consists of seven strands twisted together.

The advantages of solid wire include rigidity and efficiency at higher frequencies. Solid wire may be used for jumpers up to 3 in. long, and for longer lengths where leads are securely mounted and not subject to vibration. When used, the bare wire can be insulated by using external sleeving. Untinned solid copper wire is the most efficient for high frequencies, since tinned or stranded wires exhibit greater losses. A major disadvantage of solid wire is its susceptibility to stress concentrations. A very slight nick in the conductor, which may occur as the insulation is stripped, will become a breaking point when the wire is subjected to flexing.

Multiconductor cables are selected according to the same factors governing the selection of individual wires, with special consideration given to the interwire insulation. Special wire and insulation are formed into cables used in specific application such as low-, medium-, or high-frequency applications. A variation of the multiconductor is coaxial cable, which is used where the distributed capacity must be constant over the entire length of the line. In the coaxial cable, one conductor follows a precise concentric path through another, and the space between is filled with an insulating material. It is important that the concentricity be maintained; if the space relationship is permitted to vary, circuit efficiency will be affected. When coaxial cable is bent, the minimum radius should be no less than 10 times the outside diameter of the cable; otherwise, cold flow can cause creeping of the inner cable at the bend.

The typical cable consists of a solid or stranded inner conductor, a dielectric other than air, outer conductor of braided shielding, and a protective, insulating material covering the braid.

SIZE

Wire size is most commonly designated by American Wire Gauge (AWG), by circular mils, or by the diameter of the wire in mils. The size to be selected depends on the current-carrying capacity (Table 13-1), permissible temperature rise, and the mechanical requirements, such as limited space and strength.

AWG

Wire sizes 22 through 24 are suitable for general chassis wiring. Filament wiring, particularly when heaters are wired in parallel, should use AWG 20 or larger, depending on the current requirements. Conductors intended to carry only audio-frequency or direct currents are chosen primarily on current and voltage requirements. In choosing conductors for radio frequency (RF), size must be correlated with the cable impedance.

Table 13-1. Wire Data Chart.

Size	Diameter	Area	Area	Weight	SOFT OR ANNEALED			HARD DRAWN		
					Tensile Strength Maximum	Breaking Strength Minimum	Maximum DC Resistance at 68°F.	Tensile Strength Minimum	Breaking Strength Maximum	Maximum DC Resistance at 68°F
Awg	Inch	Cir. Mils	Sq. In.	Lbs. per 1,000 Ft.	Lbs. per Sq. In.	Lbs.	Ohms per 1,000 Ft.	Lbs. per Sq. In.	Lbs.	Ohms per 1,000 Ft.
12	.0808	6,530	.00513	19.77	38,500	197.5	1.588	65,700	336.9	1.652
13	.0720	5,180	.00407	15.68	38,500	156.6	2.003	65,900	268.0	2.083
14	.0641	4,110	.00323	12.43	38,500	124.2	2.525	66,200	213.5	2.626
15	.0571	3,260	.00256	9.858	38,500	98.48	3.184	66,400	169.8	3.312
16	.0508	2,580	.00203	7.818	38,500	78.10	4.016	66,600	135.1	4.176
17	.0453	2,050	.00161	6.200	38,500	61.93	5.064	66,800	107.5	5.266
18	.0403	1,620	.00128	4.917	38,500	49.12	6.385	67,000	85.47	6.640
19	.0359	1,290	.00101	3,899	38,500	38.95	8.051	67,200	67.99	8.373
20	.0320	1,020	.000804	3.092	38,500	30.89	10.15	67,400	54.08	10.56
21	.0285	812.0	.000638	2.452	38,500	24.50	12.80	67,700	43.07	13.31
22	.0253	640.0	.000503	1.945	38,500	19.43	16.14	67,900	34.26	16.79
23	.0226	511.0	.000401	1.542	38,500	15.41	20.36	68,100	27.25	21.17
24	.0201	404.0	.000317	1.223	40,000	12.69	25.67	68,300	21.67	26.69
25	.0179	320.0	.000252	.9699	40,000	10.07	32.37	68,600	17.26	33.66
26	.0159	253.0	.000199	.7692	40,000	7.983	40.81	68,800	13.73	42.44
27	.0142	202.0	.000158	.6100	40,000	6.331	51.47	69,000	10.92	53.52
28	.0126	159.0	.000125	.4837	40,000	5.020	64.90	69,300	8.698	67.49
29	.0113	128.0	.000100	.3836	40,000	3.981	81.84	69,400	6.908	85.10
30	.0100	100.0	.0000785	.3042	40,000	3.157	103.2	69,700	5.502	107.3
31	.0089	79.2	.0000622	.2413	40,000	2.504	130.1	69,900	4.376	135.3
32	.0080	64.0	.0000503	.1913	40,000	1.986	164.1	70,200	3.485	170.6
33	.0071	50.4	.0000396	.1517	40,000	1.575	206.9	70,400	2.772	215.2
34	.0063	39.7	.0000312	.1203	40,000	1.249	260.9	70,600	2.204	271.3
35	.0056	31.4	.0000246	.09542	40,000	.9904	329.0	70,900	1.755	342.1
36	.0050	25.0	.0000196	.07567	40,000	.7854	414.8	71,100	1.396	431.4
37	.0045	20.2	.0000159	.06001	40,000	.6228	523.1	71,300	1.110	544.0
38	.0040	16.0	.0000126	.04759	40,000	.4939	659.6	71,500	.8829	686.0
39	.0035	12.2	.00000962	.03774	40,000	.3917	831.8	71,800	.7031	865.0
40	.0031	9.61	.00000755	.02993	40,000	.3106	1,049	72,000	.5592	1,091
41	.0028	7.84	.00000616	.02374	40,000	.2464	1,323	72,000	.4434	1,375
42	.0025	6.25	.00000491	.01882	40,000	.1954	1,668	72,000	.3517	1,734
43	.0022	4.84	.00000380	.01493	40,000	.1549	2,103	72,000	.2789	2,187
44	.0020	4.00	.00000314	.01184	40,000	.1229	2,652	72,000	.2212	2,758

Table 13-2. Properties of Insulation Material.

Insulation	Notes	Applications
Cellulose acetate	Moisture and abrasion resistant. Combustible, poor flexibility.	Solid insulation
Cotton	Low resistivity in humid environments.	Cable interwire filler
Cotton, impregnated	Relatively high dielectric strength	Braided insulation
Fluorocarbon (Teflon, Kel-F)	High melting point, no appreciable moisture absorption; excellent flexibility despite low temperatures.	Solid insulation for use up to 200° C (Teflon) and 135° C (Kel-F)
Glass fiber	Nonflammable; fungi- and heat-resistant. Tendency to fray and absorb moisture at ends; subject to abrasion.	Braided insulation
Nylon	Excellent abrasion, flame, solvent resistance. High surface resistivity.	Braided insulation
Polyethylene	Low loss at high frequencies, chemically stable, moisture-resistant, excellent flexibility. Subject to abrasion, softens at comparatively low temperatures.	Solid insulation
Rubber, high temperature	Good dielectric strength, moisture- and abrasion-resistant. Subject to aging; impairment upon contact with oils.	Cable jacket; primary solid insulation
Vinyl	Chemically stable, moisture- and abrasion-resistant, flameproof. Fair dielectric.	General-purpose solid insulation for use up to 100° C

INSULATION

Electrically, insulating materials are rated according to their dielectric strength, dielectric constant, resistance, and capacity-to-Q ratio. Physically, insulating materials are rated according to the permissible operating temperatures, mechanical strength, ease of stripping, effects of aging, and resistance to abrasion, vibration, moisture, flame, oils, alkalies, and fungi.

Insulation normally consists of a solid waterproof material, which may be covered with a braid. The basic solid insulation is usually made from rubber, vinyl, polyethylene, or fluorocarbon. The braid provides additional protection against abrasion and is used to carry the color code or identification. Polyethylene and similar materials are used extensively without braided coverings.

To establish the correct wire size when the current and voltage requirements

Preferred

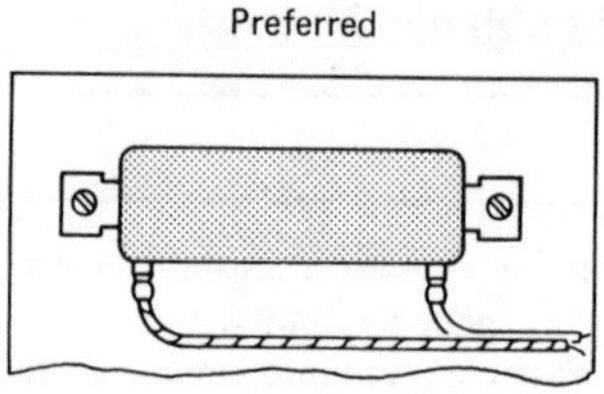

Figure 13-1. Lead dressed away from hot component, preferably $\frac{1}{2}$ inch or more.

for the equipment are known, the wire resistance is computed using Ohm's law:

$$R \text{ (ohms)} \frac{E \text{ (volts)}}{I \text{ (amperes)}}$$

E (volts) is the voltage requirement for the equipment at the using end of the cable or wire, I (amperes) is the current requirement for the equipment, and R (ohms) is the wire resistance to be selected from Table 13-1.

The outstanding properties of commonly used insulation materials are listed in Table 13-2. To a greater extent, the thermal properties of the material will determine the application of the insulation, since some materials deteriorate rapidly at high temperatures and others soften and lose their shape. At low temperatures, many insulation materials become brittle and are easily damaged by flexing. Temperature limitations must be carefully reviewed, especially in cases where wiring must be routed close to heat-generating components (Figures 13-1 and 13-2). A wide variety of insulating materials, each characterized by individual properties, may be used as insulation covers.

CODING OF CONDUCTORS

Color coding facilitates wiring, testing, and localizing faults, and is an advantage in field servicing. A single color code should be used continuously throughout a series of equipment models. Solid colors should be used wherever possible to facilitate lead tracing. When the number of circuits required exceeds ten, multicolor wires should be used.

Avoid

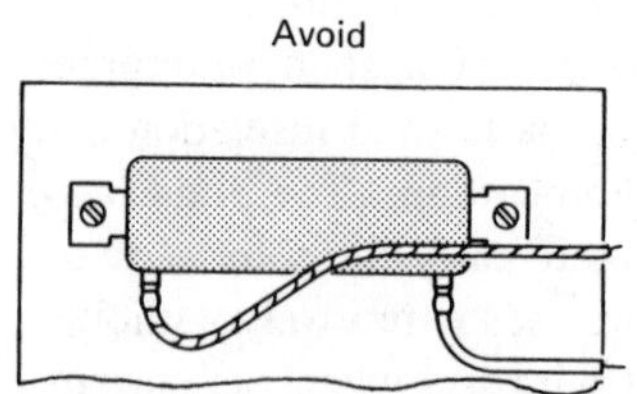

Figure 13-2. Lead near hot component.

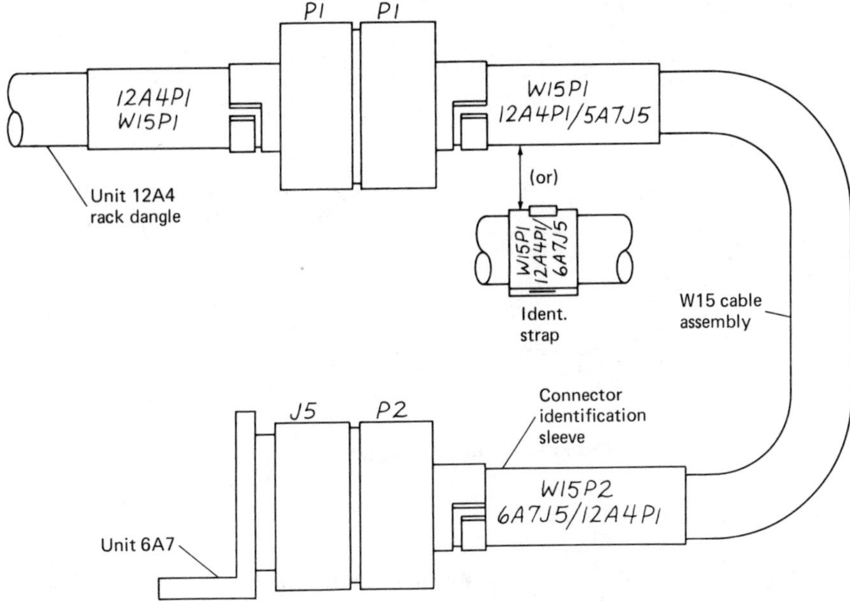

Figure 13-3. Typical cable assembly.

The colors selected should be easily distinguishable under the available lighting (normally incandescent), and should be resistant to fading, running, or discoloring due to exposure to heat. Violet is not recommended since it is difficult to distinguish in certain adverse lighting conditions.

Noninsulated leads may be color-coded by using lacquers and placing a color bead on the wire near the connecting terminals. On leads that are no more than 4 in. long or where the placement is obvious, no color coding is required.

Interconnecting cable leads are readily coded by using commercially available adhesive wire markers (Figures 13-3 through 13-6). Permanent thermosetting adhesive tapes are also available for marking heat-generating components.

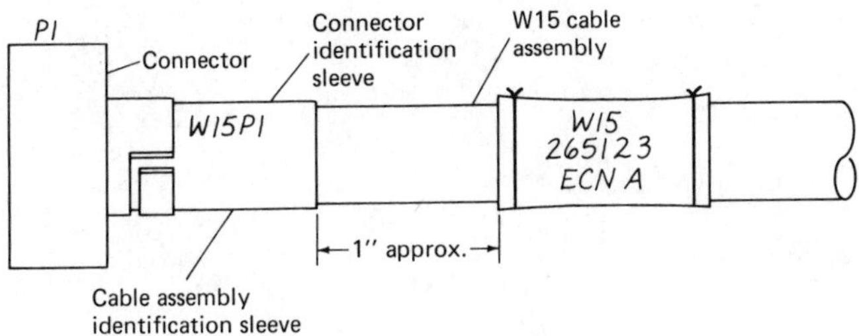

Figure 13-4. Typical connector identification.

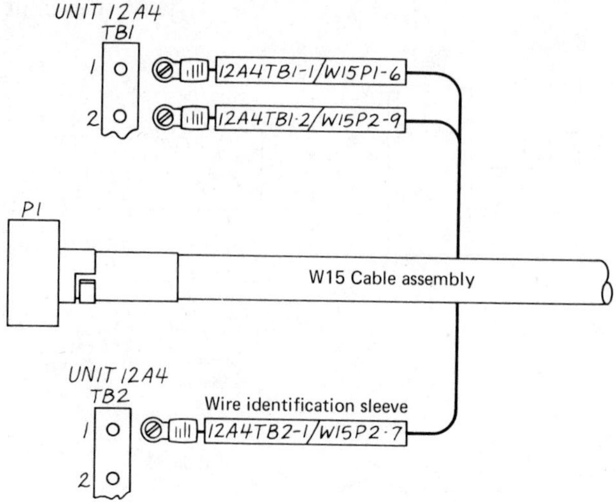

Figure 13-5. Identifying cable leads.

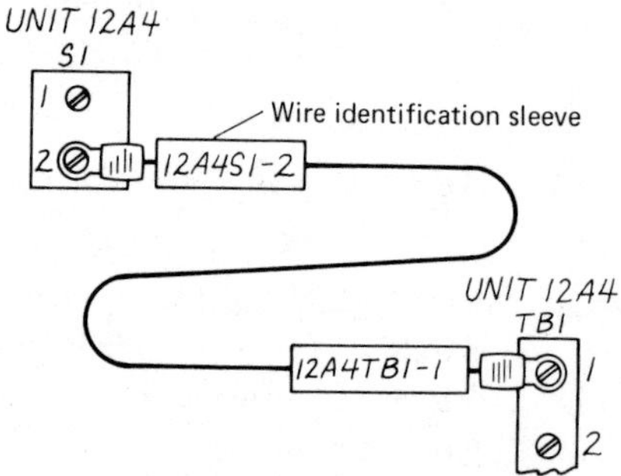

Figure 13-6. Identifying jumpers.

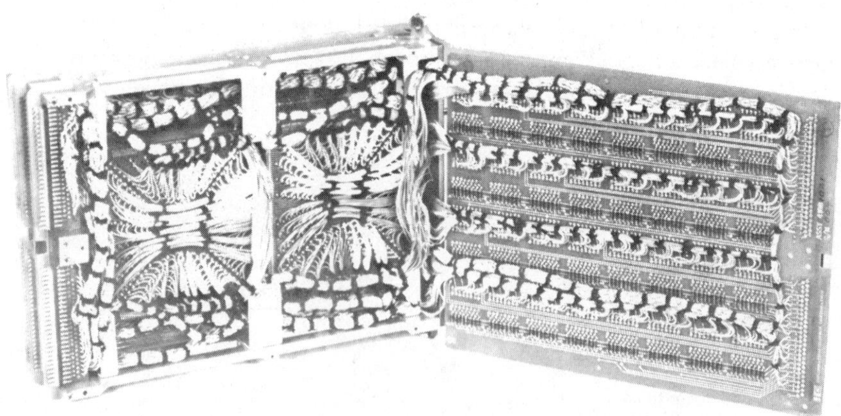

Figure 13-7. Routing.

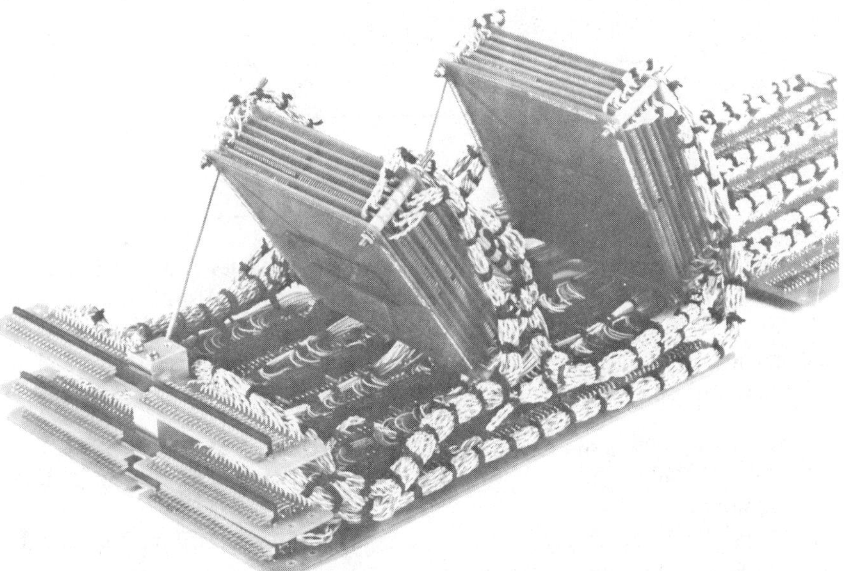

Figure 13-8. Dressing.

WIRING PROCEDURE

Whether an assembly is a single unit or a composite of many subassemblies, the basic steps of wiring are essentially the same. This includes routing, dressing, harnessing, cabling, and preparing the wires for connecting. Using the proper procedure insures that the wiring will withstand any severe environmental conditions and will facilitate maintenance.

Routing is the layout of the wiring to provide the most efficient and direct order throughout the assembly (Figure 13-7). *Dressing* is the arrangement of the wiring within any localized area to secure a neat and orderly appearance (Figures

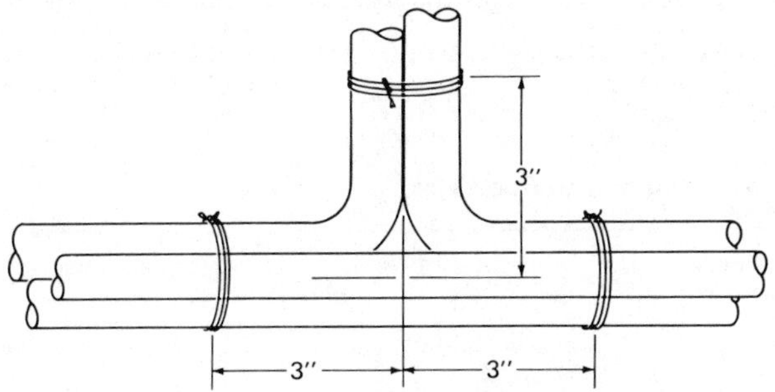

Figure 13-9. Spot ties at junction.

Table 13-3. Spacing of Spot Ties.

Cable Diameter	Spacing
Up to $\frac{1}{2}$ in.	3 in. + $\frac{1}{2}$ in.
Over $\frac{1}{2}$ in.	4 in. + $\frac{1}{2}$ in.

13-8 and 13-9, Tables 13-3 and 13-4). Wiring and cables should never cross sockets or openings provided for access to adiustments. Also, it should never interfere with the normal operation or maintenance, or the replacement of components and parts. The proper routing and dressing of wiring assemblies should always attempt to accomplish the following:

- Wiring and cabling should be neat, sturdy, and as short as practical.
- Wiring should be arranged to permit easy inspection and test at the final assembly level.
- Wiring should be arranged to prevent damage to assembled parts by the addition of cabling and wiring requirements.
- A good general practice in dressing wires is to locate insulated wires flat against the chassis or base to provide maximum compactness. Bare wire or bus wire should be placed in such a manner as to prevent shorts to the chassis.
- The proper location of the wire connections facilitates fabrication and replacement of components. Removable subassemblies, such as component boards, should always be interconnected with the main assembly after the subassembly components have been mounted.
- The interconnecting wire ends should be installed over the component leads. When the harness or cable assembly is an integral part of the subassembly, the interconnecting wire leads should be made below the component leads to permit component replacement and any required subsequent wiring. All wiring should be firmly supported to prevent any undue strain on the conductor or terminals and to eliminate possible changes in the equipment performance due to the shifting of conductors (Figure 13-8).

In making a layout of the placement of wires and cables, the following factors must be considered:

Table 13-4. Cord Size of Spot Ties.

Cable Diameter	Spot Ties Used	Lacing Cord
Up to $\frac{3}{4}$ in.	Single spot tie	No. 18 ($\frac{3}{32}$-in. width)
$\frac{3}{4}$ in. to $1\frac{1}{2}$ in.	Double spot tie	No. 18 ($\frac{3}{32}$-in. width)
$1\frac{1}{2}$ in. to 2 in.	Double spot tie	No. 26 ($\frac{1}{4}$-in. width)

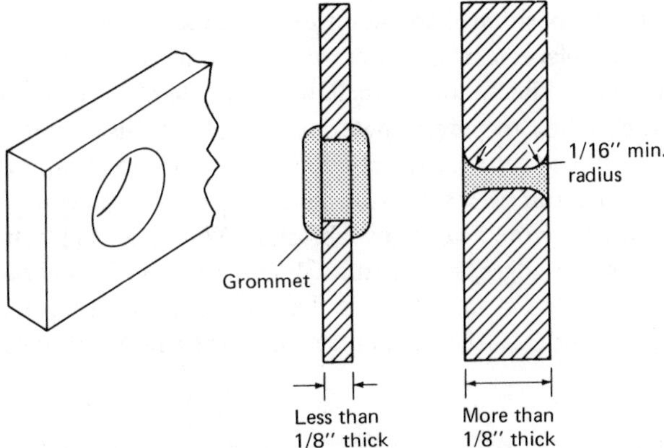

Figure 13-10. Use of grommets or rounded edges.

- The size, type, and number of wires routed between the various components.
- The location of the wire runs with respect to the framework for clamping.
- The isolation of critical wiring to reduce the intercircuit interference.
- The protection of personnel against radio frequency and high voltages.
- When conductor pairs are carrying alternating current, they should be twisted whenever possible to minimize undesired magnetic flux. Wires and

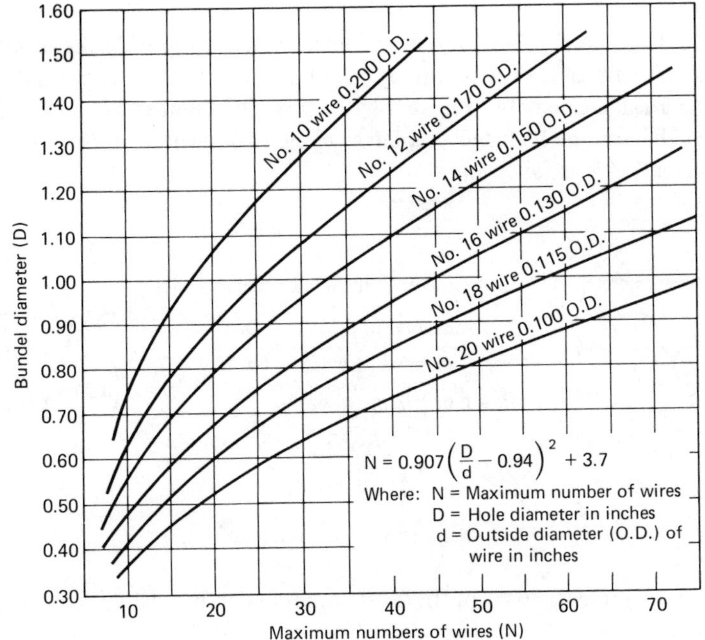

Figure 13-11. Wire bundle diameter.

cables should be positioned to minimize the inductive and capacitive effects. The improper placement of wires or cables may cause serious operating difficulty due to the spurious oscillations or other electrical interferences.

- Care must be taken to prevent deterioration of wire insulation by providing sufficient clearance between the conductors and heat-generating components such as resistors, transistors, transformers, and vacuum tubes.
- All wire must be protected against abrasion. Wires and cables should not be routed over sharp edges, screws, nuts, lugs, or terminals, and should not be routed or bent around sharp details that may cause abrasion under normal operating conditions. Also, clearance for moving mechanical parts must be provided.

When it is required to pass wiring through holes in thin metal, the conductors should be protected by using grommets (Figure 13-10). Ceramic, nylon, Teflon, or similar plastic grommets should always be used to protect wires carrying RF potentials except when coaxial cables are employed. When the metal is thicker than $\frac{1}{8}$ in. and the electrical requirements permit, the edges of the holes may be rounded to a radius of one-half the metal thickness.

The designer must always remember that harnesses and cables must always be located such that circuit tracing can be easily accomplished, and accessibility to parts and components that may require replacement is not limited. Individual wires of a harness should run parallel or at right angles to each other whenever possible.

Figures 13-11 and 13-12 have been included here to provide the engineer or designer with a means of determining the characteristics of a wire bundle when designing a cable assembly. Figure 13-12 shows the number of wires of a given diameter that can be included in a larger circle of known diameter. The attendant calculation is as follows:

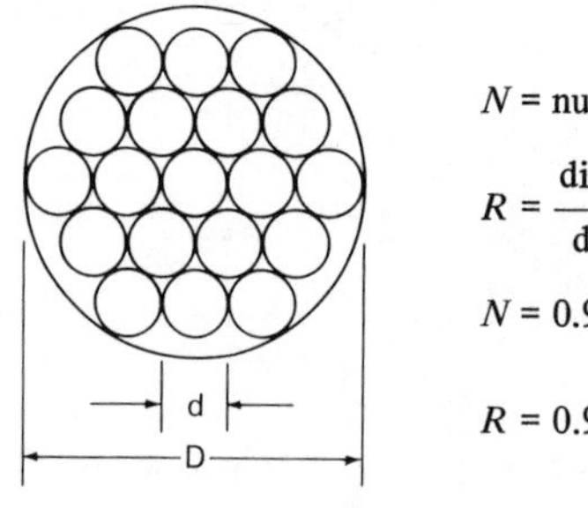

N = number of included wires

$$R = \frac{\text{diameter of bundle}}{\text{diameter of wires}} = \frac{D}{d}$$

$$N = 0.907\,(D/d - 0.94)^2 + 3.7$$

$$R = 0.94 + \sqrt{\frac{N - 3.7}{0.907}} = \frac{D}{d}$$

$$\frac{D}{d} = 0.94 + \sqrt{\frac{N - 3.7}{0.907}}$$

$$D = d\left(0.94 + \sqrt{\frac{N - 3.7}{0.907}}\right)$$

Figure 13-12. Wire bundles.

Figure 13-13 presents some of the most common commercially available devices for terminating wires.

TERMINALS

RING TONGUE

This is the basic terminal, and enjoys the widest over-all usage. Its main features are safety and reliability, since it cannot be disconnected from the component without completely removing the stud. None of the other four tongue types will lend themselves as easily to stacking (placing two or more terminals upon each other in a pile or stack over the same stud).

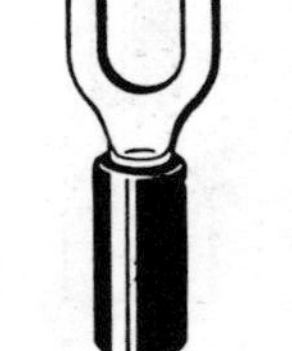

SPADE TONGUE

With its open end, the spade tongue offers an advantage in the way of reduced assembly time and effort. To make the connection, it is only necessary to loosen the stud a few turns until the tongue can be slipped under the head. However, this tongue type lacks the reliability and safety of the ring type.

FLANGED-SPADE TONGUE

This tongue has the basic features of the spade type, but the ends of the prongs, or tines, have been turned up, or flanged. It is safer and more reliable than the spade type, since the stud has to be loosened more to insert or remove the terminal. The flanges give a locking and locating action. This tongue type is most frequently used on small wires in conjunction with terminal blocks or barrier strips.

CONNECTORS

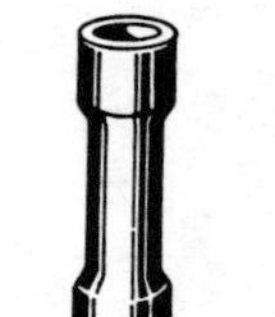

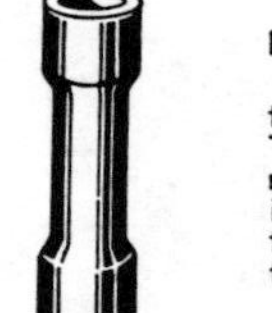

BUTT

This type is used principally to hook-up additional equipment and to repair a line or harness. The wires, inserted into both ends of the connector, meet in the middle of the barrel. A stop device is used in the center of the barrel to prevent over-travel of the wires. Each end of the connector is then crimped to give an in-line splice.

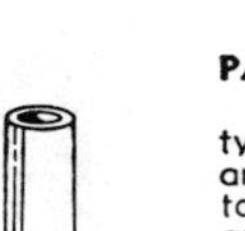

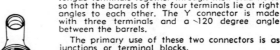

PARALLEL

This connector is shorter in length than the butt type. The wires enter the barrel from both ends and lie side by side. Only one crimp is required to secure both wires, but care must be taken to avoid pushing the wires off center during insertion. Also, a parallel connector will splice only wires whose combined total cross sectional area is within the wire range of the connector size. This type of connector is widely used where space is a factor.

X and Y

The X connector is made by riveting four ring tongue terminals together through their stud holes so that the barrels of the four terminals lie at right angles to each other. The Y connector is made with three terminals and a ~120 degree angle between the barrels.

The primary use of these two connectors is as junctions or terminal blocks.

Figure 13-13. Standard terminals and connectors.

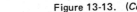

HOOK TONGUE

Resembling both the ring and the spade type, this tongue differs in that it is slotted to the side. The rearward angle of the slot gives it some of the advantages of both the ring and the spade types. It is difficult to pull off of the stud but can still be installed and removed by sliding it onto the stud.

A combined sideways and forward movement is required for installation and removal rather than an in-line motion. Hence, space limitations may prevent assembly or disassembly of this terminal without removing the screw.

RECTANGULAR TONGUE

This tongue is basically a ring type with parallel sides. It is most commonly used on terminal blocks or barrier strips where it is important that these terminals do not shift or move from side to side and come into short-circuit contact. The barrels are generally insulated also to prevent short circuits. This tongue helps to keep the general appearance of the whole assembly neat and uniform.

FLAG

The barrel of this terminal is at right angles to the tongue. It is used where space is limited, and is similar to a ring or rectangular tongue terminal in function.

CLOSED-END

This type has a fully insulated copper barrel, similar to a parallel connector. One end is closed off by the insulation, and the other end is open and flared. The closed-end connector will accommodate a wide range of wire size combinations. It makes a strong splice with only one crimp. Generally, this type is used to tie together (pig-tail) several wires.

SPECIAL DEVICES

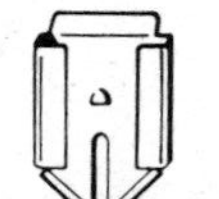

QUICK DISCONNECTS

This device is a two-piece assembly consisting of a blade and a receptacle. The advantage of this connector is its ease of connection and disconnection. It is used primarily for applications that require frequent assembly and disassembly.

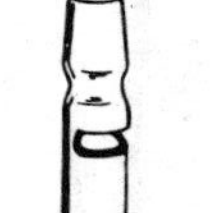

SNAP PLUGS

This connector is designed so that when the plug is pushed into the spring-tensioned receptacle, it is held firmly in place. It is used principally for easy connection and disconnection of one or more wires from a switch or other electrical component.

Figure 13-13. *(Continued)*

14

MATERIALS AND PROCESSES

The tables presented in this chapter are provided as an aid to the designer and engineer and as a ready reference on standard materials, applications, and applicable finishes. They present the common engineering materials and their recommended applications. Also, the applicable finishes are presented with their recommended use and their compatibility with various base materials.

The designer and the design engineer should be well versed in this chapter since the selection of the proper materials and finishes for any specific part or assembly will be a major factor in a product's reliability and survival in severe environments.

Table 14-1. Aluminum Alloys Guide.

Alcoa Alloy	General Characteristics	Typical Uses
1100	Excellent forming qualities, resistance to corrosion, weldability, electrical conductivity.	Chemical equipment, tank cars, heat exchangers, storage tanks, sheet-metal work, dials and nameplates, cooking utensils, decorative parts, giftware, and reflectors. 1100-0 wire stock is often used as general-purpose welding wire and where the assembly is Alumilite-treated.
2011	Good machinability, unexcelled for free-cutting qualities, good mechanical properties.	Screw-machine products, machine parts, atomizer and hose parts, pipe stems, cigarette holders, tube fittings.
2017	Relatively high strength, combined with fair workability and good machinability.	Screw-machine products, tube fittings, pulleys, gauges, coat hangers, crochet and knitting needles.
2024	A high-strength material of adequate workability. Has largely superseded 2017 for structural applications. Use of 2024-0 is not recommended unless subsequently heat-treated.	Aircraft parts, truck wheels, and caul plates, piano hinges, luggage, scientific instruments, ski poles, fastening devices, veterinary and orthopedic braces and equipment.
ALCLAD 2024	Combines high strength of 2024 with excellent resistance to corrosion in T3 and T4 temper. Its appearance is good.	Aircraft frames and skins, venetian blind slats, railroad-car roofs and sides, truck bodies, caul plates.
2017	Highest strength rivet alloy requiring no further heat treatment by user.	Rivets for general machine work, structural use, aircraft and truck bodies.
3003	Similar characteristics to 1100 but with slightly higher strength, good workability, weldability, resistance to corrosion. Low cost. 3003-H112 Plate: ASME Unfired Pressure Vessel Code Approved.	Ductwork, cooking utensils, ice-cube trays, garage doors, awning slats, trailer and truck panels, refrigerator panels, gas lines, gasoline tanks, heat exchangers, pressure vessels, storage tanks, chemical equipment, drawn and spun parts, general sheet-metal work.
ALCLAD 3003	Same as 3S, except has higher resistance to perforation under severe corrosive conditions.	Heat exchanger tubes, chemical equipment, swimming pools, tea kettles.
5005	Similar characteristics to 3003, but with finer grain structure. Good finishing characteristics.	Same as 3003. Useful where excessive finishing costs are encountered in the use of 3003 alloys due to surface roughness upon drawing.

5050	Intermediate strength, good finishing characteristics. 5050-0 Alcoa Utilitube can be used with compression or flare fittings, comparable to annealed copper.	Decorative refrigerator parts, cosmetic cases, general-purpose tubing (Alcoa® Utilitube), for instrumentation lines, fuel, lubricant and gas lines. 5050 Flat Sheet is a good general-purpose sheet-metal alloy where strengths greater than 3003 are required.
5052	Excellent resistance to corrosion, especially marine environment; good workability, higher strength than 1100 or 3003. Good finishing characteristics. 5052-H112 Plate: ASME Unified Pressure Vessel Code Approved.	Kitchen cabinets, small boats, home freezers, milk crates, bus and truck bodies, refrigerator trays, aircraft tube, fencing, fan blades, shoe eyelets. Sheet-metal parts and home appliances. 5052-H112 Plate is often used in tankage, high-temperature vessels.
5056	Highest strength aluminum alloy requiring no heat treatment. Excellent corrosion resistance.	Rivets.
5154	Excellent strength and ductility for welding alloys 3004, 5052 (and 5154 sheet and plate). Used in the consumable electrode welding process.	Often used where Alumilite finishing is required. For welds of maximum strength and ductility, particularly in thickness above $\frac{1}{4}$ in.
5356	A special welding rod for specific uses where strengths greater than 5154 are required.	Consult your Aluminum Manual.
5357	A grade of sheet especially suited for Alumilite finishing.	Ornamental trim, giftware.
6053	Medium-strength rivet alloy requiring no heat treatment by user. High corrosion resistance.	Rivets for general machine work, structural, ornamental architectural work, small boats, and truck bodies.
6061	Combines relatively high strength, good workability and high resistance to corrosion; widely available. 6061-T6 Plate: ASME Unfired Pressure Vessel Code Approved.	Sailboats, canoes, truck and bus bodies, scaffolding, transmission towers, mine skips, furniture, chemical equipment, awnings, marine equipment, fire ladders, moldings, pipe. Uses for 6061-T6 Plate include tankage, tank fittings, and flanges. Uses for 6061-T6 Tube and Pipe include general structural and high-pressure applications, paper and textile rolls.

Table 14-1. *(Continued)*

Alcoa Alley	General Characteristics	Typical Uses
6062	Equivalent to 6061 alloy in strength, better formability, lower cost.	Alternate for 6061 alloy, but with better finishing characteristics where Alumilite treatment is required.
6063	High resistance to corrosion. Pleasing natural finish greatly enhanced by the Alumilite process. Adequate strength, low cost.	Irrigation pipe, awning supports, windows, storefronts, architectural trim, storm sash, thresholds, stair rails, general utility pipe.
7075	Very high strength and hardness.	Aircraft, keys. Used where higher strengths than 2024 are required.
ALCLAD 7075	Very high strength, excellent resistance to corrosion in the T6 temper. Used for highly stressed structural parts. The 0 temper combines formability with high strength after heat-treating.	Aircraft. Used where higher strengths than Alclad 2024 are required.

Table 14-2. Properties of Aluminum Alloys.

Alcoa Alloy	Typical Mechanical Properties[1]											Joining[7]					Available Forms							
	Temper	Tensile Strength[8]	Yield Strength[8]	Elongation Sheet	Elongation Rod	Relative Resis. Outdoor Exp.[3]	Cold Forming[5]	Machinability	Brinell Hardness 500-kg Load, 10-mm Ball	Shearing Strength (psi)	Endurance Limit (psi)	Gas Welding	Inert Gas Arc Welding	Resistance Welding	Brazing	Soldering	Sheet	Plate	Rod-Bar	Tube-Pipe	Extruded Shapes	Structurals	Wire	Fasteners
1100	0	13	5	35	45	A	A+	D	23	9,000	5,000	A	A	B	A	A	×						×	
	H14	18	17	9	20	A	B	C	32	11,000	7,000	A	A	A	A	A	×						×	
	H18	24	22	5	15	A	B–	C	44	13,000	9,000	A	A	A	A	A							×	
	P[2]	–	–	–	–	A	[6]	D	–	–	–	A	A	B	A	A		×	×					×
2011	T3	55[9]	43[9]	–	15	D	C–	A	95	32,000	18,000	D	D	D	D	D			×					
	T8	59	45	–	12	C[4]	D	A	100	35,000	18,000	D	D	D	D	D			×					
2014	0	27	14	–	18	–	–	–	45	18,000	13,000	–	–	–	–	–								
	T4	62	42	–	20	C	C	B	105	38,000	20,000	D	B	B	D	–			×		×			
2017	T4	62	40	–	22	C[4]	C	B	105	38,000	18,000	D	B	B	D	D			×				×	
2024	0	27	11	20	22	–	B	C	47	18,000	13,000	D	C	C	D	D	×	×					×	
	T3	70	50	18	–	C[4]	C–	B	120	41,000	20,000	D	B	B	D	D	×			×				
	T4	68[10]	47[10]	20	19	C[4]	C	B	120	41,000	20,000	D	B	B	D	D		×	×				×	×
ALCLAD 2024	0	26	11	20	–	–	B	C	–	18,000	–	D	C	C	D	D	×	×						
	T3	65[10]	45[10]	18	–	A	C	B	–	40,000	–	D	B	A	D	D	×							
	T4	64	42	19	–	A	B–	B	–	40,000	–	D	B	A	D	D		×						
2117	T4	43	24	–	27	C	B–	B	70	28,000	14,000	–	–	–	–	–								×

Table 14-2. (*Continued*)

	Typical Mechanical Properties[1]											Joining[7]					Available Forms							
Alcoa Alloy	Temper	Tensile Strength[8]	Yield Strength[8]	Elongation Sheet	Elongation Rod	Relative Resis. Outdoor Exp.[3]	Cold Forming[5]	Machinability	Brinell Hardness 500-kg Load, 10-mm Ball	Shearing Strength (psi)	Endurance Limit (psi)	Gas Welding	Inert Gas Arc Welding	Resistance Welding	Brazing	Soldering	Sheet	Plate	Rod-Bar	Tube-Pipe	Extruded Shapes	Structurals	Wire	Fasteners
3003	0	16	6	30	40	A	A+	D	28	11,000	7,000	A	A	B	A	A	×							
	H14	22	21	8	16	A	B	C	40	14,000	9,000	A	A	A	A	A	×			×				
	H112[2]	17	12	20	30	A	6	D	–	–	–	A	A	B	A	A		×		×				
ALCLAD 3003	H14	22	21	8	16	A	B	C	–	14,000	–	A	A	A	A	A	×			×				
5005	H34	23	20	8	–	A	B	C	41	14,000	–	A	A	A	B	B	×							
5050	0	21	8	24	–	A	A+	D	36	15,000	12,000	A	A	B	B	B	×							
	H34	28	28	8	–	A	B	C	53	18,000	13,000	A	A	A	B	B				×				
5052	0	28	13	25	30	A	A+	D	47	18,000	16,000	A	A	B	C	C	×			×				
	H32	33	28	12	18	A	B+	C	60	20,000	17,000	A	A	A	C	C	×							
	H34	38	31	10	14	A	B	C	68	21,000	18,000	A	A	A	C	C	×							
	H112[2]	30	20	18	24	A	6	D	–	–	–	A	A	B	C	C		×						
5083	0	42	21	22	–	–	–	–	–	25,000	–	–	–	–	–	–		×						
	H113	46	33	16	–	–	–	–	–	–	23,000													
5086	0	38	17	22	–	A	A	D	–	23,000	–	C	A	B	D	–	×	×						
5154	0	35	17	27	–	A	A	D	58	22,000	17,000	C	A	B	D	D			×					

6053	T-61	Shear Strength 23[10]				A	C+	C	–	–	–	–	–	–	–	–							X	X
6061	0	18	8	25	30	–	A	D	30	12,000	9,000	A	A	C	A	B	X	X						
	T4	35	21	22	25	A	B	C	65	24,000	14,000	A	A	A	A	B	X			X				
	T6	45	40	12	17	A	C	C	95	30,000	14,000	A	A	A	A	B	X	X	X	X		X	X	
6062	T6	45	40	–	17	A	C	C	95	30,000	14,000	A	A	A	A	B					X			
6063	T42	22	13	20	–	A	A	C	48	14,000	9,000	A	A	A	A	B				X	X			
	T5	27	21	12	–	A	B	C	60	17,000	10,000	A	A	A	A	B				X				
	T832	42	39	12	–	A	C	C	90	27,000	–	A	A	A	A	B				X				
7075	T6	83	73	11	11	C	D	B	150	48,000	22,000	D	D	B	D	D		X	X					
ALCLAD 7075	0	32	14	17	–	A	B	C	–	22,000	–	D	D	C	D	D	X							
	T6	76	67	11	–	A	D+	B	–	46,000	–	D	D	B	D	D	X							

[1] Alcoa ratings for wrought aluminum alloys, "A," "B," "C," and "D" are relative ratings in decreasing order of merit. An "A" rating is highest.
[2] This is not a controlled temper and the mechanical properties will vary with thickness of the plate.
[3] "A" rating indicates paint protection not generally required. Paint protection advisable or necessary, in corrosive atmospheres, for "C" or lower ratings.
[4] Rating changes from "C" to "D" for sections about 0.125 in. or more in thickness.
[5] Hot forming permissible for some materials under carefully controlled conditions.
[6] No ratings assigned, as properties are not controlled in this temper. See note 2.
[7] Weldability ratings "A," "B," "C," and "D" are relative ratings as follows: (a) Generally weldable by all commercial procedures and methods. (b) Weldable with special technique or on specific applications which justify preliminary trials or testing to develop welding procedure and performance. (c) Limited weldability because of crack sensitivity or loss in resistance to corrosion and mechanical properties. (d) No commonly used welding methods have so far been developed.
[8] In thousands of pounds per square inch.
[9] Sizes greater than $1\frac{1}{2}$ in. will have strengths slightly lower than these values.
[10] Sheet less than 0.040 in. thick will have strengths lower than these values.

Table 14-3. Aluminum Casting Alloys.

Commercial Designation	Military Designation	Availability		
		Size (in.)	Form	
Aluminum 356-T6	QQ-A-601 Alloy 356 Temper T6	Not applicable	Sand casting	High-strength, dimensionally stable casting alloy. Add note on drawing. "Stress relieve before final machining for 1 hour at 440 ± 10° F, and air cool."
Aluminum 356-T6	QQ-A-596 Comp 356 Temper T6	Not applicable	Permanent mold castings	Heat-treated. Welding is permissible, but results in lower properties in the weld and heat affected zone. High strength properties, corrosion resistance, and good machinability.
Aluminum 356-T51	QQ-A-601 Alloy 356 Temper T51	Not applicable	Sand casting	–
Aluminum 356-T7	QQ-A-601 Alloy 356 Temper T7	Not applicable	Sand casting	High strength, dimensionally stable casting alloy. Add note on drawing. "Stress relieve before final machining for 1 hour at 440 ± 10° F, and air cool."
Aluminum 356-T51	QQ-A-596 Alloy 356 Temper T51	Not applicable	Permanent mold castings	Heat-treated. Welding is permissible but results in lower properties in weld and heat-affected zone. High strength properties, corrosion resistance, and good machinability.
Aluminum 356-T7	QQ-A-596 Alloy 356 Temper T7			

Aluminum A356-T4	QQ-A-596 Alloy A356 Temper T4	Not applicable	Investment and permanent mold castings	Heat-treated. Welding is permissible but results in lower properties in weld and heat-affected zone. High strength properties, corrosion resistance, and good machinability.
Aluminum A356-T51	QQ-A-596 Alloy A356 Temper T51			
Aluminum A356-T6	QQ-A-596 Alloy A356 Temper T6	Not applicable	Investment and permanent mold castings	High strength. Good machinability and corrosion resistance.
Aluminum A356-T61	QQ-A-596 Alloy A356 Temper T61			
Aluminum A356-T7	QQ-A-596 Alloy A356 Temper T7	Not applicable	Investment and permanent mold castings	High strength, dimensionally stable. Good corrosion resistance and machinability.
Aluminum A356-T61	MIL-A-21180 Alloy A356 Temper T61	Not applicable	All castings	Use where high strength, elongation, ductility, and sound castings are required.

Table 14-4. Typical Physical Properties of Wrought Aluminum Alloys.[a]

Alloy and Temper	Specific Gravity	Weight (lb/in.^{3})	Melting Range Approx. (°F)	Electrical Conductivity at 20°C (68°F) % of International Annealed Copper Standard	Thermal Conductivity at 24°C (77°F) CGS Units[b]
2EC-T61	2.70	0.098	–	59	0.53
1100-O	2.71	0.098	1190–1215	59	0.53
1100-H18	2.71	0.098	1190–1215	59	0.53
2011-T3	2.82	0.102	995–1190	39	0.36
2011-T8	2.82	0.102	995–1190	45	0.41
2014-O	2.80	0.101	950–1180	50	0.46
2014-T6	2.80	0.101	950–1180	40	0.37
2017-O	2.79	0.101	955–1185	50	0.46
2017-T4	2.79	0.101	955–1185	34	0.32
2024-O	2.77	0.100	935–1180	50	0.46
2024-T3, T36, T4	2.77	0.100	935–1180	30	0.29
2024-T6, T81, T86	2.77	0.100	935–1180	38	0.35
2117-T4	2.74	0.099	950–1200	40	0.37
3003-O	2.73	0.099	1190–1210	46	0.42
3003-H18	2.73	0.099	1190–1210	46	0.42
3004-O	2.72	0.098	1165–1205	42	0.39
3004-H38	2.72	0.098	1165–1205	42	0.39
5005-O	2.70	0.098	1170–1205	52	0.48
5005-H38	2.70	0.098	1170–1205	52	0.48
5050-O	2.69	0.097	1160–1205	50	0.46
5050-H38	2.69	0.097	1160–1205	50	0.46
5052-O	2.68	0.097	1125–1200	35	0.33
5052-H38	2.68	0.097	1125–1200	35	0.33
5056-O	2.64	0.095	1055–1180	29	0.28
5056-H38	2.64	0.095	1055–1180	29	0.28
5083-O	2.66	0.096	1075–1185	29	0.28
5083-H34	2.66	0.096	1075–1185	29	0.28
5083-H113	2.66	0.096	1075–1185	29	0.28
5086-O	2.66	0.096	1085–1185	32	0.30
5086-H34	2.66	0.096	1085–1185	32	0.30
5154-O	2.66	0.096	1100–1190	32	0.30
5154-H38	2.66	0.096	1100–1190	32	0.30
5254-O	2.66	0.096	1100–1190	32	0.30
5254-H38	2.66	0.096	1100–1190	32	0.30
5356-O	2.64	0.095	1055–1180	29	0.28
5356-H38	2.64	0.095	1055–1180	29	0.28
5357-O	2.70	0.098	1165–1210	43	0.40
5357-H38	2.70	0.098	1165–1210	43	0.40
5454-O	2.68	0.097	1115–1195	34	0.32

Table 14-4. (*Continued*)

Alloy and Temper	Specific Gravity	Weight (lb/in.^{3})	Melting Range Approx. (°F)	Electrical Conductivity at 20°C (68°F) % of International Annealed Copper Standard	Thermal Conductivity at 24°C (77°F) CGS Units[b]
5454-H38	2.68	0.097	1115–1195	34	0.32
5456-O	2.66	0.096	1055–1180	29	0.28
5458-H38	2.66	0.096	1055–1180	29	0.28
6061-O	2.70	0.098	1100–1205	45	0.41
6061-T4	2.70	0.098	1100–1205	40	0.37
6061-T6	2.70	0.098	1100–1205	43	0.40
6062-O	2.70	0.098	1100–1205	45	0.41
6062-T4, T6	2.70	0.098	1100–1205	40	0.37
6063-O	2.70	0.098	1140–1205	58	0.52
6063-T42	2.70	0.098	1140–1205	50	0.46
6063-T5	2.70	0.098	1140–1205	55	0.50
6063-T6	2.70	0.098	1140–1205	53	0.48
7075-T6	2.80	0.101	890–1180	33	0.31
7079-T6	2.74	0.099	900–1180	32	0.30
7178-T6	2.81	0.102	890–1165	31	0.29

[a]Wrought alloys of aluminum fall into two classes:

1. Alloys whose harder tempers are developed by strain hardening (cold work): Alloys 1100, 3003, 5052.
2. Alloys whose harder tempers are produced by heat treatment: Alloys 2011, 2017, 2024, 6061, 6053, 7075.

A wide range of tensile properties is found in both classes. However, the highest combination of strength and ductility appears in the heat-treated group.

Various tempers are indicated by letter following alloy designation: H18, full hard; H14, half hard; O, soft annealed; T, heat-treated; T4, quenched or (normally) aged.

[b]CGS units = cal/cm/cm/cm^2/°C/sec.

Table 14-5. Aluminum Sheets, Approximate Weights.

B.&S. Gauge No.	Dec. Inch	Wt. Lb. ft^2	B.&S. Gauge No.	Dec. Inch	Wt. Lb. ft^2	B.&S. Gauge No.	Dec. Inch	Wt. Lb. ft^2
4–0	0.4600	6.394	12	0.08081	1.123	27	0.01420	0.197
3–0	0.4096	5.694	13	0.07196	1.000	28	0.01264	0.176
2–0	0.3648	5.070	14	0.06408	0.891	29	0.01126	0.156
0	0.3249	4.516	15	0.05707	0.793	30	0.01003	0.139
1	0.2893	4.021	16	0.05082	0.706	31	0.008928	0.124
2	0.2576	3.581	17	0.04526	0.629	32	0.007950	0.111
3	0.2294	3.189	18	0.04030	0.560	33	0.007080	0.098
4	0.2043	2.840	19	0.03589	0.499	34	0.006304	0.088
5	0.1819	2.529	20	0.03196	0.444	35	0.005614	0.078
6	0.1620	2.252	21	0.02846	0.396	36	0.005000	0.069
7	0.1443	2.006	22	0.02535	0.352	37	0.004453	0.062
8	0.1285	1.786	23	0.02257	0.314	38	0.003965	0.055
9	0.1144	1.591	24	0.02010	0.279	39	0.003531	0.049
10	0.1019	1.416	25	0.01790	0.249	40	0.003144	0.044
11	0.09074	1.261	26	0.01594	0.222			

1 in. thick, 13.9. Weight/in.^{3} 0.0965.
Specific gravity 2.67 = 0.0936 lb/in.^{3}

Table 14-6. Aluminum Specifications.

Alloy	Commodity	Federal	A.M.S.
1100	Sheet and plate	QQ-A-561c	4001B 4003B
1100	Rod and wire	QQ-A-411e	4102
2017	Rod and wire	QQ-A-351d	4118C
2024	Sheet and plate bare	QQ-A-355e	4035D 4037D
2024	Sheet and plate alclad	QQ-A-362b	4040E 4041F 4042E
2024	Bar, rod and wire drawn	QQ-A-268-1	4120E
2024	Bar, rod and shapes extruded	QQ-A-267-2	4152F
2024	Tube, drawn	WW-T-785a-1	4087B 4088E
3003	Sheet and plate	QQ-A-359d	4006B 4008B
3003	Tubing, drawn	WW-T-788b-1	4065B 4067B
5052	Sheet and plate	QQ-A-318e	4015D 4016D 4017D
5052	Tubing, drawn	WW-T-787a	4069 4070E 4071E
6061	Sheet and plate	QQ-A-327b	4025C 4026C 4027C
6061	Bar, rod and wire drawn	QQ-A-325a-2	–
6061	Tubing, drawn	WW-T-789a-2	4079 4080E 4082E
6061	Tubing, extruded	–	4150C
7075	Sheet and plate bare	QQ-A-283a	4044B 4045B
7075	Sheet and plate alclad	QQ-A-287a-1	4048C 4049C
7075	Bar, rod and wire drawn	QQ-A-282-1	4112C
7075	Bar, rod and shapes extruded	QQ-A-277-1	4154D

Table 14-7. Copper and Copper Alloys.

Commercial Designation	Military Designation	Form Available	Remarks
Electrolytic tough pitch copper ASTM B152 Type ETP Temper: cold rolled, annealed	Copper QQ-C-576 Temper: cold rolled, soft annealed (ASTM B152, Type ETP)	Sheet or strip	General-purpose copper sheet for parts requiring high electrical conductivity.
Electrolytic tough pitch copper ASTM B133, Type A Temper: Soft	Copper QQ-C-502 Temper: soft (ASTM B133, Type A)	Rod or bar	For parts requiring high electrical conductivity.
Phosphorized copper ASTM B88, Type N Temper: annealed	Copper tubing WW-T-775 Temper: annealed (ASTM B88, Type N)	Tubing	Use for fuel, lubricating evacuating tubes, etc. AB No. 103
Oxygen-free copper ASTM B152	Copper QQ-C-576	Sheet, pipe, rod, tube, wire, shapes	Used for brazing or high-temperature applications where embrittlement or oxidation must be avoided.
Beryllium copper ASTM B194 Condition $\frac{1}{2}$ H	Beryllium copper QQ-C-533 Alloy 172 Condition $\frac{1}{2}$ H (ASTM B194)	Sheet or strip	Heat-treatable alloy for springs and parts requiring high strength, hardness and wear resistance.
Beryllium copper ASTM B196 Condition $\frac{1}{2}$ H	Beryllium copper QQ-C-530 Condition $\frac{1}{2}$ H (ASTM B196)	Rod or bar	Specify on drawing, "Heat-treat after forming to condition $\frac{1}{2}$ HT (Rockwell C39 min: tensile strength, 185,000 psi min)."
Beryllium copper ASTM B197 Condition $\frac{1}{2}$ H	Beryllium copper QQ-C-530 Condition $\frac{1}{2}$ H (ASTM B197)	Wire	
Cartridge Brass ASTM B36, Alloy 6 Temper: annealed	Brass QQ-B-613 Composition 2 Temper: annealed (ASTM B36 Aly 6)	Sheet or strip	General-purpose brass sheet for deep drawing, spinnings. AB Number 42
Cartridge brass ASTM B36, Alloy 6 Temper: $\frac{1}{2}$ hard	Brass QQ-B-613 Composition 2 Temper: $\frac{1}{2}$ H (ASTM B36 Aly 6)	Sheet or strip	General-purpose brass sheet for stamping, forming, etc. AB Number 42

Free Cutting Brass ASTM B16 Temper: $\frac{1}{2}$ hard	Brass QQ-B-626 Composition 22 Temper: $\frac{1}{2}$ H (ASTM B16)	Rod or bar	General-purpose brass rod for screw machine parts. AB Number 271
Cartridge Brass ASTM B134, Alloy 6 Temper: $\frac{1}{8}$ hard	Brass QQ-W-321 Composition 6 Temper: $\frac{1}{8}$ H (ASTM B134 Aly 6)	Wire	Use for forming terminals, etc. AB Number 42
Leaded tube brass ASTM B135, Alloy 3 Temper: annealed	Tubing brass WW-T-791 Grade 2, Type A (ASTM B135 Aly 3)	Tubing	General-purpose brass tubing. AB Number 218
Brass casting ASTM B146, Alloy 6A	Brass casting QQ-B-621 Class C	Casting	General-purpose brass casting alloy with good machining properties.
Leaded Brass ASTM B121, Alloy 5 Temper: $\frac{1}{2}$ hard	Clock brass QQ-B-613 Composition 24 Temper: $\frac{1}{2}$ H (ASTM B121 Aly 5)	Sheet	Mechanism plates. Brass sheet gears. AB Number 235
Phosphor-Bronze ASTM B103, Alloy A Temper: spring	Phosphor-Bronze QQ-B-750 Composition A Temper: spring (ASTM B103 Aly A)	Sheet or strip	Flat springs, shims, cams, etc. AB Number 351
Phosphor-Bronze ASTM B139, Alloy A Temper: hard	Phosphor-Bronze QQ-B-750 Composition A Temper: H (ASTM B139 Aly A)	Rod or bar	High-strength screw machine parts and bearing bushings. AB Number 351
Phosphor-Bronze ASTM B159, Alloy A Temper: spring	Phosphor-Bronze QQ-W-401 Spring wire (ASTM B159 Aly A)	Wire	General-purpose springs. AB Number 351

Table 14-8. Steel Materials.

Commercial Designation	Military Designation	Form Available	Remarks
Steel, low carbon, $\frac{1}{4}$-hard ASTM 109T	Steel QQ-S-698 Condition C, $\frac{1}{4}$ hard	Sheet or strip	General-purpose stampings, formed parts having bends with and across grain on same part.
Steel, low carbon, $\frac{1}{2}$-hard ASTM 109T	Steel QQ-S-698 Condition B, half hard	Sheet or strip	Stampings having bends across grain only where rigidity requirements are higher than for $\frac{1}{4}$ hard
Steel, low carbon, dead soft ASTM 109T	Steel QQ-S-698 Condition E, dead soft	Sheet or strip	For parts having coining, embossing, or extruded portions.
Steel, low carbon, dead soft (A1 killed) ASTM 109T	Steel QQ-S-698 Condition E (A1 killed)	Sheet or strip	For parts formed by spinning, deep drawing, or hydroforming.
Steel AISI C1095 Annealed	Steel MILS-7947 Class A (1095)	Sheet or strip	General-purpose material for small parts which are to be hardened.
Steel, high carbon Music wire	Steel QQ-W-470 (Music wire)	Wire, Cold drawn	General-purpose springs. Stress relieve after forming.
Steel AISI C1095 Annealed	Steel QQ-S-634 C1095	Rod or bar	General-purpose material for small parts which are to be hardened for heat-treatment procedures.
Steel AISI C1117	Steel QQ-S-637 C1117	Rod or bar	General-purpose screw machine stock in diameters greater than 0.25. Do not use in joining applications such as soldering or brazing.

Steel AISI C1212	Steel QQ-S-637 C1212	Rod or bar	General-purpose screw machine stock in diameters less than 0.25.
Steel AISI 4130	Steel MIL-S-6758 Condition C4 (4130)	Rod or bar	For parts requiring high strength and toughness and deep hardening.
Steel AISI E52100	Steel QQ-S-624 (AISI E52100)	Rod or bar	General-purpose AISI E52100 used for bearing retainers and other parts in contact with E52100 bearings.
Steel AISI E52100	Steel Vac Melted 52100 IAW C190000201	Rod or bar	Special-purpose AISI E52100 to be used only for bearing parts.
Steel tubing AISI C1015 Annealed	Steel tubing QQ-T-830 Type SMLS Composition MT1015 Condition ANL (1015)	Seamless tubing	May be used where joining methods such as soldering, brazing, or welding are necessary.
Steel casting ASTM A27 GR 65-35 (Similar to AISI C1025)	Steel casting QQ-S-681 Class 65-35 (1025)	Investment casting	General-purpose material. May be used in welding applications if stress relief annealed after welding.
Steel casting AMS 5334	Steel casting AMS 5334	Investment casting	For high-strength, heat-treatable castings. Heat-treatment procedure should be detailed on drawing (composition similar to AISI 8730).

Table 14-9. Corrosion-Resistant Steel.

Commercial Designation	Military Designation	Form Available	Remarks
Stainless steel AISI 302 Annealed	CRES QQ-S-766 Class 302 Condition A (AISI 302)	Sheet or strip	General-purpose, nonmagnetic, not hardenable. Good for soft soldering. Must be followed by a corrective anneal if heated above 800°F.
Stainless steel AISI 301 $\frac{1}{2}$-hard	CRES QQ-S-766 Class 301, $\frac{1}{2}$ hard Temper (AISI 301)	Shim stock	Good for blanking only. To be used for shims. Available only up to 0.031-in. thickness. Residual magnetism created by blanking may be removed by annealing.
Stainless steel AISI 303 Annealed	CRES QQ-S-764 Class 303, Condition A (AISI 303)	Rod or bar	General-purpose, free-machining, nonmagnetic, not hardenable. Not good for hard or soft soldering.
Stainless steel AISI 302 or 304 Temper: spring	QQ-W-423 Form 1 Comp 302 or 304 Condition B	Wire	General-purpose springs.
Stainless steel AISI 304 Annealed	MIL-T-8506 Type I	Tubing	General-purpose seamless tubing. See MS 33533 for tolerances. Good corrosion resistance.
Stainless steel AISI 316 Annealed	CRES QQ-S-766 Class 316 Condition A (AISI 316)	Sheet or strip	Good strength and corrosion resistance.
Stainless steel AISI 316 Annealed	CRES QQ-S-763 Class 316 Condition A (AISI 316)	Rod or bar	

Stainless steel casting ASTM A296 GR CF-8 (Similar to AISI 304)	CRES casting MIL-S-867 Class I (Similar to AISI 304)	Investment casting	Intended for use where austenitic corrosion resisting steel is required.
Stainless steel casting ASTM A296 GR CF-8C (Similar to AISI 347)	CRES casting MIL-S-867 (Similar to AISI 347)	Investment casting	Class II castings to be used in applications over 800°F.
CRES AMS 5643 Cond A (17-4PH)	CRES AMS 5643 Condition A (17-4PH)	Rod or bar	Special-purpose, high-strength, corrosion-resistant alloy. Hardenable with very low distortion. Higher cost.
Stainless steel AISI 410 Annealed Stainless steel AISI 410 Annealed	CRES QQ-S-766 Class 410 Condition A CRES QQ-S-763 Class 410 Condition A	Strip Rod or bar	Good for deep drawing, spinning, stamping, cold-heading riveting, soldering, brazing, and welding. Magnetic and hardenable. Poorest corrosion resistance of CRES types. Lowest cost.
Stainless-steel casting ASTM A296 GR CA-15 (Similar to AISI 410)	CRES casting MIL-S-16993, Class I (Similar to AISI 410)	Investment casting	Intended for use where hardenable corrosion resisting steel is required.
Stainless steel AISI 416	CRES QQ-S-764 Class 416 Condition A (AISI 416)	Rod or bar	Free machining, magnetic stock. Not good for cold-heading riveting, soldering, brazing, or welding. When used for pinion shafts, add note to material box: "Preharden to RC 28-38."
Stainless steel AISI 440C Annealed	CRES QQ-S-763 Class 440C Condition A	Rod or bar	Use for parts requiring high hardness. Can be heat-treated to RC 56-60. Not recommended for use at lower hardness. Poor corrosion resistance.

Table 14-10. Special-Purpose Alloys and Miscellaneous Materials.

Commercial Designation	Military Designation	Form Available	Remarks
Magnesium AZ91C Temper T6	Magnesium casting QQ-M-56 Composition AZ91C Temper T6	Sand, permanent mold, and investment casting	Intended for use where maximum strength and hardness consistent with good ductility and light weight are required.
Magnesium AZ61A	Magnesium QQ-M-31 Composition AZ61A	Bars or rods	Good mechanical properties with moderate ductility.
Magnesium AZ31B Condition 0, H24	Magnesium QQ-M-44 Composition AZ31B Condition 0, H24	Sheet and plate	General-purpose sheet stock alloy. Condition 0: Used for forming applications. Condition H24: Used for blanking applications.
Pure titanium	Titanium MIL-T-9047, class I Titanium MIL-T-9046, type I Composition B	Bars and forgings Sheet and plate	Special-purpose material having high strength, low density, good corrosion resistance. Good for high temperature (to 1000° F) applications. Difficult to weld.
Beryllium	–	Sintered bar or ingot	Special-purpose material having high modules, very light weight (comparable to magnesium), good corrosion resistance, and high temperature properties.
Tungsten alloy 90% W	–	Sintered bar and shapes	High-density alloy (specific gravity 16.8 min) for balance weights, flywheels, etc. Does not resist high

			humidity. List manufacturer's designation on drawing. Mallory 1000, Fansteel 77, Densalloy No. 3, Kennertium W-2, GE Hevimat. Selection of specific type on basis of evaluation by project engineer.
Elgiloy or equivalent	–	Bars, rods, and wire	Nickel cobalt alloy having excellent corrosion, wear, and fatigue resistance. Can be heat-treated to RC 60. Used for some pivot and stop applications. See C100990041.
Nickel alloy AMS 5665 Annealed (Inconel)	Nickel alloy MIL-N-6710 Condition A (Inconel)	Rod or bar	High temperature, corrosion-resistant alloy for structural applications to 1500°F.
Nickel alloy AMS 5667 Specify annealed or age-hardened (Inconel-X)	Nickel alloy MIL-N-8550 Cond A, E (Inconel-X)	Rod or bar	Condition A: Hot finished (annealed). Condition E: Age hardened. High temperature, corrosion-resistant alloy for structural applications to 1500°F and for springs to 1000°F.
Free-Cut Ivar 36 (Specify hot finished, cleaned, or centerless ground)	Nickel alloy MIL-S-16598 Finish	Bars rounds, squares, hexagons, octagons, flats	Nickel/iron alloy (36% nickel) with selenium. Used where machining is a problem. Low thermal coefficient of expansion (70°F to 400°F). Finish: Hot finished-forged or rolled, scale present. Cleaned-pickled or blast cleaned, scale removed. Centerless-ground, round bars only, bright and free from injurious defects.

Table 14-10. *(Continued)*

Commercial Designation	Military Designation	Form Available	Remarks
Synthetic rubber (Neoprene)	Synthetic rubber MIL-R-6855 (Neoprene)	Sheet	General-purpose gasket material.
Hot oils resistant rubber AMS 3226 and 3227	–	Sheet	Packings, seals.
Synthetic oil-resistant rubber	MIL-R-7362	Sheet	Packings, seals, O-rings.
Synthetic rubber	Synthetic rubber ZZ-R-765	Sheet	High-temperature gasket material. Not for oil and oxygen system.
Synthetic rubber AMS 3195 (medium) and AMS 3196 (firm) (silicone sponge)	Synthetic rubber (Silicone sponge)	Sheet	For high-temperature gaskets where high compressibility is required.
Synthetic sponge rubber AMS 3197 (soft), 3198 (medium) and 3199 (firm)	Synthetic sponge rubber (chloroprene)	Sheet	Vibration isolators, pads, and weather seals.
Synthetic rubber, heat-resistant AMS 3201 and 3202	Synthetic rubber	–	Seals, grommets.

Table 14-11. General Characteristics of Thermoplastics.

Item	Material	Key Advantages	Notable Limitations	Typical Trade Names	Maximum Tensile Strength (psi)	Maximum Service Temperature (°F)	Fabrication Methods
1	ABS	Balanced combinations of good mechanical properties and heat and chemical resistance. Good durability, high impact strength and mar resistance. Dimensionally stable, including creep resistance. Relatively low in price. Versatility and ease of processing, fabrication, and finishing	Low weather resistance. Some self-extinguishing grades display low heat resistance.	Cycolac, Kralastic, Abson, Dylel, Luran, Lustran, Tybrene	4,000 to 8,000	175 to 212	Injection molding, extrusion, calendaring, vacuum forming
2	Acetals	Good creep resistance, fatigue resistance, and dimensional stability. Heat resistance to over 200° F for extended use. Very good hydrolytic stability. Low coefficient of friction. Excellent resiliency	Not self-extinguishing. Limited resistance to acids and alkalis.	Delrin, Celcon	9,000 to 10,000	185 to 220	Forming, molding, extrusion, machining
3	Acrylics	Excellent optical properties, including high transparency. Excellent long-term resistance to sunlight and weathering	Low scratch resistance compared to glass. Limited resistance to alkalis and solvents.	Plexiglas, XT-Polymer, Acrylite, Implex, Lucite, Swedcast	6,000 to 10,000	140 to 200	Extrusion, casting, injection molding, compression
4	Cellulose acetates	Good optical properties, including high transparency. Good toughness. Low cost	Low heat resistance. High moisture absorption. Not resistant to alkalis and strong acids. Not flame retardant.	–	6,000 to 8,000	120 to 170	Injection, compression, and blow molding, vacuum forming, machining
	Cellulose acetate butyrates	Very good optical properties, including high transparency. Good toughness. Good weathering and aging resistance	Low tensile strength. Not resistant to strong acids and solvents.	Uvex	–	–	Injection, compression, and blow molding, extrusion, machining, drawing
	Ethyl cellulosics	Good toughness (even at low temperature). Good sunlight and weathering resistance. High transparency.	Low strength. Readily oxidized. Not flame retardant.	–	–	–	Injection and compression molding, extrusion, machining, drawing

Table 14-11. *(Continued)*

Item	Material	Key Advantages	Notable Limitations	Typical Trade Names	Maximum Tensile Strength (psi)	Maximum Service Temperature (°F)	Fabrication Methods
	Cellulose propionates	Good toughness; slightly better than cellulose acetate. High transparency. Excellent molding characteristics.	Low tensile strength. Not flame retardant.	Forticel	–	–	Injection and compression molding, extrusion, machining
5	Ethylene copolymers	Very flexible, good toughness and resiliency. Excellent chemical resistance. Tasteless, odorless. Good flex cracking and environmental stress cracking resistance. High filler loading capacity.	Low tensile strength, Some loss of inherent ethylene solvent resistance.	Ultrathene, Zetafax, Zetafin, Co-Mer	–	–	–
6	Ethylene vinyl acetate	Flexible, high impact strength. Useful to −150°F.	–	3,600	150	–	Injection and blow molding, extrusion
7	PTFE fluoropolymer	Outstanding chemical resistance, even at high temperature. Excellent sunlight and weathering resistance. Temperature resistant up to 500°F in continuous use. Coefficient of friction lowest of any solid material. Very good dielectric properties	Very high cost. Low strength. Cannot be processed by standard plastic methods.	Teflon, Halon, Tetran	6,500 to 7,000	400 to 500	Molding, extrusion, hot forming, cold forming, machining
	FEP fluoropolymer	Excellent chemical resistance, even at high temperature. Excellent sunlight and weathering resistance. Temperature resistant up to 400°F in continuous use. Very good dielectric properties.	Very high cost. Low strength.	Teflon	6,500 to 7,000	400 to 500	Molding, extrusion, hot forming, cold forming, machining
	CTFE fluoropolymer	Very good chemical resistance. Excellent sunlight and weathering resistance. Temperature resistant up to 390°F in continuous use. Stronger and stiffer than PTFE or FEP fluoropolymers.	Very high cost. Dielectric properties inferior PTFE or FEP.	Halon, Kel-F			
	ETFE &	Continuous use up to 300°F. Excellent weath-	High cost.	Tefzel, Halar			

	ECTFE fluoro-polymers	ering and chemical resistance. Good injection molding and extrusion characteristics.					
	PVF_2 fluoro-polymer	Very good chemical resistance. Excellent sunlight and weathering resistance. Strongest and most rigid fluorocarbon-type plastic	High cost. Lower heat resistance than other fluoropolymers.	Kynar			
8	Ionomer	Good toughness and barrier resistance	Not self-extinguishing. Not weather resistant. Attacked by acids.	Surlyn, Bakelite	–	–	Injection or blow moldings extrusion thermoforming
9	Nylon, Type 6	Good creep resistance, high fatigue endurance limit. Good toughness and resiliency. Resistance to petroleum oils and greases and many chemical solvents. Good abrasion resistance.	High moisture absorption. Low impact strength in dry environments.	Weldamid, Plaskon	8,500 to 12,500	250 to 300	Injection, compression, and blow molding, cold forming, extrusion, machining
	Nylon, Type 6/6	Very good creep resistance, high fatigue endurance limit. Good toughness and tensile strength. Resistant to petroleum oils and greases and many chemical solvents. Good abrasion resistance.	High moisture absorption. Low impact strength in dry environments.	Weldamid, Xylon, Zytel			
	Nylon, Type 6/10	More rigid and tougher than Types 6 and 6/6	Lower tensile strength than Types 6 and 6/6.	Zytel	8,500 to 12,500	250 to 300	Injection, compression, and blow molding, cold forming, extrusion, machining
	Nylon, Type 8	High impact strength. Low moisture absorption compared to other nylons.	Low tensile strength. High cost.	–			
	Nylon, Type 11	Low moisture absorption. Good impact strength.	High cost.	Rilsan			
	Nylon, Type 12	Low moisture absorption	High cost. Some grades not self-extinguishing.	–			
	Nylon copolymers	Good flexibility. Very tough	Not self-extinguishing.	Plaskon			
10	Parylene	Good for insulation and protective coating	–	–	–	200 to 240	Coating
11	Phenoxy	Clear, tough, rigid, hard. Impact strength: 2 ft-lb/in.	High cost.	–	–	–	–

Table 14.11. *(Continued)*

Item	Material	Key Advantages	Notable Limitations	Typical Trade Names	Maximum Tensile Strength (psi)	Maximum Service Temperature (°F)	Fabrication Methods
12	Phenylene oxide based materials	Good engineering properties, including creep and dimensional stability. Excellent hydrolytic stability. Temperature resistant to 160°F in continuous use. Good self-extingiushing and nondripping characteristics. Relatively low cost.	Attacked by many solvents.	Noryl	11,600	225	All conventional techniques.
13	Polyallomer methyl penetene	Very light weight	Low tensile strength. Not self-extinguishing or weather-resistant.	TBX	4,000	200	Injection molding, extrusion, vacuum forming
14	Polyamide-imide	Heat resistant to 500°F	Difficult to process.	Al Polymers, Torlon	–	–	–
15	Polyaryl ether	Heat resistant to 250°F. Good creep and fatigue resistance. Good chemical resistance. Excellent dimensional stability.	Not self-extinguishing. Poor weathering resistance. Difficult to mold.	Arylon	–	–	–
16	Polybutylene	Light weight. Continuous use to 225°F. Very high impact strength. Good chemical resistance.	Not weather-resistant. Not self-extinguishing.	–	–	–	–
17	Polycarbonates	Good creep and fatigue resistance and dimensional stability. Excellent toughness. Heat resistant to 250°F in continuous use. Very good optical properties, including transparency. Very good dielectric properties.	Poor resistance to solvents. Low scratch resistance.	Lexan, Merlon	–	250	All molding techniques, machining
	Polycarbonate ABS alloy	Good creep and fatigue resistance and dimensional stability. Very high impact strength. Good chemical resistance. Temperature resistant to 190°F in continuous use.	Attacked by solvents. Not self-extinguishing.	Cycoloy			
18	Polyethylenes low density	Excellent chemical resistance. Very flexible with good fatigue resistance and toughness.	Low tensile strength. Very difficult to bond	Marlex, Petrothene, Dylan, Hi-fax,	1,000 to 5,000	175 to 220	Molding, extrusion, calender-

		Very good dielectric properties. Tasteless, odorless. Low cost.	or print on. Susceptible to environmental stress cracking.	Norchem, Bakelite, Alathon			ing, coating, casting, vacuum forming
	Polyethylenes, high density	Excellent chemical resistance. Very good dielectric properties. Highest rigidity of ethylene plastics. Good toughness. Readily molded and extruded. Low cost.	Very difficult to bond or print on. Self-extinguishing grades have low properties.	Marlex, Petrothene, Dylan, Hi-fax, Norchem, Bakelite, Forteflex, Microthene, Alathon			
	Polyethylenes, high molecular weight	Excellent chemical resistance. Very good toughness. Low coefficient of friction. High dielectric strength. Good creep resistance.	High cost. Difficult to process	Marlex, Hi-fax	1,000 to 5,000	175 to 220	Molding, extrusion, calendering, coating, casting, vacuum forming
19	Polymides	Retention of high mechanical strength and dielectric properties up to about 500°F. Highly resistant to damage from ionizing radiation. Very good wear resistance.	Very high cost. Very difficult to process and fabricate.	Gemon, Meldin, Pyralin, Pyre-ML. Skybond, Vespel, Novimide	10,000 to 12,000	425 to 450	Machining
20	Polyphenylene sulfide	Continuous use to 500°F. Good creep and fatigue resistance. Good dimensional stability. Very good chemical resistance.	High cost. Difficult to process. Low impact strength.	Ryton	–	–	–
21	Polypropylenes	Strong, relatively rigid, tough, and light weight. Retain most mechanical properties at elevated temperatures. Outstanding chemical and stress cracking resistance. Excellent dielectric properties. Low cost.	Embrittles below 0°F. Not weather-resistant.	Oleform, Olemer, Olefil, Marlex, Moplen, Pro-fax	5,000	230	Injection and blow molding, extrusion
22	Polystyrenes	Good transparency. High hardness and rigidity. Readily molded and extruded. Low cost	Brittle. Low heat resistant to solvents.	Duratron, Dylene, Lustrex, Sytron, Bakelite, Fostrene	3,000 to 8,000	125 to 165	Injection and compression molding, extrusion, machining
23	Polysulfones	Good creep resistance and dimensional stability. Temperature resistant to 300°F in continuous use. Very good dielectric properties. Good chemical resistance and hydrolytic stability. Self-extinguishing.	Not resistant to polar solvents.	Ucardel	10,000	300	Injection and blow molding, extrusion, thermoforming

Table 14-11. *(Continued)*

Item	Material	Key Advantages	Notable Limitations	Typical Trade Names	Maximum Tensile Strength (psi)	Maximum Service Temperature (°F)	Fabrication Methods
	Polyaryl sulfone	Continuous use of 500°F. Good creep and fatigue resistance. Good dimensional stability. Excellent chemical resistance.	–	Astrel 360			
24	Vinyls, rigid	Excellent corrosion resistance. High dielectric strength. Good toughness and abrasion resistance. Self-extinguishing. Good weatherability.	Susceptible to staining.	Dacovin, Plaskon, Marvinol, Opalon, Rucon, Vyram, Bakelite, Pliovic	–	130	All molding techniques, extrusion, casting, calendering (machining for PVC)
	Vinyls (chlorinated). rigid	Excellent corrosion resistance. Heat resistance 50°F higher than rigid vinyls. Good toughness and abrasion resistance. Self-extinguishing. Good weatherability.	Difficult to process.	Geon			
	Vinyls, flexible	Very flexible. Good chemical and weatherability resistance. Inherently self-extinguishing. High dielectric strength. Low cost.	Stiffen at low temperatures. Susceptible to straining. Some plasticizers migrate to surface.	Dacovin, Plaskon, Marvinol, Opalon, Rucon, Vyram, Bakelite, Pliovic			
25	PVC/ABS alloy	Good creep and fatigue resistance and dimensional stability. High impact strength. Good chemical resistance. Self-extinguishing.	Attacked by solvents.	Cycovin	–	–	–
	PVC/acrylic alloy	High rigidity and toughness. Long-term resistance to sunlight and weathering. Self-extinguishing.	Attacked by solvents	Kydene, Kydez			
	Cross-linked PVC	Good heat resistance. Self-extinguishing in flexible grades.	Fabrication limited to extrusion.	–			
26	Cross-linked polyethylene	Good heat resistance. Self-extinguishing. Low temperature toughness. Good stress cracking resistance.	Low impact strength.	Flamovin, Blanex, Cab	–	–	–
	Cross-linked polypropylene	Good heat resistance. Self-extinguishing.	Some loss of inherent PP mechanical properties chemical resistance.	Flamovin			
	Cross-linked acrylic	Good heat resistance and transparency. Can be thermoformed.	Available only in sheet.				

Table 14-12. General Characteristics of Thermosetting Plastics.

Item	Material	Key Advantages	Notable Limitations	Typical Trade Names	Maximum Tensile Strength (psi)	Maximum Service Temperature (°F)	Fabrication Processes
1	Alkyds	Good dimensional stability. Very good dielectric properties. Temperature resistant to 300°F in continuous use.	Low impact strength. No resistant to high humidities.	Plenco, Glaskyd, Plaskon, Chempol, Duraplex	8,000	300	Compression molding
2	Allyl diglycol carbonates	Very high transparency. Stability of optical properties under load, heat, and many chemical environments. Good dimensional stability and radiation resistance.	Available only in cast stock shapes or machined parts. High cost.	CR 39	6,000	350	Injection, compression, and transfer molding, extrusion
3	Diallyl phthalates	Excellent dimensional stability. Retention of electrical properties at high temperature and humdity. Excellent resistance to moisture, acids, alkalis, and solvents. Self-extinguishing. Colors remain stable at high temperatures.	High cost.	Diall, Durex, Dapon	–	–	–
4	Epoxies	Excellent strength and toughness. Outstanding adhesion to many other materials. Low power factor and high dielectric strength. Good resistance to many acids, alkalis, and solvents. Versatility and ease of processing.	High cost.	Epon, Bakelite. Araldite, Epiall, Fiberite, Plenco, Epi-Rex, Polyox	12,000	500	Molding, extrusion, casting, potting
5	Melamines	Very high hardness. Resistant to detergents, water, and staining. High arc tracking resistance. Permanency of color and molded in designs. Self-extinguishing.	Fair dimensional stability. Low impact strength.	Admino, Fiberite, Cymel, Diaron, Melmac	–	–	–
6	Phenolics	Temperature resistant to 300°F (some to 600°F). Outstanding resistance to deformation under load. Dimensional stability over wide temperature range. Resistant to common solvents, weak acids, and many detergents. Low cost and ease of processing.	Color limitations. Low impact strength.	Genal, Durez, Bakelite, Phenall, Fiberite	4,000 to 10,000	250 to 450	Injection, compression, transfer, and plunger molding, casting

Table 14-12. *(Continued)*

Item	Material	Key Advantages	Notable Limitations	Typical Trade Names	Maximum Tensile Strength (psi)	Maximum Service Temperature (°F)	Fabrication Methods
7	Polybuta-dienes	Low dielectric constant and dissipation factor. High dielectric strength. Good temperature stability and chemical resistance.	Elevated processing temperature required.	Hystl, Budene	–	–	–
8	Polyesters	Good strength and rigidity. Versatility and ease of processing. Special grades display good weather resistance, very good chemical resistance, and flame retardancy. Good dielectric properties. Low cost.	–	Hetron, Atlac, Derakane, Cyglas, Dion, Genpol, Selectron	1,000 to 17,000	150 to 300	Molding, casting
9	Polyurethanes	Very flexible and fatigue resistant. Excellent abrasion resistance. Very high tear strength. Good chemical and solvent resistance. Very resistant to oxygen aging. Temperature resistant to 300°F in continuous use.	–	Roylar, Texin, Estane, Pellethane	5,000	–	Molding, extrusion, casting, calendering
10	Silicones	Retention of mechanical and dielectrical properties at very high temperature and humidity. Self-extinguishing.	Very high cost.	Silastic	4,000 to 6,000	500 and up	–
11	Ureas	Very high mechanical strength and wear resistance. Very low cost. Available in wide range of colors and finishes. Resistant to many chemicals and solvents. Self-extinguishing.	Heat sensitive and relatively brittle.	Beetle, Amaform, Arodure, Plaskon	–	–	–
12	Urethanes	Retention of properties over wide temperature range. Very good wear resistance and toughness. Good dielectric properties.	–	Scotchcast	–	–	–

Table 14-13. Commonly Used Plastics.

Commercial Designation	Description and Military Designation	Form Available	Remarks
Laminated, nylon base, phenolic NEMA N-1	Laminated plastic, MIL-P-15047, Type NPG	Sheet	Good mechanical and electrical properties, (Not good over 165° F continuous operation.) Fungus non-nutrient, can be used outside hermetic seal.
Melamine glass MEGA G-5	Laminated plastic, MIL-P-15037, Type GMG (TP1 Min Fill 50-Warp 50)	Sheet	Good mechanical and electrical properties. High-strength material for higher temperature applications. (Not good over 300° F continuous operation.) Fungus non-nutrient, can be used outside hermetic seal. Cannot be used for electrical insulation outside hermetic seal without moisture proofing.
	Laminated plastic, MIL-P-79, Type GMG, form R (TP1 Min Fill 50-Warp 50)	Rod	
	Laminated plastic, MIL-P-79, Type GMG, form Tr (TP1 Min Fill 50-Warp 50)	Tubing	
Plastic sheet, acrylic. Plexiglass II, UVA, or equivalent	Plastic sheet, acrylic, MIL-P-5425, finish A	Sheet	General-purpose transparent acrylic sheet. Meets requirements of MIL-P-7788 for lighting panels (80° C max). Transparent thermoplastic used for enclosures deicer shields, edge-lighted panels, etc.
Epoxy glass NEMA G-10	Laminated plastic, MIL-P-18177, Type GEE (TP1 Min Fill 50-Warp 50)	Sheet	Excellent electrical values. Low moisture absorption and dissipation factor over a wide range of humidities and temperatures. Good dimensional stability and excellent mechanical properties. Fungus non-nutrient. Can be used outside hermetic seal. Continuous operation limit: 300° F for G-10: 350° F for G-11.

Table 14-13. *(Continued)*

Commercial Designation	Description and Military Designation	Form Available	Remarks
Melamine resin	Mineral-filled melamine, MIL-M-14, Type MME	Molded parts	For use where good dielectrical properties and arc and flame resistance are required. Most stable of melamines.
	Glass fiber-filled, melamine molding compound, MIL-M-14, Type MM1-5		Lower impact strength and higher dielectric constant and dissipation factor at 1 megahertz than Type MM1-30. Superior moldability. Notch impact strength; approximately 0.5 ft-lb/in.
	Glass fiber-filled melamine, MIL-M-14, Type MM1-30		High impact strength. For use where heat arc, and flame resistance are required.
Polyester resin	Mineral-filled polyester, MIL-M-14, Type MAG	Molded parts	For use where good dielectric properties and arc resistance are required.
	Glass fiber-filled polyester, MIL-M-14, Type MAI-60		For use where high impact strength, good dielectric properties, and good arc resistance are required.
	Mineral-filled, glass fiber reinforced, alkyd resin, MIL-M-14, Type MAI-30		Excellent handling and molding characteristics. Arc, heat, and flame resistant. Good mechanical and excellent electrical characteristics.
Diallyl phthalate resins	Mineral-filled diallyl phthalate, MIL-M-14, Type MDG	Molded parts	For use where good dielectric properties and low shrinkage are required.
	Glass-filled diallyl phthalate resin, MIL-M-14, Type SDG		Low loss, high dielectric strength, low shrinkage, good moisture resistance. Relatively low impact strength.

	Acrylic polymer fiber-filled, diallyl phthalate resin. MIL-M-14, Type SDI-5		Low loss, high dielectric strength, low shrinkage, excellent moisture resistance. Moderate impact strength.
	Polyethylene terephthalate fiber-filled diallyl phthalate resin, MIL-M-14, Type SDI-30		Low loss, high dielectric strength, low shrinkage, very good moisture resistance, and high impact strength.
	MIL-P-19833, Type GDI-30-1		Flame retardancy is at a maximum when plastic material is burned.
Silicone resin	Mineral-filled silicone, MIL-M-14, Type MSG	Molded parts	Low loss, high dielectric strength, and excellent heat resistance.
	Glass fiber-filled silicone, MIL-M-14, Type MSI-30		High impact strength and heat resistance. Somewhat poorer electrical properties than Type MSG.
TFE (Teflon)	Teflon L-P-403, Type I, Class 1	Molding	Used for high-temperature insulation and molded parts. Excellent electrical properties, moisture resistance, and chemical resistance.
	MIL-P-22241 Teflon	Film	
	MIL-P-19468 Teflon	Rod	
	MIL-P-22242 Teflon	Sheet	
Nylon 6/10	L-P-410 Nylon	Extruded, molded rods, tubes and sheets	Jacketing for wire and cable; special molded parts. Low moisture absorption, heat resistance, and yield strength. Good flexibility. Resists alkalis and salt solutions. Self-extinguishing.
Nylon 6/6			Used for mechanical parts where lubrication is difficult or undesirable. High rigidity and surface temperature limits. Resists alkalis and salt solutions. Self-extinguishing.

Table 14-13. (*Continued*)

Commercial Designation	Description and Military Designation	Form Available	Remarks
Nylon 6 (poly-caprolactam)	L-P-410 Nylon	Extruded, molded rods, tubes and sheets	Used for bearings, gears, bushings. tape, and coil forms. Excellent frictional characteristics. High resiliency, impact strength, and heat resistance. Self-extinguishing. Attacked by mineral acids.
Nylon 6 (monomer)	–	Cast parts	Application and characteristics same as Nylon 6 (polycaprolactam).
Acetal (Delrin)	L-P-392	Molded, extruded	Thermoplastic with excellent abrasion resistance and dimensional stability. Low coefficient of friction. Good retention of mechanical and electrical properties. Maximum continuous service temperature is 250° F.
Acrylic sheet	MIL-P-8184	Sheet	Transparent thermoplastic similar to MIL-P-5425. Better resistance to heat and crazing. Used for enclosures and various formed panels. Maximum continuous service temperature is 215° F.
Epoxy glass laminate, ('u-('lad	Laminated plastic sheet, copper-clad to conform to MIL-P-13949	Sheet	Used for printed circuit boards. Extremely smooth copper surface exceeds NEMA standards.

Silicone, glass-base	MIL-P-997, Type GSG	Sheet	Fine-weave laminate. Good electrical insulating properties. Maximum continuous service temperature is 480°F. Attacked by hydrocarbon oils, greases, and solvents.
Teflon AMS 3651	Polytetrafluoroethylene	All forms	Extreme chemical inertness. Water absorption; 0.005 percent by weight. Very good dimensional stability. Nonflammable. Continuous operating temperature is –90°F to 500°F. Excellent machinability. Can be cemented, if etched. Cannot be rubber-stamped.

Table 14-14. Finishing Processes and Applicable Specifications.

Process	Metal	Applicable Specification	Normal Coating Thickness and Color	Remarks
Anodize (Chromic acid)	Aluminum	MIL-A-8625 Type I Class 1: Non-dyed Class 2: Dyed	0.0001 Maximum. Clear to dark gray depending on alloy.	Not to be applied to alloys with copper content in excess of 5.0% silicon in excess of 7.0%, or when total alloying elements exceed 7.5%. Nonconductive, good paint base, poor abrasion resistance, not good for dyes except black. Should be used for close tolerance parts.
Anodize (Sulfuric acid)	Aluminum	MIL-A-8625 Type II Class 1: Non-dyed Class 2: Dyed	0.0001–0.0004 Clear	Not to be applied to assemblies or parts with joints or recesses which might entrap solution. Good electrical barrier and paint base, may be dyed in all colors. Dichromate seal will impart yellow color. Produces excellent decorative finishes when buffed, satined, or bright dipped prior to anodizing. Good abrasion resistance.
Anodized (Hard) (Includes Alumilite, Martin Hardcoat, or Surge Hardcoat)	Aluminum	MIL-A-8625 Type III Class 1: Non-dyed Class 2: Dyed	Unless otherwise specified, 0.002 ± 0.0002 will be applied. Color will vary with alloy and thickness.	Dense, hard wear resistance coating. Coatings are approximately 50% penetration and 50% buildup. Excellent dielectric and heat absorption properties. Should be considered for salvage applications. Sealing greatly increases corrosion resistance, but slightly reduced wearing qualities.
Anodize with Duralox seal	Aluminum	–	All colors, no thickness increase.	The ultimate seal for salt-spray resistance on all alloys. Independent tests have withstood 8000 hr + without pitting on 6061 and 2024 aluminum alloys.
Alodine	–	*See* Chemical Films	–	–
Baking (Hydrogen embrittlement or stress relief)	All	In accordance with applicable plating specifications	–	General specifications may require this operation before and/or after processing for hydrogen embrittlement relief, improved adhesion, or hardness. Slight staining may result.
Black oxide	Iron or steel (Including stainless steel)	MIL-C-13924 Class 1: Iron and steel Class 2: Stainless steel Class 3: Fused salt process Class 4: 300 series stainless	No buildup of coating, color black.	Poor corrosion and abrasion resistance. Supplementary wax or oil dip will improve appearance and corrosion resistance. Good for decorative purposes when buffed or polished before processing.
	Copper alloys	MIL-F-495		

Brass	All	None	0.0002–0.0005. Matte to lustrous.	Normally preceded by bright nickel plate and post-treated with a coat of clear lacquer or baked enamel.
Bright dips	Aluminum Brass Copper	None	Removes metal.	Clean for welding, pressure test, or spotwelding improves appearance of aluminum, brass, and copper after processing.
Cadmium	All	QQ-P-416 Type I: Without supplementary chromate treatment Type II: With supplementary chromate treatment Type III: With supplementary phosphate treatment	Class 1: 0.0005 Class 2: 0.0003 Class 3: 0.0002 Note: embrittlement relief required on metals above Rockwell C 40 metals	Most commonly used plating. High density of plate affords excellent corrosion resistance. Type I coating (bright) highly susceptible to stains and fingerprints. Type II coating (golden) is excellent for resistance to moisture and humidity, excellent paint base may be chromate treated black, olive drab, or clear if specified as such, but is normally golden. Type III (gray) good paint base; final appearance is not attractive.
Chemical films (Alodine, Iridite Chromicoat, Kenvert, etc.)	Aluminum	MIL-C-5541 Class 1A for maximum protection against corrosion on painted or unpainted surfaces Class 3 for protection against corrosion where low electrical resistance is required	No dimensional change. Golden or clear as specified.	Used mainly as a paint base. Good corrosion resistance. Poor abrasion resistance. Electrically conductive. Normally gold unless otherwise specified.
Chromium	All	QQ-C-320 Class 1: Decorative Type I: Bright Type II: Satin Class 2 see specifications for requirements of Class 2a, 2b, 2c, 2d, and 2e.	Class 1: Decorative 0.00001–0.00005 (plus under plate) Class 2: Engineering 0.002 unless otherwise specified	Decorative chrome is normally applied over copper and nickel plate. Hard chrome is plated directly to the basis metal. Parts requiring heavy deposits should be overplated and ground to the finished dimension.
Chrome pickle	Magnesium	*See* Dow #1.	–	–
Chromic acid (Touch-up)	Magnesium	*See* Dow #19.	–	–
Copper	All	MIL-C-14550	Class 0: 0.001–0.005 Class 1: 0.001 Class 2: 0.0005 Class 3: 0.0002 Class 4: 0.0001	Stop off for carburizing, shielding for brazing. As an undercoat for other plating, increases electrical conductivity to prevent base metal migration into tin (prevents poisoning solderability).

Table 14-14. (*Continued*)

Process	Metal	Applicable Specification	Normal Coating Thickness and Color	Remarks
Continuous wire and strip-plating	Most	As specified	As specified	Continuous plating is a newly perfected method of continuous reel-to-reel plating on blank or prestamped strip. This method allows one to plate rapidly and accurately on almost all base metals in matte or bright tin. 60–40 solder, silver or gold. Underplates of copper or nickel are also possible if required.
Degreasing	All	TT-C-490	–	Although degreasing will remove oils, grease, etc., it will not always remove solids, additional cleaning may be necessary.
Dichromate	Magnesium	*See* Dow #7	–	–
Dow #1 (Chrome pickle)	Magnesium	MIL-M-3171 Type I	Removes metal, gray to brown.	For temporary corrosion resistance. Affords such protection as may be required during shipment, storage, or machining. May be used as a paint base.
Dow #7 (Dichromate treatment)	Magnesium	MIL-M-3171 Type III	No dimensional change, brassy to dark brown.	Good paint base, best corrosion protection of chemical coatings. May be applied to all alloys except EK30A, EK41A, EZ33A, HK31A, M1A, HM31A, HM21A, and LA141A. Good paint base. May be used on all other alloys.
Dow #9 (Galvanic anodize)	Magnesium	MIL-M-3171 Type IV	No dimensional change, dark brown to black.	Good paint base. May be used on all alloys. May be used for close tolerance and optical parts.
Dow #17 (Anodize)	Magnesium	MIL-M-45202 Type I: Light coat Type II: Heavy coat	Class C 0.0001–0.0005 light green Class D 0.0009–0.0016 dark green	Machining allowances must be made. Good paint base and corrosion protection. Hard nonconductive coating castings should be pickled prior to machining to remove scale and insure uniformity of coating. The coating thickness is approximately 40% penetration and 60% buildup.
Dow #19 (Chromic acid touch-up treatment)	Magnesium	MIL-M-3171 Type VI	No dimensional change, gray to brown.	Commonly used to touch up rack marks, surface scratches, reworked areas and surfaces that have been remachined or abraded.

Dow #21 (Ferric nitrate bright pickle)	Magnesium	None	Removes metal satin to bright metallic.	May be lacquered to preserve brightness, good paint base, poor corrosion resistance.
Dow #23	Magnesium	–	Under 0.0001.	Immersion tin deposition on magnesium.
Dry film lubrication	–	*See* Solid Film Lubricant	–	–
Electroless nickel	All	MIL-C-26074 Class I: As coated no heat treat. Class II: Metals not adversely affected by heat treat. Class III: Aluminum not heat treatable and beryllium alloys. Class IV: Heat treatable aluminum alloys.	Grade A: 0.001 min. Grade B: 0.0005 min. Grade C: 0.0015	100% uniformity of plate thickness on all accessible internal and external areas. Exceptionally good for salvage purposes. Heat treatment at 500°F for 1 hr, should result in coatings with a Vickers hardness of 850 ± 150 for Class 2 applications. Note: Class 2 coating should not be applied to those alloys that would be affected by this heat conditioning.
Cadmium	All	QQ-P-416 Type I: Without supplementary chromate treatment Type II: With supplementary chromate treatment Type III: With supplementary phosphate treatment	Class 1: 0.0005 Class 2: 0.0003 Class 3: 0.0002 *Note:* embrittlement relief required on metals above Rockwell C 40 metals	Most commonly used plating. High density of plate affords excellent corrosion resistance. Type I coating (bright) highly susceptible to stains and fingerprints. Type II coating (golden) is excellent for resistance to moisture and humidity, excellent paint base may be chromate treated black, olive drab, or clear if specified as such, but is normally golden. Type III (gray) good paint base, final appearance is not attractive.
Chemical films (Alodine, Iridite Chromicoat, Kenvert, etc.)	Aluminum	MIL-C-5541 Class 1A for maximum protection against corrosion on painted or unpainted surfaces Class 3 for protection against corrosion where low electrical resistance is required	No dimensional change, golden or clear as specified.	Used mainly as a paint base. Good corrosion resistance. Poor abrasion resistance. Electrically conductive. Normally gold unless otherwise specified.
Chromium	All	QQ-C-320 Class 1: Decorative Type I: Bright Type II: Satin Class 2: see specifications for requirements of Class 2a, 2b, 2c, 2d, and 2e.	Class 1: Decorative 0.00001–0.00005 (plus underplate) Class 2: Engineering 0.002 unless otherwise specified	Decorative chrome is normally applied over copper and nickel plate. Hard chrome is plated directly to the basis metal. Parts requiring heavy deposits should be overplated and ground to the finished dimension.

Table 14-14. (*Continued*)

Process	Metal	Applicable Specification	Normal Coating Thickness and Color	Remarks
Chrome pickle	Magnesium	*See* Dow #1	–	–
Chromic acid (Touch-up)	Magnesium	*See* Dow #19	–	–
Copper	All	MIL-C-14550	Class 0: 0.001–0.005 Class 1: 0.001 Class 2: 0.0005 Class 3: 0.0002 Class 4: 0.0001	Stop off for carburizing, shielding for brazing. As an undercoat for other plating, increases electrical conductivity to prevent base metal migration into tin (prevents poisoning solderability).
Continuous wire and strip-plating	Most	As specified	As specified	Continuous plating is our newly perfected method of continuous reel-to-reel plating on blank or prestamped strip. This method allows us to plate rapidly and accurately on almost all base metals in matte or bright tin. 60–40 solder, silver or gold. Underplates of copper or nickel are also possible if required.
Etching	Most	None	Removes metal, matte to bright.	To improve appearance, cleaning for spotwelding, etc.
Galvanic anodize	–	*See* Dow #9	–	–
Gold	All	MIL-G-45204 Type I: 99.7% gold Type II: 99.0% gold Type III: 99.9% gold	Class 00: 0.00002 Class 0: 0.00003 Class 1: 0.00005 Class 2: 0.0001 Class 3: 0.0002 Class 4: 0.0003 Class 5: 0.0005 Class 6: 0.0015 Matte to bright yellow or orange dependent on thickness and surface finish	Good solderability and corrosion resistance. Normally used over nickel or silver base. Consult military specifications for proper class for intended use. Improve tarnish resistance.
HAE (A superior type of anodic coating	Magnesium (all alloys)	MIL-M-45202 Type I: Light coating	Type I Class A: 0.0001–0.0003, Tan	It is generally agreed by most authorities, that this treatment is the best all-around coating for magnesium in existence.

		Type II: Heavy coating Type I, Grade 1: Without posttreatment (dyed) Grade 2: With Chromate posttreatment Type II, Grade 1: Without posttreatment Grade 3: With Bifluoride Dichromate posttreatment Grade 4: With Dichromate posttreatment including moist heat aging Grade 5: Same as Grade 4 except double Biflouride Dichromate posttreatment	Type II Class A: 0.0013–0.0017, Oatmeal to dark brown	The coating thickness may be considered as approximately 35% to 40% penetration and 55% to 60% buildup. When selected organic posttreatments are applied, a salt-spray test of up to 1000 hr may be expected. The coating is nonconductive. Type II coatings have extremely good resistance to abrasion. Type I light coatings are normally tan in color. Type II heavy coating will range from an oatmeal color to dark brown, dependent on the thickness applied. HAE coatings are not affected by extreme temperature variations. A heavy coated panel heated to 1075°F and plunged into ice water will show no ill effects to the coating. Recently developed techniques have now made it possible to mask off selective areas for electrical grounding, close tolerances or heavy buildups for salvage purposes.
Iridite	–	*See* Chemical Film	–	–
Iridite #15	Magnesium	MIL-M-3171 Type VIII	Brassy to dark brown. Experience shows it may remove metal	A good paint base on all alloys. Limited corrosion protection. Used on alloys, not adaptable to Dow #7. Careful control necessary to avoid etching.
Magnesium alloy processes	Magnesium	MIL-M-3171 Type I (*see* Dow #1) Type III (*see* Dow #7) Type IV (*see* Dow #9) Type VI (*see* Dow #19) Type VII (*see* Fluoride Anodizing) Type VIII (*see* Iridite #15)	–	This is a general specification that spells out in detail the processes referred to under Applicable Specification. *Also see* Dow #17 and HAE processes.
Magnetic particle inspection	Steel and other magnetic materials	MIL-I-6868	No dimensional or appearnace change.	A method for detecting flaws in steel, welds, and magnetic materials.
Nickel (Electrodeposited)	All	QQ-N-290 Class 1: corrosion protective Class 2: engineering plating to specified thickness and dimensions (*see* Sulfamate Nickel)	Class 1 Grade A: 0.0016 Grade B: 0.0012 Grade C: 0.001 Grade D: 0.0008 Grade E: 0.0006 Grade F: 0.0004 Grade G: 0.0002 Class 2 as specified	There is a nickel finish for almost any need. Nickel can be deposited soft or hard, dull or bright. Corrosion resistance is related to thickness applied. Low coefficient or thermal expansion. Slightly magnetic. Class 2 plating thickness will be 0.003 unless otherwise specified, but may be controlled to fit any engineering application. *Note:* Class 1 coatings that include copper undercoat are used for decorative chromium systems.

Table 14-14. (*Continued*)

Process	Metal	Applicable Specification	Normal Coating Thickness and Color	Remarks
Painting	All	As specified (MIL-F-14072 may be consulted for references)	Thickness as specified. Colors per FED-STD-595.	Include hammertone, wrinkles, enamels, epoxys, urethanes, polyvinyls, zinc chromate, etc.
Passivate	Stainless steels	MIL-S-5002 QQ-P-35 (Fastening device) Type I: low temperature Type II: medium temperature Type III: high temperature (For 300 series, all types are useable; for 400 series, Types II and III only.)	No dimensional change. No appearance change.	Removes all traces of foreign metals such as pieces of steel wool, tool scrapings, chips, etc. The presence of heavy scale may necessitate a pre-pickle. A coating of light oil is necessary sometimes to prevent rusting. For more magnetic series of stainless steel, proper heat-treat is important.
Penetrant inspection (Zyglo, a registered trade name)	All	MIL-6866 Type I: Fluorescent Type II: Visible dye Method A: Water washable Method B: Postemulsified Method C: Solvent removed	No dimensional change. No appearance change.	This process is generally used on aluminum, magnesium, and stainless steels. It can, however, be useful on other metals and materials. Detects cracks, discontinuities, and welding flaws.
Phosphate (Heavy)	Iron and steel	MIL-P-16232 Type M: Manganese base Type Z: Zinc base	0.0002–0.0004 Black 0.0002–0.0006 Gray	Good paint base for primers and enamels. Some corrosion resistance. Type M assists in "breaking in" of bearing surfaces. Type Z may be used to prevent galling in cold extrusion and deep drawing applications.
Pickling	All	None	Removes metal. Cleaner and brighter surface.	Generally used as a cleaner to remove scale, corrosion or rust. The hazards involved in this process should be investigated before using. Removes heat treat or welding scale from stainless steel and titanium.
Polishing	All	None	Removes metal. Satin to bright.	Decorative applications. Removes surface imperfections. Bright or satin finishes.
Rhodium	All	MIL-R-46085	Class 1: 0.000002 min. Class 2: 0.00001 min. Class 3: 0.00002 min. Class 4: 0.00010 min. Class 5: 0.00025 min.	Excellent abrasion resistance. Good electrical conductivity, hard as chromium. Will not tarnish. Excellent wear in contact areas.

Sandblast (Includes vaporblast, glass bead, liquid hone)	All	MIL-S-177726	Removes metal.	For removing scale and rust, etc., and blending imperfections. Amount of metal removed will depend on media selected. May be used for decorative and optical finishes. Provides uniform matte finish before decorative processing.
Silkscreen printing	All	As specified	–	An excellent means to label parts for identification and information.
Silver	All	QQ-S-365 Type I: Matte Type II: Semibright Type III: Bright Grade A: With supplementary tarnish-resistant treatment Grade B: Without supplementary tarnish-resistant treatment	0.0005 unless otherwise specified.	Excellent conductivity. Application of light water dip lacquer of chromate treatment per Grade A does not impair solderability. Greatly increases conductivity of lesser metals. An attractive decorative finish, particularly when base metal polished or satined.
Solder plate	All	MIL-P-81728	0.0003–0.0005 unless otherwise specified.	60/40 tin-lead alloy plating. Excellent solderability. Should not be used on part of irregular configuration or with deep recesses. This process is also available for wire and strip.
Solid film lubricant	All	MIL-L-8937	0.0002–0.0005 Dull gray to black	Pretreatment compatible with parent material is normally necessary (anodize, cadmium plate, etc.). Low coefficient of friction, reduces wear, prevents galling and seizing. Excellent fluid and corrosion resistance when used in conjunction with pretreatment.
Stress relief	–	*See* Baking	–	–
Sulfamate nickel	All	QQ-N-290 Class 2 AMS 2424	0.002–0.003 or as specified	Low stressed nickel deposit. Used for wear and abrasion resistance. Also, for heavy buildups on worn parts and for salvage purposes.
Tin	All	MIL-T-10727 Type I: Electrodeposited Type II: Hot-dipped	Type I: No thickness is specified. Suggested thicknesses are: 0.0001–0.00025 for soldering 0.0002–0.0004 to prevent galling or seizing	Excellent solderability with fair corrosion resistance. Oxides on Type I parts must be removed before soldering or parts must be protected by lacquer or similar finish immediately after plating. Type I parts may be fused but thickness will vary after treatment. A maximum of 0.0002 must be main-

Table 14-14. *(Continued)*

Process	Metal	Applicable Specification	Normal Coating Thickness and Color	Remarks
			0.0002–0.0006 to prevent formation of case during nitriding Type II: No thickness is specified. Average will be approximately 0.0005. Gray to semibright.	tained to permit satisfactory fusing. Fused tin is not covered in this specification. This process is also availble for wire and strip.
Tin, bright	All	MIL-T-10727	0.0001–0.0005	Also known as acid tin plating. Good solderability. Excellent shelf life.
Tin, lead plate	All	MIL-P-81728	0.003–0.0005 unless otherwise specified	60/40 tin-lead alloy plating. Excellent solderability. Should not be used on part of irregular configuration or with deep recesses. This process is also available for wire and strip.
Zinc	All	QQ-Z-325 Type I: Without supplementary chromate treatment Type II: With supplementary chromate treatment Type III: With supplementary phosphate treatment	Class 1: 0.001 Class 2: 0.0005 Class 3: 0.0002	Gives galvanic protection to base metal. Untreated (Type I) zinc does not maintain its bright surface for a very long period of time. Weather exposure will change zinc to a dull gray.
Zyglo	–	*See* Penetrant Inspection	–	–

Table 14-15. Properties of Common Electroplated Coatings.

Plate	Melting Point, °F/ft	Thermal Conductivity (Btu/hr, ft^2/°F/ft)	Electric Resistance (μohm)	Reflectance (Polished) % at 5000 A	Hardness	Abrasive Resistance
Cadmium	610	5.3	7.5	–	30–50 Brn	Fair
Chromium	2939	–	14–66	High	1000–1100 Brn	Excellent
Copper	1981	222	3.8	44	60–150 Brn	Poor
Gold	1944	169	2.4	47	65–325 Knoop	Poor–Good
Nickel	2651	34.4	7.4–10.8	61	150–500 Brn	Very good
Rhodium	3553	50.9	4.7	76	400–800 Brn	Good
Silver	1760	244	1.6	91	50–150	Good
Lead-Tin	460	–	–	–	Soft	Poor
Tin	448	36.3	11.5	54	5	Poor unalloyed
Zinc	786	64.2	5.8	55	40–50	Good

15

SAFETY

Design of any equipment must employ features to protect personnel from electrical and mechanical hazards, and those dangers that may arise from fire, elevated operating temperatures, and toxic fumes.

There are numerous methods of incorporating adequate safeguards, many of which are implicit in routine design procedures. However, certain design practices are of such importance as to warrant special attention.

In the design, attention must be given to the protection of both operating and maintenance personnel. Operating personnel must not be exposed to any mechanical or electrical hazards, nor should operation of the equipment necessitate any unusual precautions; safeguards must be provided in the equipment to protect maintenance personnel working on energized circuits.

Operating personnel must be safeguarded from hazards that may cause physical injury during either normal operation or malfunctioning of the equipment. The design must minimize the possibility of the operator's clothing becoming caught or entangled in the equipment. Handles and knobs should be so arranged that clothing will not catch, and corners should be rounded. Potentials greater than 70 V must be physically shielded or removed by the action of interlock switches. Despite safety regulations, the ever-present possibility exists that operating personnel may attempt to service equipment in a nonapproved manner.

Safeguarding maintenance personnel is more difficult; tests and repairs are often made with the apparatus exposed. It may be necessary to short out interlock switches and to remove covers that shield high voltages or moving parts.

Every effort should be made in the design to protect maintenance personnel against contact with dangerous voltages. Controls for adjustment should be located away from high voltages and moving parts. Danger labels next to dangerous parts or on protective covers should be used to alert maintenance personnel.

ELECTRIC SHOCK

Potentials exceeding 70 V are considered to be possible electric shock hazards. Research reveals that most deaths result from contact with the relatively low potentials ranging from 70 to 500 V, although under extraordinary circumstances, even lower potentials can cause injury.

Three factors determine the severity of electric shock: (1) the quantity of current flowing through the body, (2) the path of current through the body, and (3) the duration of time during which current flows through the body. Relatively small currents can cause death if the path includes the heart or lungs. Electric burns are usually of two types: those produced by the heat of the arc, which occurs when the body touches a high-voltage circuit, and those caused by passage of electric current through skin and tissue. Sufficient current passing through any part of the body will cause severe burns and hemorrhages. In cases on record, potentials below 10 V have proved fatal when points of contact have pierced the skin.

Table 15.1. Probable Effects of Shock.

Current Values (mA)	Effects
0–1	Perception
1–4	Surprise
4–21	Reflex action
21–40	Muscular inhibition
40 and above	Respiratory block

WARNINGS

Warnings signs marked *Caution: High Voltage* or *Caution: XXX Volts* should be placed in prominent positions on covers, access doors, and inside equipment wherever danger may be encountered. These signs should be durable, easily read, and so placed that dust or other foreign deposits will not obscure them in time.

PERSONNEL PROTECTION

Grounding

Various grounding techniques are used to protect personnel from dangerous voltages in equipment. All enclosures, exposed parts, and chassis should be maintained at ground potential. Reliable grounding systems should be incorporated

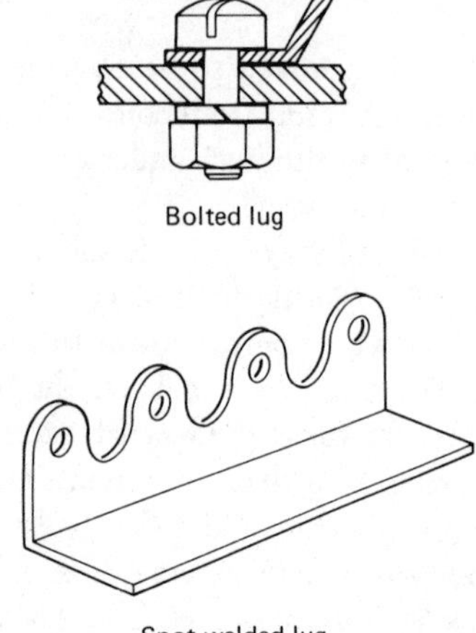

Figure 15-1. Ground connections.

in all electronic equipment. Enclosures and chassis should not be used as electric conductors to complete a circuit because of possible intercircuit interference. A terminal spot welded to the chassis provides a reliable ground connector. For chassis where welding is not feasible, a terminal properly secured by a machine screw, lockwasher, and nut is satisfactory (Figure 15-1). The machine screw used should be of sufficient size so that eventual relaxation will not result in a poor connection. A lockwasher is necessary to maintain a secure connection. All nonconductive finishes of the contacting surfaces should be removed prior to inserting the screw.

The common ground of each chassis should connect to a through-bolt, mounted on the enclosure and clearly marked *Enclosure Ground*, which in turn should connect to an external, safety ground strap. The external ground conductor should be fabricated from suitably plated, flexible copper strap, capable of carrying at least twice the current required for the equipment (Figure 15-2).

Electronic test equipment must be furnished with a grounding pigtail at the end of the line cord. Signal generators, vacuum tube voltmeters, amplifiers, oscilloscopes, and tube testers are among the devices so equipped. These leads are to be used for safety grounding purposes. Should a fault inside the portable instrument connect a dangerous voltage to the metal housing, the dangerous current is bypassed to ground without endangering the operator.

Power Lines

Designers are often inclined to confine their safety considerations to high-voltage apparatus. However, considerable attention should be devoted to the

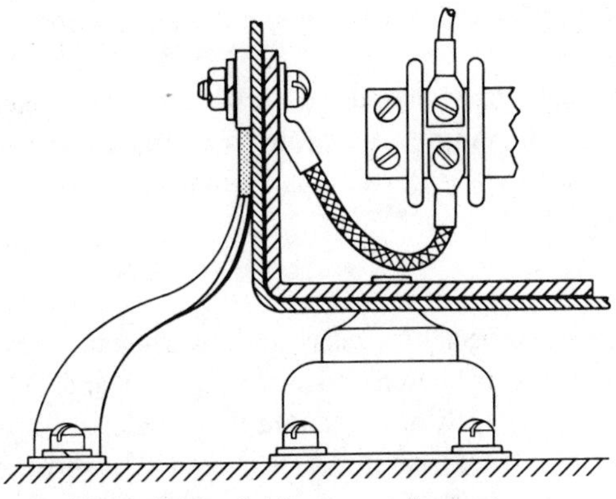

Figure 15-2. Cabinet grounding system.

hazards of power lines. Fires, severe shocks, and serious burns have resulted from personnel contacting, short-circuiting, or grounding the incoming lines.

Fusing

All leads from the primary service lines should be protected by fuses. Fuses should be connected to the load side of the main power switch. Holders for branch-line fuses should be such that when correctly wired, fuses can be changed without the hazard of accidental shock. At least one of the fuse-holder connections should be inaccessible to bodily contact; this terminal should be connected to the supply main power switch (Figure 15-3).

Main power switches should be equipped with safety devices that afford protection against possible heavy arcing. Devices that prevent opening of the

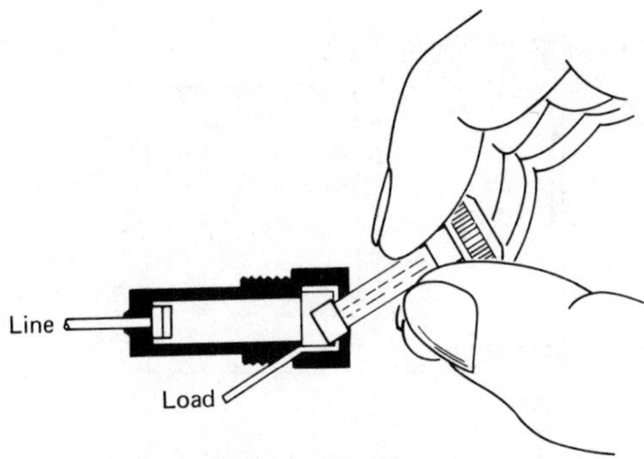

Figure 15-3. Fuse holder wiring.

switchbox when the switch is closed should be provided as protection for personnel.

Each piece of equipment should be furnished with a clearly labeled main power switch that will remove all power from the equipment by opening all leads from the primary power service connections.

Panel-Mounted Parts

Panel-mounted parts, especially jacks, are occasionally employed in power circuits for the insertion of meters, output lines, test apparatus, and other supplementary equipment. Such items should be connected to the grounded leg of the monitored circuit, rather than in the ungrounded, high-voltage line.

Shields and Guards

Safety enclosure covers should be anchored by means of screws or screwdriver-operated locks. They should be plainly marked by warnings.

Hinged covers, hinged doors, and withdrawable chassis should be counterbalanced or provided with other means to retain them in their open position, thus preventing accidental closing.

Terminal boards carrying hazardous voltages above 500 V should be protected by means of a cover provided with holes for the insertion of test probes. Terminal numbers should be plainly marked on the external side of the cover.

High-voltage meters should be recessed and shatterproof windows used (Figure 15-4).

Housings, cabinets, or covers may require perforations to provide air circulation. The area of a perforation should be limited to that of a $\frac{1}{2}$-in. square or round hole. High-voltage components within should be set back far enough to

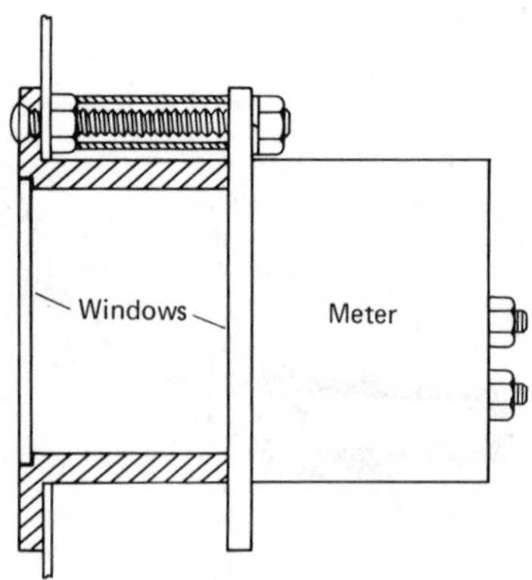

Figure 15-4. High-voltage panel meter.

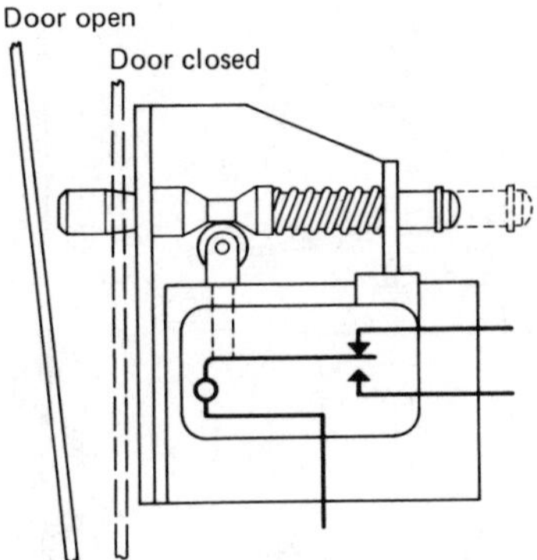

Figure 15-5. Door interlock switch.

prevent accidental contact. If this cannot be done, the size of the openings should be reduced.

Interlocks

Interlock switches are used to remove power during maintenance and repair operations. Each cover and door providing access to potentials greater than 70 V should be equipped with interlocks.

An interlock switch is ordinarily wired in series with one of the primary service leads to the power supply unit. It is usually actuated by the movable access cover, thus breaking the circuit when the enclosure is entered (Figure 15-5).

The selection of a type of interlock switch must be based upon its reliable operation. The so-called self-aligning switch seems most reliable, but in actual usage, the door interlock switch, although it contains moving parts, has proven most satisfactory.

Since electronic equipment must often be serviced with the power on, a switch enabling maintenance personnel to bypass the interlock system should be mounted inside the equipment. The switch should be so located that reclosing of the access door or cover automatically restores interlock protection. Also a panel-mounted visual indicator, such as a neon lamp, and a suitable nameplate should be used to warn personnel when interlock protection is removed.

Discharging Devices

Since high-grade filter capacitors can store lethal charges over relatively long periods of time, adequate discharging devices must be incorporated in all medium- and high-voltage power supplies. Such devices should be used wherever the time constant of capacitors and associated circuitry exceeds 5 sec; they should

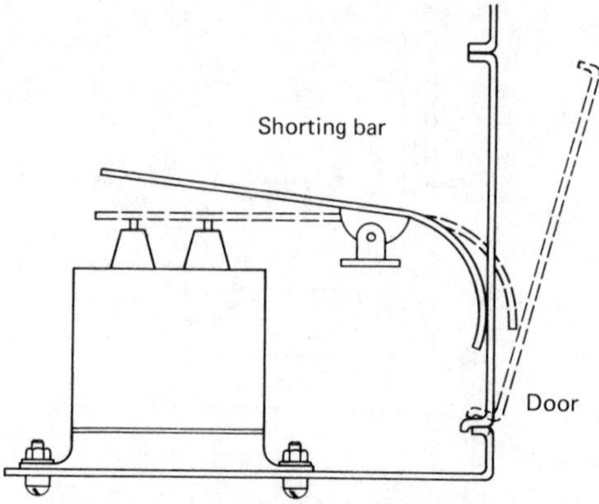

Figure 15-6. Shorting bar actuation.

be positive acting and reliable, and should be automatically actuated whenever the enclosure is opened (Figure 15-6). Shorting bars should be actuated either by mechanical release or by an electrical solenoid when the cover is opened.

Parts Safety

Protective devices should be incorporated in the design for all parts carrying hazardous voltages. Wherever possible, such components should be mounted beneath the chassis. Ventilation requirements must always be considered.

If it is impracticable to mount parts below the chassis and thus reduce the hazard to maintenance personnel when replacing above-chassis parts, protective housings having ventilating holes or louvers should be provided. If such housings cannot be used, exposed terminals of the parts should be oriented away from the direction of easy contact. These expedients lessen the possibility of accidental shock and arcing.

MECHANICAL AND OTHER HAZARDS

Thin edges should be avoided, and chassis construction should be such that the chassis may be carried without danger of cutting the hands on the edges. To minimize the possibility of physical injury, all enclosure edges and corners should be rounded to maximum practical radii.

Recessed mountings are recommended for small projecting parts such as toggle switches and small knobs located on front panels (Figure 15-7). To prevent hazardous protrusions on panel surfaces, flathead screws should be used wherever sufficient panel thickness is available; otherwise, panhead screws should be used.

In their normal installed positions, chassis should be securely retained in enclosures. Stops should be provided on chassis slides to prevent inadvertent re-

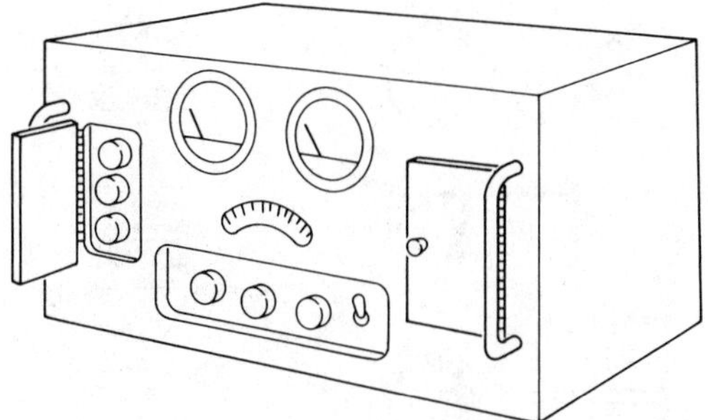

Figure 15-7. Recessed controls.

moval. Provision for firmly holding the chassis handles while releasing the equipment from the cabinet should also be incorporated. In the tilt-up position, a secure latch should support the equipment firmly.

All reasonable precautions should be taken to minimize fire, high temperature, and toxic hazards. In particular, any capacitors, inductors, or motors involving fire hazards should be enclosed by a noncombustible material having minimum openings. As stated previously, ventilation requirements and elevated operating temperatures are primary considerations in personnel protection. Since many equipments are installed in confined spaces, materials that may produce toxic fumes must not be employed. Finished equipment should be carefully checked for verification of protective features in the design.

Push buttons and toggle switches that project from a console surface are much more susceptible to accidental actuation than flush-mounted devices. Toggles are particularly dangerous. Individual controls can be recessed, or a group of controls can be mounted on a subpanel, so that their top surfaces are flush with the panel. Push buttons designed for flush mounting must be large enough for the operator's finger to clear the panel opening (Figure 15-8).

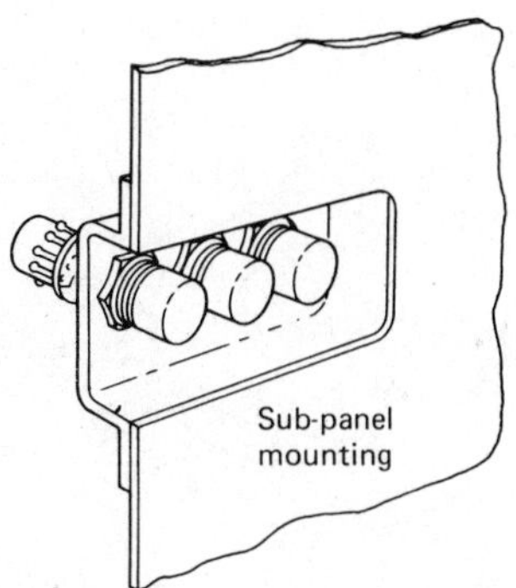

Figure 15-8. Recessed controls.

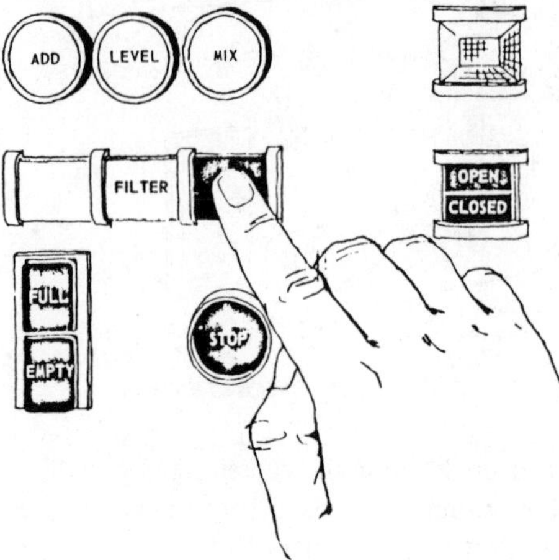

Figure 15-9. Raised barriers.

The converse to recessing is to provide raised barriers around individual controls. The barrier should extend as far as the maximum projection of the control above the panel surface. Push buttons, switch-lights, and toggle switches are easily guarded in this way, and the use of the control is not slowed to any significant extent. However, it is not absolutely foolproof; an elbow or a screwdriver can still cause accidents (Figure 15-9).

Controls that are not manipulated very often, such as those used in setup or calibration of equipment, should be actuated by a screwdriver or other special tool to prevent accidental disturbance as well as tampering (Figure 15-10).

Critical controls can be safeguarded by built-in two-step operation. Pull-before-push, push-before-turn, or pull-before-turn features are available for knobs and selector switches. Whenever such out-of-the-ordinary operations are required,

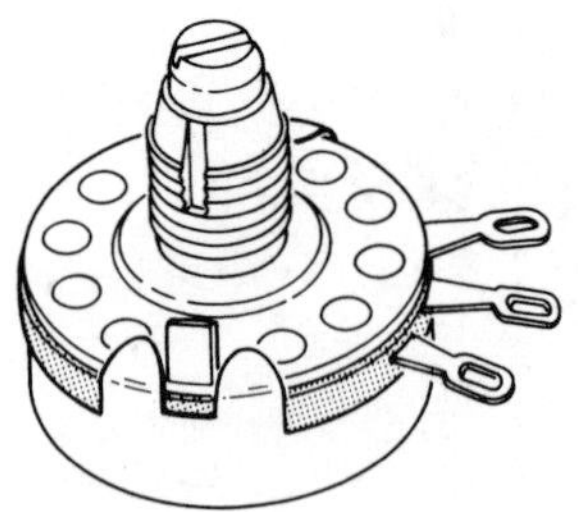

Figure 15-10. Tool-actuated controls.

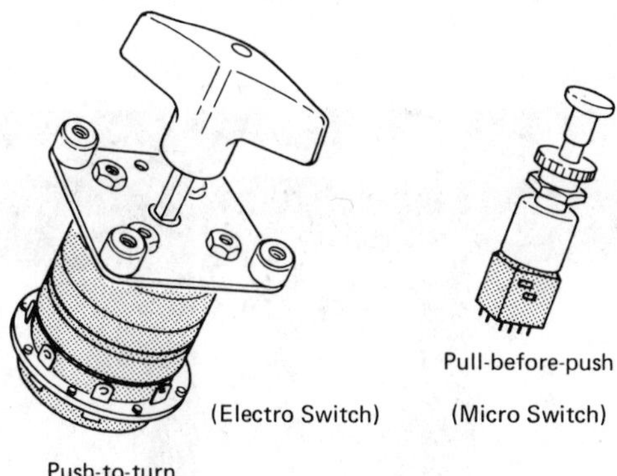

Figure 15-11. Two-step controls.

however, instructions for control operation should be placed on the panel (Figure 15-11).

Controls actuated by a linear motion should be oriented so that chance contacts by the operator or others will not disturb their settings. Toggles, levers, and joysticks fall in this category. Control consoles that must be placed in narrow passageways are particularly prone to this danger, and toggles or levers oriented in the up-down direction are less likely to be tripped than if mounted in the left-right position (Figure 15-12).

Large numbers of controls and displays should be broken into smaller, functionally related groups, each differentiated by colored panel areas or demarcation lines. This clarifies relationships between the areas and reduces confusion

Figure 15-12. Orientation of controls.

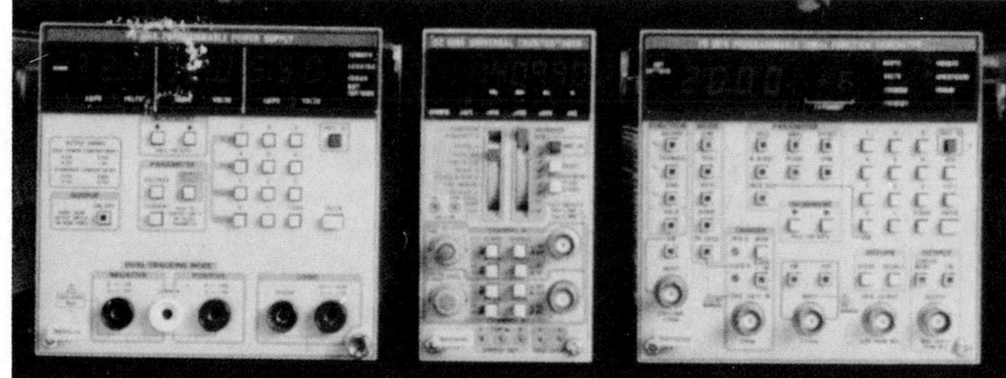

Figure 15-13. Functional marking of panel areas.

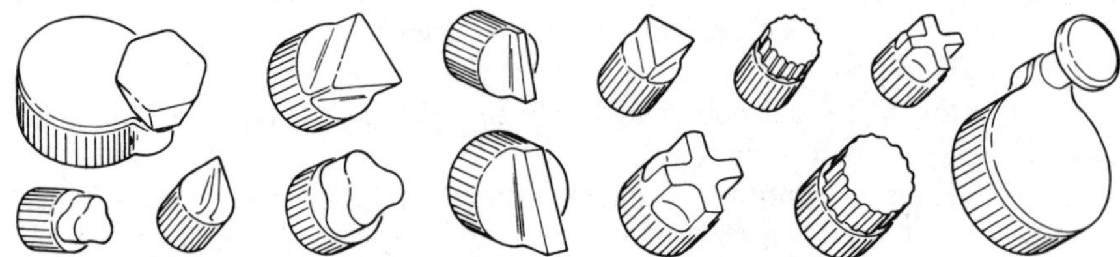

Figure 15-14. Code by size, shape, and texture.

and misidentifications. The photograph shows three basic colors: a general background color, a darker tone for controls and displays in one operating mode, and the darkest tone for panel elements used in another (Figure 15-13).

When the operator's visual sense is overburdened, he or she may reach "blind" for controls. If controls are closely spaced, misidentification and incorrect actuation is possible. Tactual clues such as size, shape, and texture of controls can minimize these mistakes. Several standard control shape codes have been developed for military applications, with the shape representing the function, such as fuel or power. Knob diameters can vary from $\frac{3}{8}$ to 4 in. Gross differences in surface texture are also effective (Figure 15-14).

Table 15-2. How the Safeguarding Techniques Compare.

Technique	Panel Space Increased	Response Time Increased	Degree of Security: Minimum	Average	Maximum	Most Suited to
Recessing	•			•		Push button, switch-light, entire panel or subpanel
Raised barriers	•			•		Push button, switch-light, toggle, thumbwheel
Built-in lock		•			•	Knobs, toggle, lever, selector switch pointer
Tool actuation		•			•	Slotted calibration control shaft
Two-step operation		•			•	Knob, selector switch pointer, lever, handwheel, switch-light (some)
Mechanical and electrical interlocks				• (Mechanical)	• (Electrical)	Series of push buttons; any control in an electrical system
Protective cover	•	•			•	Switch-light, toggle, push button, thumbwheel
Detachable controls		•			•	Knob, selector switch pointer, handwheel
Key-actuated controls		•			•	Selector switch
Control resistance			•			All controls
Orientation			•			Toggle, lever, joystick
Spacing	•		•			All controls
Positioning			•			All controls

16

PRINTED CIRCUITS

Printed wiring can lead to a better and lower cost product, and it can be accomplished in a number of different ways with widely different design techniques and varying levels of performance. To standardize the design areas that affect the quality and the reliability of the printed circuit, the military has set up certain minimum standards that can also be effectively used in commercial products. The design standards presented herein attempt to achieve the combined minimum requirements of all of the military and commercial standards in use today (Table 16-1).

All of the printed circuit boards discussed in this chapter are produced by the etch-and-plated-through hole process. This is the most widely used method, and it provides the greatest design flexibility at the lowest cost. The etch-and-plated-through hole process has a history of reliability and excellent pattern definition. The process is limited to the use of low-moisture-absorption materials and conductor and insulation combinations that provide good bond strength. The two most commonly used printed circuit board types in use are: (1) two-sided printed circuits and (2) multilayer printed circuits.

TWO-SIDED PRINTED CIRCUIT BOARDS

The two-sided printed circuit board consists of two layers of conductor (normally copper). These conductive layers are bonded on both sides of an insulation material. The most common insulation materials are epoxy-glass and paper-filled phenolic resin. The two-sided printed circuit board should be used whenever the design will permit because it costs approximately one-fifth as much as a multilayer printed circuit (Figure 16-1).

MULTILAYER PRINTED CIRCUIT BOARDS

Multilayer printed circuit boards are made up of three or more layers of single- or two-sided printed circuit boards, which have been etched to produce a circuit pattern. The individual printed circuits are bonded together using a layer of

Table 16-1. Printed Wiring Materials that Meet Military Specifications.

Material or Function	Description of Material	Commercial Designation	Military or Federal Specification	Referenced Printed Wiring Specification
Base material (unclad)	Epoxy resin			
	Glass fabric filler	GEC	MIL-P-18177	–
Base material (unclad)	Phenolic resin	–	MIL-P-31158	MIL-P-21193 BuOrd
	Paper filler	XXXP	MIL-P-13949	SCL-6225 Sig. Corps
Conductor	Copper 99.5% pure	ASTM-53-48	–	MIL-P-21193 BuOrd
Solder	60% tin, 40% lead	–	QQ-S-571	–
	63% tin, 37% lead	–	–	–
Flux	Noncorrosive	–	MIL-F-14256	SCL-6225 Sig. Corps
	Moderately active	–	–	–
Moisture protection	Varnish	–	MIL-V-173	–
Corrosion protection and soldering aid	Gold flash immersion	–	–	MIL-STD-275A
	0.00008 in. min			
Noncorrosive surface for wiping contacts	Nickel-rhodium plating	–	–	MIL-P-21193 BuOrd
	0.00025 to 0.0005 nickel	–	–	–
	0.00002 to 0.00005 rhodium	–	–	–
Noncorrosive surface for nonwiping service	Nickel-gold plating	–	–	MIL-P-21193 BuOrd
	0.00025 nickel min	–	–	–
	0.00002 gold min	–	–	–
Clad base material	Plastic material, foil clad	–	MIL-P-13949	MIL-P-21193 Bu Ord

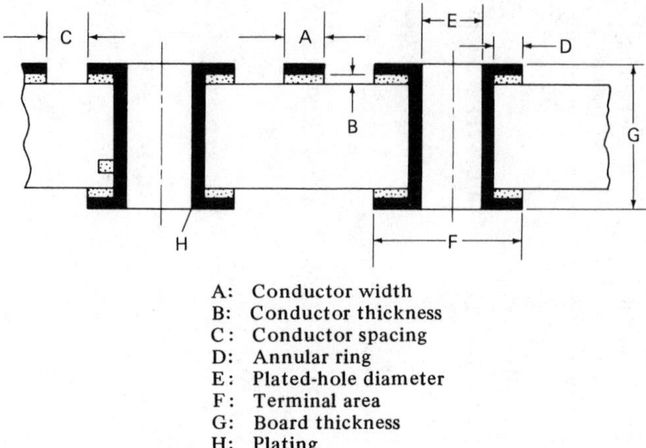

A: Conductor width
B: Conductor thickness
C: Conductor spacing
D: Annular ring
E: Plated-hole diameter
F: Terminal area
G: Board thickness
H: Plating

Figure 16-1. Printed circuit characteristics.

epoxy-impregnated fiberglas (prepreg) between each layer. The conductors of each layer are interconnected by plated-through holes (Figure 16-2). The multilayer printed circuits are used when very dense and complex circuits are required in a limited space.

DESIGNING THE PRINTED CIRCUIT BOARD

Using the input from the mechanical and electrical engineers, a layout is generated. The designer prepares a master layout on a translucent material, such as Mylar, consisting of at least two sheets. A modular grid is taped to the designer's board as a guide for the positioning of the components in the layout. The components should be positioned on one side of the printed circuit board, which is referred to as the "component side"; the conductor should be placed on the back side or "circuit side" of the board, which should be free of components. Occasionally, space limitations require that components are mounted on both sides. But the mounting of components on both sides should be avoided whenever possible. Sheet 1 of the layout should be the component side of the board and should be drawn in green. Sheet 2, the circuit side, should be drawn in red. When the two sheets are placed one on top of the other, a view of both sides of the board provides the designer with a view of the total circuitry so the interconnects from side to side can be aligned and circuit interferences can be avoided.

The printed circuit layout must be carefully drawn to scale. The completed layout will be used to prepare a "tape-up" to be photoreduced and used to process the printed circuit board. The master layout should establish the size and shape of the board, the size and location of all holes, and the location of all elec-

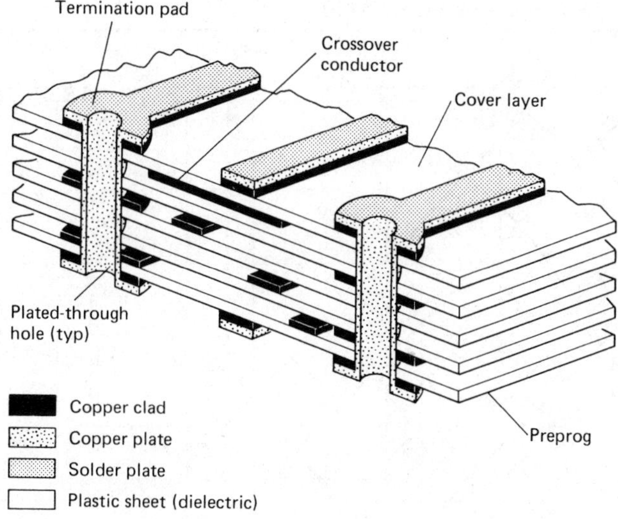

Figure 16-2. Multilayer printed circuit board.

trical components. All of the pattern features that are not controlled by the artwork should be dimensioned by notes, or directly on the face of the artwork. All of the plated-through holes, component mounting holes, and circuit should be dimensioned using a grid system. The basic grid unit applied in the x- and y-axes of the Cartesian coordinates should be 0.100, 0.050, or 0.025 to conform to the industry standard. The grid system of layout assures the designer that the board could be mass-produced using commercially available component insertion equipment.

Design Requirements

In the design of two-sided boards for both military and commercial products, MIL-STD-275 and MIL-P-55110 should be used as design guides.

Dimensions and Tolerances

All of the dimensions given herein will refer to the finished product. The artwork and design considerations must include the proper compensation for the photography and the etching processes.

Conductor Width

The etched copper conductor width for a particular thickness, based on the allowable current-carrying capacity, is specified in Table 16-2.

Table 16-2. Minimum Conductor Width.

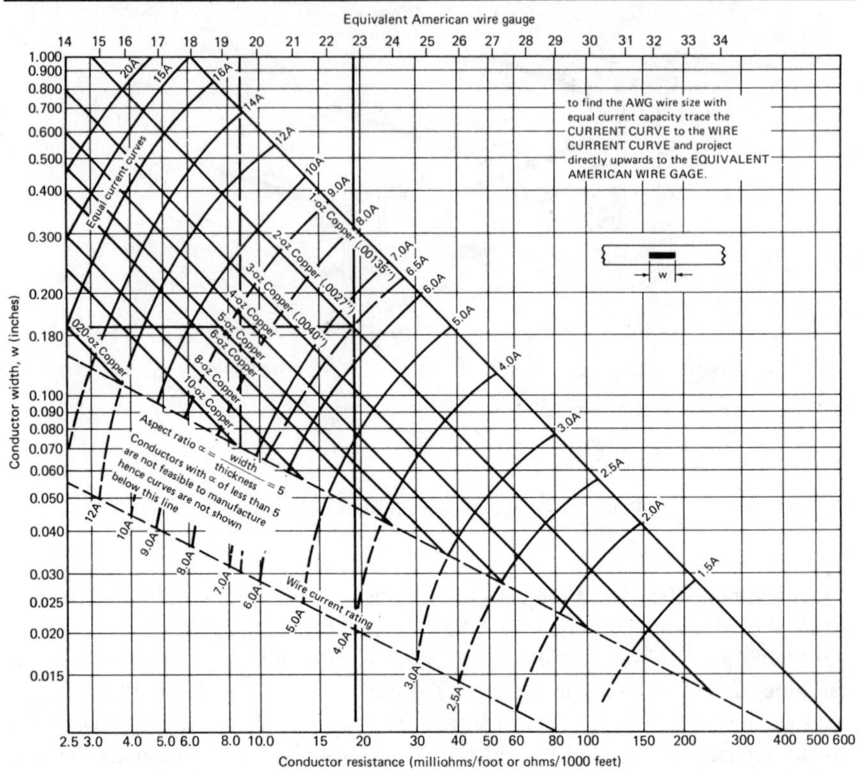

Table 16-3. Conductor Spacing for Uncoated Boards.

DC or Peak AC Voltage Between Conductors (V)	Minimum Spacing (in.), Sea Level to 10,000 ft	Minimum Spacing (in.), Above 10,000 ft
0 to 50	0.025	0.025
51 to 100	0.025	0.060
101 to 150	0.025	0.125
151 to 170	0.050	0.125
171 to 250	0.050	0.250
251 to 300	0.050	0.500
301 to 500	0.100	0.500
Greater than 500	0.0002 per volt	0.001 per volt

Table 16-4. Conductor Spacing for Conformal Coated Boards.

DC or Peak AC Voltage Between Conductors (V)	Minimum Spacing (in.)
0 to 30	0.010
31 to 50	0.015
51 to 150	0.020
151 to 300	0.030
301 to 500	0.060
Greater than 500	0.00012 per volt

Conductor Thickness

Whenever possible, all designs should employ 1-oz copper. Thicker and heavier copper should be used when line widths are restricted due to spacing or when unwanted coupling occurs.

Conductive Areas

Conductive areas larger than an 0.5-in. square should be avoided where practical. When such conductive areas are necessary, such as for shielding purposes, these areas should be broken up by a design that will leave the conductive pattern electrically continuous, or other means should be used to prevent blistering or warpage resulting from the soldering operation.

Conductor Spacing

The minimum spacing from the edge of the printed circuit board should be 0.010 in. When space allows, the preferred spacing is two times the board thickness. The minimum spacing between conductors is specified in Tables 16-3 and 16-4.

Hole Dimensions

The hole dimensions recommended herein apply to the finished printed circuit board. Table 16-5 applies to plated-through holes used for electrical communication from the circuit on the component side of the printed circuit board to the circuit side. Table 16-6 applies to the plated-through holes for the insertion of the component leads.

Table 16-5. Plated Feed-Through Holes.

Drilled Hole Size (in.)		Plated-Through Hole Size (in.)	
Min	Max	Min	Max
0.012	0.018	0.006	0.012
0.013	0.019	0.007	0.013
0.014	0.020	0.008	0.014
0.015	0.021	0.009	0.015
0.016	0.022	0.010	0.016
0.017	0.023	0.011	0.017
0.018	0.024	0.012	0.018
0.019	0.025	0.013	0.019
0.020	0.026	0.014	0.020
0.021	0.027	0.015	0.021
0.022	0.028	0.016	0.022
0.023	0.029	0.017	0.023
0.024	0.030	0.018	0.024
0.025	0.031	0.019	0.025
0.026	0.032	0.020	0.026
0.027	0.033	0.021	0.027
0.028	0.034	0.022	0.028
0.029	0.035	0.023	0.029
0.030	0.036	0.024	0.030
0.031	0.047	0.025	0.031
0.032	0.038	0.026	0.032
0.033	0.039	0.027	0.033
0.034	0.040	0.028	0.034
0.035	0.041	0.029	0.035
0.036	0.042	0.030	0.036
0.037	0.043	0.031	0.037
0.038	0.044	0.032	0.038
0.039	0.045	0.033	0.039
0.040	0.046	0.034	0.040
0.041	0.047	0.035	0.041
0.042	0.048	0.036	0.042
0.043	0.049	0.037	0.043
0.044	0.050	0.038	0.044
0.045	0.051	0.039	0.045
0.046	0.052	0.040	0.046
0.047	0.053	0.041	0.047
0.048	0.054	0.042	0.048

Table 16-6. Plated-Through Holes for Lead Insertion.

Wire Lead Diameter (in. Max)	Plated Hole Size (in.)				Minimum Master Pattern Terminal Area Diameter (in.)	
	Manual Insertion		Automatic Insertion			
	Min	Max	Min	Max	Manual Insertion	Automatic Insertion
0.012	0.016	0.024	0.030	0.038	0.060	0.070
0.013	0.017	0.025	0.031	0.039	0.060	0.070
0.014	0.018	0.026	0.032	0.040	0.060	0.070
0.015	0.019	0.027	0.033	0.041	0.060	0.075
0.016	0.020	0.028	0.034	0.042	0.060	0.075
0.017	0.021	0.029	0.035	0.043	0.060	0.075
0.018	0.022	0.030	0.036	0.044	0.060	0.075
0.019	0.023	0.031	0.037	0.045	0.064	0.080
0.020	0.024	0.032	0.038	0.046	0.064	0.080
0.021	0.025	0.033	0.039	0.047	0.064	0.080
0.022	0.026	0.034	0.040	0.048	0.064	0.080
0.023	0.027	0.035	0.041	0.049	0.064	0.080
0.024	0.028	0.036	0.042	0.050	0.070	0.080
0.025	0.029	0.037	0.043	0.051	0.070	0.085
0.026	0.030	0.038	0.044	0.052	0.070	0.085
0.027	0.031	0.039	0.045	0.053	0.070	0.085
0.028	0.032	0.040	0.046	0.054	0.070	0.085
0.029	0.033	0.041	0.047	0.055	0.080	0.085
0.030	0.034	0.042	0.048	0.056	0.080	0.090
0.031	0.035	0.043	0.049	0.057	0.080	0.090
0.032	0.036	0.044	0.050	0.058	0.080	0.090
0.033	0.037	0.045	–	–	0.080	–
0.034	0.038	0.046	–	–	0.080	–
0.035	0.039	0.047	–	–	0.080	–
0.036	0.040	0.048	–	–	0.080	–
0.037	0.041	0.049	–	–	0.080	–
0.038	0.042	0.050	–	–	0.080	–
0.039	0.043	0.051	–	–	0.090	–
0.040	0.044	0.052	–	–	0.090	–
0.041	0.045	0.053	–	–	0.090	–
0.042	0.046	0.054	–	–	0.090	–
0.043	0.047	0.055	–	–	0.090	–
0.044	0.048	0.056	–	–	0.090	–
0.045	0.049	0.057	–	–	0.090	–
0.046	0.050	0.058	–	–	0.090	–
0.047	0.051	0.059	–	–	0.090	–
0.048	0.052	0.060	–	–	0.090	–

Table 16-7. Standard Printed Circuit Board Thickness.

Standard Thickness	Tolerance (in.)	Expected Board Total Thickness Over Conductors (in. Nominal)[a]
0.093	±0.004	0.102
0.062	±0.003	0.070
0.047	±0.003	0.055
0.031	±0.003	0.039
0.030 and less in 0.001 increments	–	–

[a]Does not include effects of fusion.

Printed Circuit Board Thickness

Two-sided printed circuit boards are available in standard sizes as shown in Table 16-7. When a nonstandard thickness is required, it can be obtained as a special product from the raw material fabricator, but at a premium cost.

Warp and Twist

The maximum allowable warp and twist should be 1.5%. This is determined by measuring the maximum vertical displacement (the vertical distance from the surface to the maximum height of the concave surface). This vertical distance is then divided by the length of the longest side of the printed circuit board, and the quotient is multiplied by 100.

MULTILAYER PRINTED CIRCUIT BOARDS

In the design of multilayer printed circuit boards, MIL-P-55640 should be used as a design guide for both military and commercial products. Figure 16-3 defines the general characteristics of the multilayer printed circuit board.

Conductor Width

The conductor width for a particular thickness is specified in Table 16-2. The selection of the conductor width is dependent on the current-carrying requirements and the available space.

Conductor Thickness

Whenever practical, the conductor thickness on the external layers of all multilayer printed circuit boards should be 1-oz copper. All of the internal layers should be 2-oz copper (0.0028).

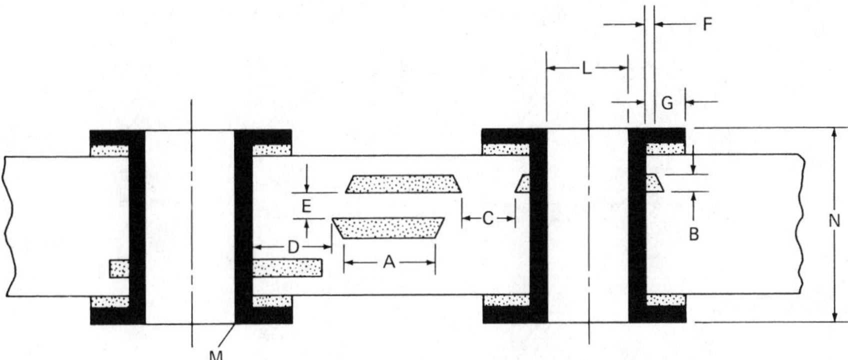

Note:
A = conductor width, B = conductor thickness, C = coplanar spacing, D = conductor-to-hole spacing, E = layer-to-layer spacing, F = internal annular ring, G = external annular ring, L = plated-hole diameter, M = plating, and N = printed circuit board thickness.

Figure 16-3. Multilayer printed circuit characteristics.

Thickness

The multilayer board thickness should be the total measured over the entire assembly. This should include the external conductors and all plating and coatings specified (Figure 16-4).

PLATING

Where plating is required, the following methods are recommended, depending on the intended application of the printed circuit board:

1. Tin-lead plating in accordance with MIL-P-81728. The tinplate will suffice as a protective finish for the copper circuitry in most instances.
2. Nickel-rhodium plating should be used for all insertion strips, for example, friction-type edge-card connectors. Rhodium plating should never be used where a soldering operation is to follow. A low-stress nickel-plating should always be used between the rhodium overplating and the copper circuitry.

Plating Thickness

1. An electroless deposition of a copper adherent should be used to provide a conductive path for the copper-plating. The copper-plating on the wall of a plated-through hole should be a minimum of 0.001.
2. The tin-lead plating should be a maximum of 0.0015 in. and a minimum of 0.0003. In areas where flat-pack devices are to be used, a plating thickness of 0.0006 should be provided for the lap-solder operation when installing the devices.

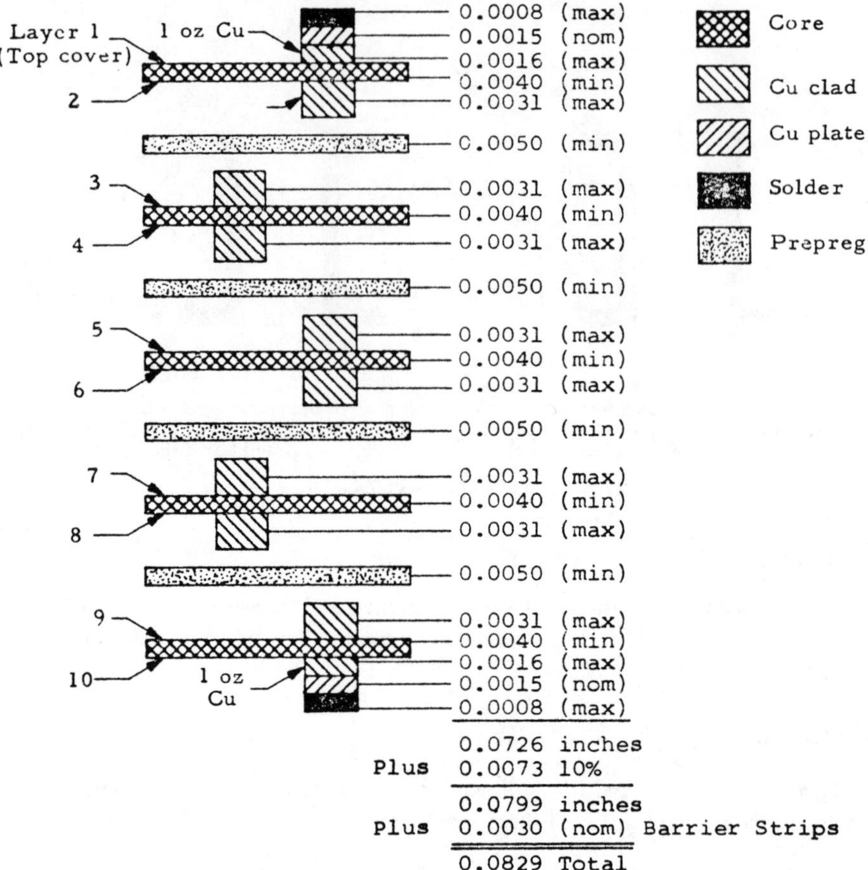

Figure 16-4. Multilayer board thickness.

3. Rhodium plating should have a minimum thickness of 0.000020 (20 millionths) and a maximum of 0.000100 (100 millionths). The rhodium should be applied using MIL-R-46085, type 1, class 3 as a guide.
4. The low-stress nickel undercoat used with the rhodium plate should be 0.0003 in. minimum and 0.0006 in. maximim. MIL-N-290, class 2 should be used as a guide.

Conformal Coating

Conformal coatings are used on printed circuit boards to provide protection against moisture, grease, high-altitude arc-over (corona), and all other detrimental contaminants.

17

REFERENCE TABLES AND FIGURES

Table 17-1. Decimal Equivalents.

	FRACTIONS OF ONE INCH				DECIMAL EQUIVALENT
	8 THS	16 THS	32 NDS	64 THS	
				1	.015625
			1	2	.031250
				3	.046875
		1	2	4	.062500
				5	.078125
			3	6	.093750
				7	.109375
1/8	1	2	4	8	.125000
				9	.140625
			5	10	.156250
				11	.171875
		3	6	12	.187500
				13	.203125
			7	14	.218750
				15	.234375
1/4	2	4	8	16	.250000
				17	.265625
			9	18	.281250
				19	.296875
		5	10	20	.312500
				21	.328125
			11	22	.343750
				23	.359375
3/8	3	6	12	24	.375000
				25	.390625
			13	26	.406250
				27	.421875
		7	14	28	.437500
				29	.453125
			15	30	.468750
				31	.484375
1/2	4	8	16	32	.500000
				33	.515625
			17	34	.531250
				35	.546875
		9	18	36	.562500
				37	.578125
			19	38	.593750
				39	.609375
5/8	5	10	20	40	.625000
				41	.640625
			21	42	.656250
				43	.671875
		11	22	44	.687500
				45	.703125
			23	46	.718750
				47	.734375
3/4	6	12	24	48	.750000
				49	.765625
			25	50	.781250
				51	.796875
		13	26	52	.812500
				53	.828125
			27	54	.843750
				55	.859375
7/8	7	14	28	56	.875000
				57	.890625
			29	58	.906250
				59	.921875
		15	30	60	.937500
				61	.953125
			31	62	.968750
				63	.984375
1"	8	16	32	64	1.000000

Table 17-2. Millimeter Conversions.

FRACTIONAL INCHES AND MILLIMETERS

Fractional	Decimal	Millimeters	Fractional	Decimal	Millimeters
1/64	.015625		33/64	.515625	
1/32	.03125			.51181	13
	.03937	1	17/32	.53125	
3/64	.046875		35/64	.546875	
1/16	.0625			.55118	14
5/64	.078125		9/16	.5625	
	.07874	2	37/64	.578125	
3/32	.09375			.59055	15
7/64	.109375		19/32	.59375	
	.11811	3	39/64	.609375	
1/8	.125		5/8	.625	
9/64	.140625			.62992	16
5/32	.15625		41/64	.640625	
	.15748	4	21/32	.65625	
11/64	.171875			.66929	17
3/16	.1875		43/64	.671875	
	.19685	5	11/16	.6875	
13/64	.203125		45/64	.703125	
7/32	.21875			.70866	18
15/64	.234375		23/32	.71875	
	.23622	6	47/64	.734375	
1/4	.250			.74803	19
17/64	.265625		3/4	.750	
	.27559	7	49/64	.765625	
9/32	.28125		25/32	.781250	
19/64	.296875			.7874	20
5/16	.3125		51/64	.796875	
	.31496	8	13/16	.8125	
21/64	.328125			.82677	21
11/32	.34375		53/64	.828125	
	.35433	9	27/32	.84375	
23/64	.359375		55/64	.859375	
3/8	.375			.86614	22
25/64	.390625		7/8	.8750	
	.3937	10	57/64	.890625	
13/32	.40625			.90551	23
27/64	.421875		29/32	.90625	
	.43307	11	59/64	.921875	
7/16	.4375		15/16	.9375	
29/64	.453125			.94488	24
15/32	.46875		61/64	.953125	
	.47244	12	31/32	.96875	
31/64	.484375			.98425	25
1/2	.500		63/64	.984375	

DECIMAL EQUIVALENTS OF MILLIMETERS

Millimeters	Inches	Millimeters	Inches
25	.98425	63	2.48031
26	1.02362	64	2.51968
27	1.06299	65	2.55905
28	1.10236	66	2.59842
29	1.14173	67	2.63779
30	1.18110	68	2.67716
31	1.22047	69	2.71653
32	1.25984	70	2.75590
33	1.29921	71	2.79527
34	1.33858	72	2.83464
35	1.37795	73	2.87401
36	1.41732	74	2.91338
37	1.45669	75	2.95275
38	1.49606	76	2.99212
39	1.53543	77	3.03149
40	1.57480	78	3.07086
41	1.61417	79	3.11023
42	1.65354	80	3.14960
43	1.69291	81	3.18897
44	1.73228	82	3.22834
45	1.77165	83	3.26771
46	1.81102	84	3.30708
47	1.85039	85	3.34645
48	1.88976	86	3.38582
49	1.92913	87	3.42519
50	1.96850	88	3.46456
51	2.00787	89	3.50393
52	2.04724	90	3.54330
53	2.08661	91	3.58267
54	2.12598	92	3.62204
55	2.16535	93	3.66141
56	2.20472	94	3.70078
57	2.24409	95	3.74015
58	2.28346	96	3.77952
59	2.32283	97	3.81889
60	2.36220	98	3.85826
61	2.40157	99	3.89763
62	2.44094	100	3.93700

Table 17-3. Weights and Measures.

Troy Weight

For gold, silver and precious metals.

Grains		Dwts.		Ozs.		Lbs.
24	=	1				
480	=	20	=	1		
5760	=	240	=	12	=	1

Pounds Avoirdupois .82286 = pounds. Troy.

Pounds Troy × 1.2153 pounds Avoirdupois.

The jewelers' carat is equal in the United States to 3.086 grains, in London to 3.0 grains, in Paris to 3.18 grains.

Avoirdupois or Commercial Weight

The grain is the same in Troy, Apothecaries and Avoirdupois weights.

The standard Avoirdupois pound is the weight of 27.7015 cubic inches of distilled water weighed in the air at 39.2 degrees Fahrenheit, barometer at 30 inches. 27.343 grains = 1 drachm.

Drachms		Oz.		Lb.		Long Qrs.		Long Cwt.		Long Ton
16	=	1								
256	=	16	=	1						
7168	=	448	=	28	=	1				
28672	=	1792	=	112	=	4	=	1		
573440	=	35840	=	2240	=	80	=	20	=	1

The above table gives what is known as the "long ton." The "short ton" weighs 2000 pounds.

Long Measure (Measures of Length)

Inch		Feet		Yards		Fath.
12	=	1				
36	=	3	=	1		
72	=	6	=	2	=	1

Inch		Feet
198	=	$16\frac{1}{2}$
7920	=	660
63360	=	5280

Feet		Yards		Fath.		Rods		Furl.		Mile
$16\frac{1}{2}$	=	$5\frac{1}{2}$	=	$2\frac{3}{4}$	=	1				
660	=	220	=	110	=	40	=	1		
5280	=	1760	=	880	=	320	=	8	=	1

6080.26 feet = 1.15 Statute Miles = 1 Nautical Mile or Knot

Equivalent Measures (Measures of Length)

1 Meter =

39.37	inches.
3.28083	feet.
1.09361	yards.
1000.	millimeters.
100.	centimeters.
10.	decimeters.
0.001	kilometers.

1 Inch =

1000.	mils.
0.0833	foot.
0.02777	yard.
25.40	millimeters.
2.540	centimeters.

1 Foot =

12.	inches.

Table 17-3. *(Continued)*

Equivalent Measures (Measures of Length) *(Continued)*

1 Centimeter =
- 0.3937 inch.
- 0.0328083 foot.
- 10. millimeters.
- 0.01 meters.

1 Millimeter =
- 39.370 mils.
- 0.03937 inch (or $\frac{1}{25}$ inch nearly).
- 0.001 meter.

1 Kilometer =
- 3280.83 feet.
- 1093.61 yards.
- 0.62137 mile.

1 Mil =
- 0.001 inch.
- 0.02540 millimeter.
- 0.00254 centimeter.

- 1.33333 yard.
- 0.0001893 miles.
- 0.30480 meter.
- 30.480 centimeters.

1 Yard =
- 36. inches.
- 3. feet.
- 0.0005681 mile.
- 0.914402 meter.

1 Mile
- 63360. inches.
- 5280. feet.
- 1760. yards.
- 320. rods.
- 8. furlongs.
- 1609.35 meters.
- 1.60935 kilometers.

Square Measure (Measures of Surface)

Sq. Inch		Sq. Ft.		Sq. Yds.		Sq. Rods		Roods		Acre
144	=	1	=							
1296	=	9	=	1						
39204	=	$272\frac{1}{4}$	=	$30\frac{1}{4}$	=	1				
1568160	=	10890	=	1210	=	40	=	1		
6272640	=	43560	=	4840	=	160	=	4	=	1

640 acres = 1 square mile

An acre = a square whose side is 69.57 yards or 208.71 feet

Measures of Weight

1 Gram =
- 15.432 grains.
- 0.022046 lb. (avoir.)
- 0.3527 oz. (avoir.)

1 Kilogram =
- 1000. grams.
- 2.20462 lb. (avoir.)
- 35.2739 oz. (avoir.)

1 Metric ton =
- 2204. 62 pounds.
- 0.984206 ton of 2240 pounds.
- 22.0462 cwt.
- 1.10231 ton of 2000 pounds.
- 1000. kilograms.

1 Gram =
- 0.064799 grains.

1 Ounce =
- 437.5 grains.
- 0.0625 pounds.
- 28.3496 grams.

1 Pound =
- 7000. grains.
- 16. ounces.
- 453.593 grams.
- 0.453593 kilograms.

1 Ton (2240 pounds) =
- 1.01605 metric tons.
- 1016.05 kilograms.

Table 17-3. *(Continued)*

Measures of Volume and Capacity

1 Cubic Meter =		1 Liter =	
61023.4	cubic ins.	1.	cubic decimeter.
35.3145	cubic feet.	61.0234	cubic inches.
1.30794	cubic yds.	0.353145	cubic foot.
1000.	liters.	1000.	cubic centimeters or centiliters.
264.170	gallons U.S. liquid = 231 cubic ins.	0.001	cubic meter.
		0.26417	U.S. gallon liquid.
		1.0567	U.S. quart.
1 Cubic decimeter =		2.202	lbs. of water at 62 degrees Fahrenheit.
61.0234	cubic ins.	1 Cubic Yard =	
0.0353145	cubic foot.	46656.	cubic inches.
0.26417	U.S. liquid gallon.	27.	cubic feet.
1000.	cubic centimeters.	0.76456	cubic meter.
0.001	cubic meter	1 Cubic foot =	
		1728.	cubic inches.
1 Cubic Centimeter =		0.03703703	cubic yard.
0.0000353	cubic foot.	28.317	cubic decimeters or liters.
0.0610234	cubic inch.		
1000.0	cubic millimeters.	0.028317	cubic meter.
0.001	liter.	7.4805	gallons.
		1 Cubic inch =	
1 Cubic Millimeter =		16.3872	cubic centimeters.
		1 Gallon (British) =	
0.000061023	cubic inch.	4.54374	liters.
0.0000000353	cubic foot.	1 Gallon (U.S.) =	
0.001	cubic centimeter.	3.78543	liters.

Table 17-4. Measurement Rules.

Length

Side of square of equal periphery as circle = diameter × 0.7854.
Diameter of circle of equal periphery as square = side × 1.2732.
Length of arc = number of degrees × diameter × 0.008727.

Area

Triangle = base × half perpendicular height.
Parallelogram = base × perpendicular.
Trapezoid - half the sum of the parallel sides × perpendicular height.
Trapezium, divide two triangles and find area of the triangles.
Parabola = base × $\frac{2}{3}$ height.
Ellipse = long diameter × short diameter × 0.7854.
Regular polygon = sum of sides × half perpendicular distance from center to sides.
Surface of cylinder = circumference × length + area of two ends.
Surface of pyramid or cone = circumference of base × $\frac{1}{2}$ of the slant height + area of the base.
Surface of a frustrum of a regular right pyramid or cone = sum of peripheries or circumferences of the two ends × half slant height + area of both ends.
Area of rectangle = length × breadth.

Solid Contents

Prism, right or oblique = area of base × perpendicular height.
Cylinder, right or oblique = area of section at right angles to sides × length of side.
Pyramid or cone, right or oblique, regular or irregular = area of base × $\frac{1}{3}$ perpendicular height.
Contents of segment of sphere = (height 2 + three times the square of radius of base) × (height × .5236).
To find the volume of a cylinder: Multiply the area of the section in square inches by the length in inches = the volume in cubic inches. Cubic inches divided by 1728 = volume in cubic feet.
Solidity of a sphere = cube of diameter × .5236; or surface × $\frac{1}{6}$ diameter.
Side of an inscribed cube = radius of a sphere × 1.1547.
Contents of frustrum of cone or pyramid. Multiply areas of two ends together and extract square root. Add to this root the two areas and × $\frac{1}{3}$ altitude.
Contents of a wedge = area of a base × $\frac{1}{2}$ altitude.

Prismoidal Formula

A prismoid is a solid bounded by six plane surfaces, only two of which are parallel.
To find the contents of a prismoid, add together the areas of the two parallel surfaces and four times the area of a section taken midway between and parallel to them, and multiply the sum by $\frac{1}{6}$ of the perpendicular distance between the parallel surfaces.

Weight

Ascertain the number of cubic inches in piece and multiply same by weight per cubic inch. Or, multiply the length by the breadth (in feet) and product by weight in pounds per square foot.

Table 17-5. Standard Gauges for Sheets.

Gauge No.	Approx. Fraction Inches	U.S. Standard Gauge For Steel Sheets: Weight Pounds Per Sq. Ft.	U.S. Standard Gauge For Steel Sheets: Approx. Decimal Thickness Inches	Galvanized Sheet Gauge Weight Pounds Per Sq. Ft.	Birmingham or Stubs' Iron Wire Gauge For H. R. Strip (Lighter than $\frac{1}{4}''$ thick), C. R. Strip, Spring Steel, Stainless Steel, Brass and Copper Tubes Decimal Thickness Inches	American or Brown & Sharpe Wire Gauge For Sheet Brass, Sheet Bronze, Brass Rods Decimal Thickness Inches
7–0	–	–	–	–	–	0.651354
6–0	–	–	–	–	–	0.580000
5–0	–	–	–	–	0.500	0.516500
4–0	–	–	–	– –	0.454	0.460000
3–0	–	–	–	–	0.425	0.409642
2–0	–	–	–	–	0.380	0.364796
1–0	–	–	–	–	0.340	0.324861
1	–	–	–	–	0.300	0.289297
2	–	–	–	–	0.284	0.257627
3	–	10.0000	0.2391	–	0.259	0.229423
4	–	9.3750	0.2242	–	0.238	0.204307
5	–	8.7500	0.2092	–	0.220	0.181940
6	–	8.1250	0.1943	–	0.203	0.162023
7	$\frac{3}{16}$	7.5000	0.1793	–	0.180	0.144285
8	$\frac{11}{64}$	6.8750	0.1644	7.03125	0.165	0.128490
9	–	6.2500	0.1495	6.40625	0.148	0.114423
10	$\frac{9}{16}$	5.6250	0.1345	5.78125	0.134	0.101897
11	$\frac{1}{8}$	5.0000	0.1196	5.15625	0.120	0.090742
12	$\frac{7}{64}$	4.3750	0.1046	4.53125	0.109	0.080808
13	$\frac{3}{32}$	3.7500	0.0897	3.90625	0.095	0.071962
14	$\frac{5}{64}$	3.1250	0.0747	3.28125	0.083	0.064084
15	–	2.8125	0.0673	2.96875	0.072	0.057068
16	$\frac{1}{16}$	2.5000	0.0598	2.65625	0.065	0.050821
17	–	2.2500	0.0538	2.40625	0.058	0.045257
18	$\frac{1}{20}$	2.0000	0.0478	2.15625	0.049	0.040303
19	–	1.7500	0.0418	1.90625	0.042	0.035890
20	$\frac{3}{80}$	1.5000	0.0359	1.65625	0.035	0.031961
21	–	1.3750	0.0329	1.53125	0.032	0.028462
22	$\frac{1}{32}$	1.2500	0.0299	1.40625	0.028	0.025346

Table 17-5. *(Continued)*

Gauge No.	Approx. Fraction Inches	U.S. Standard Gauge For Steel Sheets: Weight Pounds Per Sq. Ft.	U.S. Standard Gauge For Steel Sheets: Approx. Decimal Thickness Inches	Galvanized Sheet Gauge Weight Pounds Per Sq. Ft.	Birmingham or Stubs' Iron Wire Gauge For H. R. Strip (Lighter than $\frac{1}{4}''$ thick), C. R. Strip, Spring Steel, Stainless Steel, Brass and Copper Tubes Decimal Thickness Inches	American or Brown & Sharpe Wire Gauge For Sheet Brass, Sheet Bronze, Brass Rods Decimal Thickness Inches
23	–	1.1250	0.0269	1.28125	0.025	0.022572
24	$\frac{1}{40}$	1.0000	0.0239	1.15625	0.022	0.020101
25	–	0.87500	0.0209	1.03125	0.020	0.017900
26	–	0.75000	0.0179	0.90625	0.018	0.015941
27	–	0.68750	0.0164	0.84375	0.016	0.014195
28	$\frac{1}{64}$	0.62500	0.0149	0.78125	0.014	0.012641
29	–	0.56250	0.0135	0.71875	0.013	0.011257
30	$\frac{1}{80}$	0.50000	0.0120	0.65625	0.012	0.010025
31	–	0.43750	0.0105	0.59375	0.010	0.008928
32	–	0.40625	0.0097	0.56250	0.009	0.007950
33	–	0.37500	0.0090	–	0.008	0.007080
34	–	0.34375	0.0082	–	0.007	0.006305
35	–	0.31250	0.0075	–	0.005	0.005615
36	–	0.28125	0.0067	–	0.004	0.005000
37	–	0.265625	0.0064	–	–	0.004453
38	–	0.25000	0.0060	–	–	0.003965
39	–	–	–	–	–	0.003531
40	–	–	–	–	–	0.003144

Table 17-6. Wire Gauges.

Number of Wire Gauge	American or Brown & Sharpe, Inches	Washburn & Moen Mfg. Co., A. S. & W. Roebling, Inches	Imperial Wire Gauge, Inches	Stubs' Steel Wire, Inches	Birmingham or Stubs' Iron Wire, Inches
0000000	–	0.4900	0.5000	–	–
000000	0.5800	0.4615	0.4640	–	–
00000	0.5165	0.4305	0.4320	–	0.500
0000	0.460	0.3938	0.4000	–	0.454
000	0.40964	0.3625	0.3720	–	0.425
00	0.3648	0.3310	0.3480	–	0.380
0	0.32486	0.3065	0.3240	–	0.340
1	0.2893	0.2830	0.3000	0.227	0.300
2	0.25763	0.2625	0.2760	0.219	0.284
3	0.22942	0.2437	0.2520	0.212	0.259
4	0.20431	0.2253	0.2320	0.207	0.238
5	0.18194	0.2070	0.2120	0.204	0.220
6	0.16202	0.1920	0.1920	0.201	0.203
7	0.14428	0.1770	0.1760	0.199	0.180
8	0.12849	0.1620	0.1600	0.197	0.165
9	0.11443	0.1483	0.1440	0.194	0.148
10	0.10189	0.1350	0.1280	0.191	0.134
11	0.090742	0.1205	0.1160	0.188	0.120
12	0.080808	0.1055	0.1040	0.185	0.109
13	0.071961	0.0915	0.0920	0.182	0.095
14	0.064084	0.0800	0.0800	0.180	0.083
15	0.057068	0.0720	0.0720	0.178	0.072
16	0.05082	0.0625	0.0640	0.175	0.065
17	0.045257	0.0540	0.0560	0.172	0.058
18	0.040303	0.0475	0.0480	0.168	0.049
19	0.03589	0.0410	0.0400	0.164	0.042
20	0.031961	0.0348	0.0360	0.161	0.035
21	0.028462	0.0317	0.0320	0.157	0.032
22	0.025347	0.0286	0.0280	0.155	0.028
23	0.022571	0.0258	0.0240	0.153	0.025
24	0.0201	0.0230	0.0220	0.151	0.022
25	0.0179	0.0204	0.0200	0.148	0.020
26	0.01594	0.0181	0.0180	0.146	0.018
27	0.014195	0.0173	0.0164	0.143	0.016
28	0.012641	0.0162	0.0148	0.139	0.014

Table 17-6. *(Continued)*

Number of of Wire Gauge	American or Brown & Sharpe, Inches	Washburn & Moen Mfg. Co., A. S. & W. Roebling, Inches	Imperial Wire Gauge, Inches	Stubs' Steel Wire, Inches	Birmingham or Stubs' Iron Wire, Inches
29	0.011257	0.0150	0.0136	0.134	0.013
30	0.010025	0.0140	0.0124	0.127	0.012
31	0.008928	0.0132	0.0116	0.120	0.010
32	0.00795	0.0128	0.0108	0.115	0.009
33	0.00708	0.0118	0.0100	0.112	0.008
34	0.006304	0.0104	0.0092	0.110	0.007
35	0.005614	0.0095	0.0084	0.108	0.005
36	0.005	0.0090	0.0076	0.106	0.004
37	0.004453	0.0085	0.0068	0.103	–
38	0.003965	0.0080	0.0060	0.101	–
39	0.003531	0.0075	0.0052	0.099	–
40	0.003144	0.0070	0.0048	0.097	–

Table 17-7. Head Dimensions of Threaded Fasteners.

Head Dimensions

Decimal Inches

Finished Hexagon Head Machine Bolts and Cap Screws

Nominal Size or Basic Major Diameter of Thread		Body Dia. Minimum (Maximum Equal to Nominal Size)	F Width Across Flats			G Width Across Corners		H Height			R Radius of Fillet	
			Max.	(Basic)	Min.	Max.	Min.	Nom.	Max.	Min.	Max.	Min.
1/4	.2500	.2450	7/16	.4375	.428	.505	.488	5/32	.163	.150	.023	.009
5/16	.3125	.3065	1/2	.5000	.489	.577	.557	13/64	.211	.195	.023	.009
3/8	.3750	.3690	9/16	.5625	.551	.650	.628	15/64	.243	.226	.023	.009
7/16	.4375	.4305	5/8	.6250	.612	.722	.698	9/32	.291	.272	.023	.009
1/2	.5000	.4930	3/4	.7500	.736	.866	.840	5/16	.323	.302	.023	.009
9/16	.5625	.5545	13/16	.8125	.798	.938	.910	23/64	.371	.348	.041	.021
5/8	.6250	.6170	15/16	.9375	.922	1.083	1.051	25/64	.403	.378	.041	.021
3/4	.7500	.7410	1 1/8	1.1250	1.100	1.299	1.254	15/32	.483	.455	.041	.021
7/8	.8750	.8660	1 5/16	1.3125	1.285	1.516	1.465	35/64	.563	.531	.062	.041
1	1.0000	.9900	1 1/2	1.5000	1.469	1.732	1.675	39/64	.627	.591	.093	.062
1 1/8	1.1250	1.1140	1 11/16	1.6875	1.631	1.949	1.859	11/16	.718	.658	.093	.062
1 1/4	1.2500	1.2390	1 7/8	1.8750	1.812	2.165	2.066	25/32	.813	.749	.093	.062
1 3/8	1.3750	1.3630	2 1/16	2.0625	1.994	2.382	2.273	27/32	.878	.810	.093	.062
1 1/2	1.5000	1.4880	2 1/4	2.2500	2.175	2.598	2.480	15/16	.974	.902	.093	.062
1 3/4	1.7500	1.7350	2 5/8	2.6250	2.538	3.031	2.893	1 3/32	1.134	1.054	.093	.062
2	2.0000	1.9880	3	3.0000	2.900	3.464	3.306	1 7/32	1.263	1.175	.093	.062

Square Head Set Screws

Nominal Size		F Width Across Flats Max.	F Width Across Flats Min.	G Width Across Corners Min.	H Height of Head Nom.	H Height of Head Max.	H Height of Head Min.	X Radius of Head Nom.
#10	.1900	.1875	.180	.247	9/64	.148	.134	15/32
#12	.2160	.2160	.208	.292	5/32	.163	.147	35/64
1/4	.2500	.2500	.241	.331	3/16	.196	.178	5/8
5/16	.3125	.3125	.302	.415	15/64	.245	.224	25/32
3/8	.3750	.3750	.362	.497	9/32	.293	.270	15/16
7/16	.4375	.4375	.423	.581	21/64	.341	.315	1 3/32
1/2	.5000	.5000	.484	.665	3/8	.389	.361	1 1/4
9/16	.5625	.5625	.545	.748	27/64	.437	.407	1 13/32
5/8	.6250	.6250	.606	.833	15/32	.485	.452	1 9/16
3/4	.7500	.7500	.729	1.001	9/16	.582	.544	1 7/8
7/8	.8750	.8750	.852	1.170	21/32	.678	.635	2 3/16
1	1.0000	1.0000	.974	1.337	3/4	.774	.726	2 1/2
1 1/8	1.1250	1.1250	1.096	1.505	27/32	.870	.817	2 13/16
1 1/4	1.2500	1.2500	1.219	1.674	15/16	.966	.908	3 1/8
1 3/8	1.3750	1.3750	1.342	1.843	1 1/32	1.063	1.000	3 7/16
1 1/2	1.5000	1.5000	1.464	2.010	1 1/8	1.159	1.091	3 3/4

Cup points are standard with stock sizes of all set screws.

Hexagonal Socket Type Set Screws

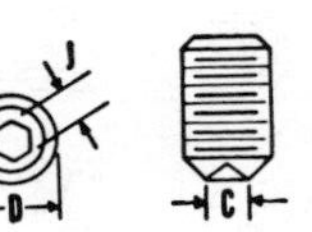

D Nominal Diameter	C Cup and Flat Point Dia. Max.	C Cup and Flat Point Dia. Min.	R Oval Point Radius	Y Cone Point Angle 118°±2° for these Lengths and Under	Y Cone Point Angle 90°±2° for these Lengths and Over	P Full Dog Point and Half Dog Point Diameter Max.	P Diameter Min.	Q Full	q Half	J Socket Width Across Flats Max.	J Socket Width Across Flats Min.
5	0.067	0.057	3/32	1/8	3/16	0.083	0.078	0.06	0.03	0.0635	1/16
6	0.074	0.064	7/64	1/8	3/16	0.092	0.087	0.07	0.03	0.0635	1/16
8	0.087	0.076	1/8	3/16	1/4	0.109	0.103	0.08	0.04	0.0791	5/64
10	0.102	0.088	9/64	3/16	1/4	0.127	0.120	0.09	0.04	0.0947	3/32

Table 17-7. (*Continued*)

12	0.115	0.101	5/32	3/16	1/4	0.144	0.137	0.11	0.06	0.0947	3/32
1/4	0.132	0.118	3/16	1/4	5/16	5/32	0.149	1/8	1/16	0.1270	1/8
5/16	0.172	0.156	15/64	5/16	3/8	13/64	0.195	5/32	5/64	0.1582	5/32
3/8	0.212	0.194	9/32	3/8	7/16	1/4	0.241	3/16	3/32	0.1895	3/16
7/16	0.252	0.232	21/64	7/16	1/2	19/64	0.287	7/32	7/64	0.2207	7/32
1/2	0.291	0.270	3/8	1/2	9/16	11/32	0.334	1/4	1/8	0.2520	1/4
9/16	0.332	0.309	27/64	9/16	5/8	25/64	0.379	9/32	9/64	0.2520	1/4
5/8	0.371	0.347	15/32	5/8	3/4	15/32	0.456	5/16	5/32	0.3155	5/16
3/4	0.450	0.425	9/16	3/4	7/8	9/16	0.549	3/8	3/16	0.3780	3/8
7/8	0.530	0.502	21/32	7/8	1	21/32	0.642	7/16	7/32	0.5030	1/2
1	0.609	0.579	3/4	1	1 1/8	3/4	0.734	1/2	1/4	0.5655	9/16

Maximum socket depth does not exceed three-fourths of minimum head height.
Cup points are standard with stock sizes of all set screws.

Headless Slotted Set Screws

Nominal Size	D Body Diameter of Screw	R Radius of Oval Point Screw	I Radius of Headless Crown	J Width of Slot	T Depth of Slot	C Diameter of Cup and Flat Points Max.	C Diameter of Cup and Flat Points Min.	P Diameter of Dog Point Max.	P Diameter of Dog Point Min.	Q Height of Dog Full	q Height of Dog Half
5	0.125	0.094	0.125	0.023	0.031	0.067	0.057	0.083	0.078	0.060	0.030
6	0.138	0.109	0.138	0.025	0.035	0.074	0.064	0.092	0.087	0.070	0.035
8	0.164	0.125	0.164	0.029	0.041	0.087	0.076	0.109	0.103	0.080	0.040
10	0.190	0.141	0.190	0.032	0.048	0.102	0.088	0.127	0.120	0.090	0.045
12	0.216	0.156	0.216	0.036	0.054	0.115	0.101	0.144	0.137	0.110	0.055
1/4	0.250	0.188	0.250	0.045	0.063	0.132	0.118	0.156	0.149	0.125	0.063
5/16	0.312	0.234	0.313	0.051	0.078	0.172	0.156	0.203	0.195	0.156	0.078
3/8	0.375	0.281	0.375	0.064	0.094	0.212	0.194	0.250	0.241	0.188	0.094
7/16	0.438	0.328	0.438	0.072	0.109	0.252	0.232	0.297	0.287	0.219	0.109
1/2	0.500	0.375	0.500	0.081	0.125	0.291	0.270	0.344	0.344	0.250	0.125
9/16	0.562	0.422	0.563	0.091	0.141	0.332	0.309	0.391	0.379	0.281	0.140
5/8	0.625	0.469	0.625	0.102	0.156	0.371	0.347	0.469	0.456	0.313	0.156
3/4	0.750	0.563	0.750	0.129	0.188	0.450	0.425	0.563	0.549	0.375	0.188

Cup points are standard with stock sizes of all set screws.

Round Head, Square Neck Carriage Bolts

Size	Body Diam. Max.	A Diameter of Head Min.	A Diameter of Head Basic		A Diameter of Head Max.	H Height of Head Min.		H Height of Head Max.	P Depth of Square Min.	P Depth of Square Max.	B Width of Square Min.	B Width of Square Max.
#10-24	.199	.438	7/16	.438	.469	3/32	.094	.114	.094	.125	.185	.199
1/4-20	.260	.563	9/16	.563	.594	1/8	.125	.145	.125	.156	.245	.260
5/16-18	.324	.688	11/16	.688	.719	5/32	.156	.176	.156	.187	.307	.324
3/8-16	.388	.782	13/16	.813	.844	3/16	.188	.208	.188	.219	.368	.388
7/16-14	.452	.907	15/16	.938	.969	7/32	.219	.239	.219	.250	.431	.452
1/2-13	.515	1.032	1 1/16	1.063	1.094	1/4	.250	.270	.250	.281	.492	.515
5/8-11	.642	1.219	1 5/16	1.313	1.344	5/16	.313	.344	.313	.344	.616	.642
3/4-10	.768	1.469	1 9/16	1.563	1.594	3/8	.375	.406	.375	.406	.741	.768
7/8-9	.895	1.719	1 13/16	1.813	1.844	7/16	.438	.469	.438	.469	.865	.895
1-8	1.022	1.969	2 1/16	2.063	2.094	1/2	.500	.531	.500	.531	.990	1.022

Lag Bolts (Screws)

Diameter of Bolt		Threads per Inch	Thread Dimensions P Pitch	Thread Dimensions B Flat at Root	Thread Dimensions T Depth of Thread	Thread Dimensions R Root Dia.	F Width Across Flats Max.	F Width Across Flats (Basic)	F Width Across Flats Min.	H Height Nom.	H Height Max.	H Height Min.
#10	.1900	11	.091	.039	.035	.120	9/32	.2810	.271	1/8	.140	.110
1/4	.2500	10	.100	.043	.039	.173	3/8	.3750	.362	11/64	.188	.156
5/16	.3125	9	.111	.048	.043	.227	1/2	.5000	.484	13/64	.220	.186
3/8	.3750	7	.143	.062	.055	.265	9/16	.5625	.544	1/4	.268	.232
7/16	.4375	7	.143	.062	.055	.328	5/8	.6250	.603	19/64	.316	.278
1/2	.5000	6	.167	.072	.064	.371	3/4	.7500	.725	21/32	.348	.308
5/8	.6250	5	.200	.086	.077	.471	15/16	.9375	.906	27/64	.444	.400
3/4	.7500	4 1/2	.222	.096	.085	.579	1 1/8	1.1250	1.088	1/2	.524	.476
7/8	.8750	4	.250	.108	.096	.683	1 5/16	1.3125	1.269	19/32	.620	.568
1	1.0000	3 1/2	.286	.123	.110	.780	1 1/2	1.5000	1.450	21/32	.684	.628
1 1/8	1.1250	3 1/4	.308	.133	.119	.887	1 11/16	1.6875	1.631	3/4	.780	.720
1 1/4	1.2500	3 1/4	.308	.133	.119	1.012	1 7/8	1.8750	1.812	27/32	.876	.812

Most lag bolts are supplied with A cut thread and full diameter unthreaded shank.

Table 17-7. *(Continued)*

Hexagonal Socket Head Cap Screws

	D Body Diameter[1]		A Head Diameter		H Head Height		S Head Side-Height			J Socket Width Across Flats	
Nom.	Max.	Min.	Max.	Min.	Max.	Min.	Nom.	Max.	Min.	Max.	Min.
2	0.0860	0.0840	0.140	0.136	0.086	0.083	0.079	0.081	0.078	0.0635	1/16
3	0.0990	0.0968	0.161	0.157	0.099	0.096	0.091	0.093	0.089	0.0791	5/64
4	0.1120	0.1096	0.183	0.178	0.112	0.109	0.103	0.105	0.101	0.0791	5/64
5	0.1250	0.1226	0.205	0.200	0.125	0.122	0.115	0.117	0.113	0.0947	3/32
6	0.1380	0.1353	0.226	0.221	0.138	0.134	0.127	0.129	0.125	0.0947	3/32
8	0.1640	0.1613	0.270	0.265	0.164	0.160	0.150	0.152	0.148	0.1270	1/8
10	0.1900	0.1867	5/16	0.306	0.190	0.185	0.174	0.176	0.172	0.1582	5/32
12	0.2160	0.2127	11/32	0.337	0.216	0.211	0.198	0.200	0.196	0.1582	5/32
1/4	0.2500	0.2464	3/8	0.367	1/4	0.244	0.229	0.232	0.226	0.1895	3/16
5/16	0.3125	0.3084	7/16	0.429	5/16	0.306	0.286	0.289	0.283	0.2207	7/32
3/8	0.3750	0.3705	9/16	0.553	3/8	0.368	0.344	0.347	0.341	0.3155	5/16
7/16	0.4375	0.4326	5/8	0.615	7/16	0.430	0.401	0.405	0.397	0.3155	5/16
1/2	0.5000	0.4948	3/4	0.739	1/2	0.492	0.458	0.462	0.454	0.3780	3/8
9/16	0.5625	0.5569	13/16	0.801	9/16	0.554	0.516	0.520	0.512	0.3780	3/8
5/8	0.6250	0.6191	7/8	0.863	5/8	0.616	0.573	0.577	0.569	0.5030	1/2
3/4	0 7500	0.7436	1	0.987	3/4	0.741	0.688	0.693	0.684	0.5655	9/16
7/8	0.8750	0.8680	1 1/8	1.111	7/8	0.865	0.802	0.807	0.797	0.5655	9/16
1	1.0000	0.9924	1 5/16	1.297	1	0.989	0.917	0.922	0.912	0.6290	5/8

[1]BODY DIAMETER (D) refers to the unthreaded portion of the screw and the maximum diameter conforms to the basic diameter or size of the screw.

Flat Head (F) Machine Screws, Tapping Screws, Wood Screws, Stove Bolts

		Standard Slotted									Phillips Recessed		
		A Head Diameter			E Height of Head		J Width of Slot		T Depth of Slot		D Diameter of Recess		C Depth of Recess
Nominal Size	Diameter of Screw Basic	Max.	Min.	Absol. Min.	Max.	Min.	Max.	Min.	Max.	Min.	Max.	Min.	Max.
2	.086	.172	.156	.147	.051	.040	.031	.023	.023	.015	.102	.089	.063
3	.099	.199	.181	.171	.059	.048	.035	.027	.027	.017	.107	.094	.068
4	.112	.225	.207	.195	.067	.055	.039	.031	.030	.020	.128	.115	.089
5	.125	.252	.232	.220	.075	.062	.043	.035	.034	.022	.154	.141	.086
6	.138	.279	.257	.244	.083	.069	.048	.039	.038	.024	.174	.161	.106
7	.151	.305	.283	.268	.091	.076	.048	.039	.041	.027	.182	.169	.114
8	.164	.332	.308	.292	.100	.084	.054	.045	.045	.029	.189	.176	.121
8											**.204**	**.191**	**.136**
9	.177	.358	.334	.316	.108	.091	.054	.045	.049	.032	.214	.201	.146
10	.190	.385	.359	.340	.116	.098	.060	.049	.053	.034	.204	.191	.136
10											**.258**	**.245**	**.146**
12	.216	.438	.410	.389	.132	.112	.067	.056	.060	.039	.268	.255	.156
12											**.283**	**.270**	**.171**
14	.242	.491	.461	.437	.148	.127	.075	.064	.068	.044	.283	.270	.171
1/4	.250	.507	.477	.452	.153	.131	.075	.064	.070	.046	.283	.270	.171
16	.268	.544	.512	.485	.164	.141	.075	.064	.075	.049	.303	.290	.191
18	.294	.597	.563	.534	.180	.155	.084	.072	.083	.054	.365	.352	.216
5/16	.312	.635	.600	.568	.191	.165	.084	.072	.088	.058	.365	.352	.216
3/8	.375	.762	.722	.685	.230	.200	.094	.081	.106	.070	.393	.380	.245
7/16	.438	.812	.771	.723	.223	.190	.094	.081	.103	.066	.409	.396	.261
1/2	.500	.875	.831	.775	.223	.186	.106	.091	.103	.065	.424	.411	.276

Edges of head may be rounded.
Dimensions shown in bold type apply to Wood Screws only.

Table 17-7. *(Continued)*

Round Head (R) Machine Screws, Tapping Screws, Wood Screws, Stove Bolts

		Standard Slotted								Phillips Recessed		
		A Head Diameter		E Height of Head		J Width of Slot		T Depth of Slot		D Diameter of Recess		C Depth of Recess
Nominal Size	Diameter of Screw Basic	Max.	Min.	Max.	Min.	Max.	Min.	Max.	Min.	Max.	Min.	Max.
2	.086	.162	.146	.069	.059	.031	.023	.048	.037	.100	.087	.053
2										**.114**	**.101**	**.064**
3	.099	.187	.169	.078	.067	.035	.027	.053	.040	.109	.096	.062
3										**.122**	**.109**	**.073**
4	.112	.211	.193	.086	.075	.039	.031	.058	.044	.118	.105	.072
4										**.130**	**.117**	**.083**
5	.125	.236	.217	.095	.083	.043	.035	.063	.047	.154	.141	.074
6	.138	.260	.240	.103	.091	.048	.039	.068	.051	.162	.149	.084
7	.151	.285	.264	.111	.099	.048	.039	.072	.055	.170	.157	.092
8	.164	.309	.287	.120	.107	.054	.045	.077	.058	.178	.165	.101
9	.177	.334	.311	.128	.115	.054	.045	.082	.062	....	....	....
10	.190	.359	.334	.137	.123	.060	.050	.087	.065	.195	.182	.119
12	.216	.408	.382	.153	.139	.067	.056	.096	.073	.249	.236	.125
14	.242	.457	.429	.170	.155	.075	.064	.106	.080	.265	.252	.142
1/4	.250	.472	.443	.175	.160	.075	.064	.109	.082	.268	.255	.147
16	.268	.506	.476	.187	.171	.075	.064	.115	.087	.281	.268	.159
18	.294	.555	.523	.204	.187	.084	.072	.125	.094	.329	.316	.167
5/16	.312	.590	.557	.216	.198	.084	.072	.132	.099	.308	.295	.187
3/8	.375	.708	.670	.256	.237	.094	.081	.155	.117	.387	.374	.228
7/16	.438	.750	.707	.328	.307	.094	.081	.196	.148	.402	.389	.241
1/2	.500	.813	.766	.355	.332	.106	.091	.211	.159	.416	.403	.256

Dimensions shown in bold type apply to Wood Screws only.

Fillister Head (P) Machine Screws, Tapping Screws

		Standard Slotted										Phillips Recessed		
		A Head Diameter		H Height of Head		J Width of Slot		T Depth of Slot		E Total Height of Head		D Diameter of Recess		C Depth of Recess
Nominal Size	Diameter of Screw Basic	Max.	Min.	Max.	Min.	Max.	Min.	Max.	Min.	Max.	Min.	Max.	Min.	Max.
2	.086	.140	.124	.062	.053	.031	.023	.037	.025	.083	.066	.104	.091	.059
3	.099	.161	.145	.070	.061	.035	.027	.043	.030	.095	.077	.112	.099	.068
4	.112	.183	.166	.079	.069	.039	.031	.048	.035	.107	.088	.122	.109	.078
5	.125	.205	.187	.088	.078	.043	.035	.054	.040	.120	.100	.148	.135	.067
6	.138	.226	.208	.096	.086	.048	.039	.060	.045	.132	.111	.166	.153	.091
8	.164	.270	.250	.113	.102	.054	.045	.071	.054	.156	.133	.182	.169	.108
10	.190	.313	.292	.130	.118	.060	.050	.083	.064	.180	.156	.199	.186	.124
12	.216	.357	.334	.148	.134	.067	.056	.094	.074	.205	.178	.259	.246	.141
1/4	.250	.414	.389	.170	.155	.075	.064	.109	.087	.237	.207	.281	.268	.161
5/16	.312	.518	.490	.211	.194	.084	.072	.137	.110	.295	.262	.322	.309	.235
3/8	.375	.622	.590	.253	.233	.094	.081	.164	.133	.355	.315	.393	.380	.233
7/16	.438	.625	.589	.265	.242	.094	.081	.170	.135	.368	.321	.413	.400	.259
1/2	.500	.750	.710	.297	.273	.106	.091	.190	.151	.412	.362	.435	.422	.280

Oval Head (O) Machine Screws, Tapping Screws, Wood Screws

Nominal Size	Diameter of Screw Basic	Standard Slotted: A Head Diameter Max.	A Head Diameter Min.	A Head Diameter Absolute Min.	H Height of Head Max.	H Height of Head Min.	J Width of Slot Max.	J Width of Slot Min.	T Depth of Slot Max.	T Depth of Slot Min.	E Total Height of Head Max.	E Total Height of Head Min.	Phillips Recessed: D Diameter of Recess Max.	D Diameter of Recess Min.	C Depth of Recess Max.
2	.086	.172	.156	.147	.051	.040	.031	.023	.045	.037	.080	.063	.112	.099	.069
3	.099	.199	.181	.171	.059	.048	.035	.027	.052	.043	.092	.073	.124	.111	.081
4	.112	.225	.207	.195	.067	.055	.039	.031	.059	.049	.104	.084	.136	.123	.094
5	.125	.252	.232	.220	.075	.062	.043	.035	.067	.055	.116	.095	.158	.145	.085
6	.138	.279	.257	.244	.083	.069	.048	.039	.074	.060	.128	.105	.178	.165	.105
7	.151	.305	.283	.268	.091	.076	.048	.039	.081	.066	.140	.116	.183	.170	.111
8	.164	.332	.308	.292	.100	.084	.054	.045	.088	.072	.152	.126	.192	.179	.119
8													**.205**	**.192**	**.131**
9	**.177**	**.358**	**.334**	**.316**	**.108**	**.091**	**.054**	**.045**	**.095**	**.078**	**.164**	**.137**	**.216**	**.203**	**.144**
10	.190	.385	.359	.340	.116	.098	.060	.050	.103	.084	.176	.148	.209	.196	.137
10													**.261**	**.248**	**.142**
12	.216	.438	.410	.389	.132	.112	.067	.056	.117	.096	.200	.169	.270	.257	.152
12													**.283**	**.270**	**.165**
14	.242	.491	.461	.437	.148	.127	.075	.064	.132	.108	.224	.190	.288	.275	.165
1/4	.250	.507	.477	.452	.153	.131	.075	.064	.136	.112	.232	.197	.290	.277	.173
16	.268	.544	.512	.485	.164	.141	.075	.064	.146	.120	.248	.212	.332	.319	.214
18	.294	.597	.563	.534	.180	.155	.084	.072	.160	.132	.272	.233	.381	.368	.226
5/16	.312	.635	.600	.568	.191	.165	.084	.072	.171	.141	.290	.249	.390	.377	.238
3/8	.375	.762	.722	.685	.230	.200	.094	.081	.206	.170	.347	.300	.410	.397	.257
7/16	.438	.812	.771	.723	.223	.190	.094	.081	.210	.174	.345	.295	.422	.409	.269
1/2	.500	.875	.831	.775	.223	.186	.106	.091	.216	.176	.354	.299	.437	.424	.283

Edges of head may be rounded.
Dimensions shown in bold type apply to Wood Screws only.

Table 17-7. (*Continued*)

Binding Head (B) Machine Screws

Standard Slotted

Nominal Size	Diameter of Screw Basic	A Head Diameter Max.	A Head Diameter Min.	E Total Height of Head Max.	E Total Height of Head Min.	F Height of Oval Max.	F Height of Oval Min.	J Width of Slot Max.	J Width of Slot Min.	T Depth of Slot Max.	T Depth of Slot Min.	U Diameter of Undercut Max.	U Diameter of Undercut Min.	X Depth of Undercut Max.	X Depth of Undercut Min.
2	.086	.181	.171	.050	.041	.018	.013	.031	.023	.030	.024	.141	.124	.010	.005
3	.099	.208	.197	.059	.048	.022	.016	.035	.027	.036	.029	.162	.143	.011	.006
4	.112	.235	.223	.068	.056	.025	.018	.039	.031	.042	.034	.184	.161	.012	.007
5	.125	.263	.249	.078	.064	.029	.021	.043	.035	.048	.039	.205	.180	.014	.009
6	.138	.290	.275	.087	.071	.032	.024	.048	.039	.053	.044	.226	.199	.015	.010
8	.164	.344	.326	.105	.087	.039	.029	.054	.045	.065	.054	.269	.236	.017	.012
10	.190	.399	.378	.123	.102	.045	.034	.060	.050	.077	.064	.312	.274	.020	.015
12	.216	.454	.430	.141	.117	.052	.039	.067	.056	.089	.074	.354	.311	.023	.018
1/4	.250	.513	.488	.165	.138	.061	.046	.075	.064	.105	.088	.410	.360	.026	.021
5/16	.312	.641	.609	.209	.174	.077	.059	.084	.072	.134	.112	.513	.450	.032	.027
3/8	.375	.769	.731	.253	.211	.094	.071	.094	.081	.163	.136	.615	.540	.039	.034

Truss Head (T) Machine Screws, Tapping Screws

		Standard Slotted								Phillips Recessed		
Nominal Size	Diameter of Screw Basic	A Head Diameter Max.	A Head Diameter Min.	E Height of Head Max.	E Height of Head Min.	J Width of Slot Max.	J Width of Slot Min.	T Depth of Slot Max.	T Depth of Slot Min.	D Diameter of Recess Max.	D Diameter of Recess Min.	C Depth of Recess Max.
2	.086	.194	.180	.053	.044	.031	.023	.031	.022	.104	.091	.059
3	.099	.226	.211	.061	.051	.035	.027	.036	.026	.110	.097	.066
4	.112	.257	.241	.069	.059	.039	.031	.040	.030	.112	.099	.069
5	.125	.289	.272	.078	.066	.043	.035	.045	.034	.128	.115	.085
6	.138	.321	.303	.086	.074	.048	.039	.050	.037	.158	.145	.084
7	.151	.352	.333	.094	.081	.048	.039	.054	.041	.165	.152	.091
8	.164	.384	.364	.102	.088	.054	.045	.058	.045	.173	.160	.099
10	.190	.448	.425	.118	.103	.060	.050	.068	.053	.188	.175	.115
12	.216	.511	.487	.134	.118	.067	.056	.077	.061	.248	.235	.128
1/4	.250	.573	.546	.150	.133	.075	.064	.087	.070	.263	.250	.143
5/16	.312	.698	.666	.183	.162	.084	.072	.106	.085	.352	.339	.193
3/8	.375	.823	.787	.215	.191	.094	.081	.124	.100	.388	.375	.226

Pan Head (D) Machine Screws, Tapping Screws

Nominal Size	Diameter of Screw Basic	Standard Slotted: A Head Diameter Max.	A Head Diameter Min.	E Height of Slotted Head Max.	E Height of Slotted Head Min.	J Width of Slot Max.	J Width of Slot Min.	T Depth of Slot Max.	T Depth of Slot Min.	Phillips Recessed: R Height of Recessed Head Max.	R Height of Recessed Head Min.	D Diameter of Recess Max.	D Diameter of Recess Min.	C Depth of Recess Max.
2	.086	.167	.155	.053	.045	.031	.023	.031	.022	.062	.053	.104	.091	.059
3	.099	.193	.180	.060	.051	.035	.027	.036	.026	.071	.062	.112	.099	.068
4	.112	.219	.205	.068	.058	.039	.031	.040	.030	.080	.070	.122	.109	.078
5	.125	.245	.231	.075	.065	.043	.035	.045	.034	.089	.079	.158	.145	.083
6	.138	.270	.256	.082	.072	.048	.039	.050	.037	.097	.087	.166	.153	.091
7	.151	.296	.281	.089	.079	.048	.039	.054	.041	.106	.096	.176	.163	.100
8	.164	.322	.306	.096	.085	.054	.045	.058	.045	.115	.105	.182	.169	.108
10	.190	.373	.357	.110	.099	.060	.050	.068	.053	.133	.122	.199	.186	.124
12	.216	.425	.407	.125	.112	.067	.056	.077	.061	.151	.139	.259	.246	.141
1/4	.250	.492	.473	.144	.130	.075	.064	.087	.070	.175	.162	.281	.268	.161
5/16	.312	.615	.594	.178	.162	.084	.072	.106	.085	.218	.203	.350	.337	.193
3/8	.375	.740	.716	.212	.195	.094	.081	.130	.113	.261	.244	.393	.380	.233

Set Screw Dimensions—Square Head Cup Point

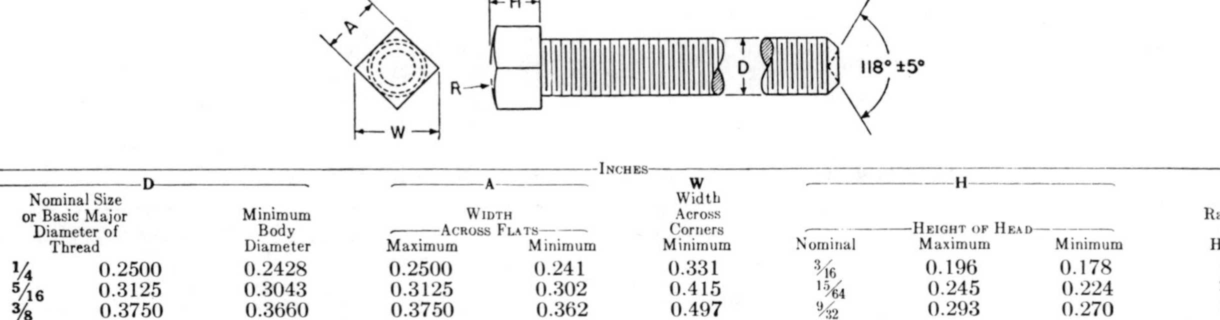

D: Nominal Size or Basic Major Diameter of Thread (Inches)		D: Minimum Body Diameter	A: Width Across Flats Maximum	A: Width Across Flats Minimum	W: Width Across Corners Minimum	H: Height of Head Nominal	H: Height of Head Maximum	H: Height of Head Minimum	R: Radius of Head
1/4	0.2500	0.2428	0.2500	0.241	0.331	3/16	0.196	0.178	5/8
5/16	0.3125	0.3043	0.3125	0.302	0.415	15/64	0.245	0.224	25/32
3/8	0.3750	0.3660	0.3750	0.362	0.497	9/32	0.293	0.270	15/16

Table 17-7. *(Continued)*

Set Screw Dimensions—Square Head Cup Point *(Continued)*

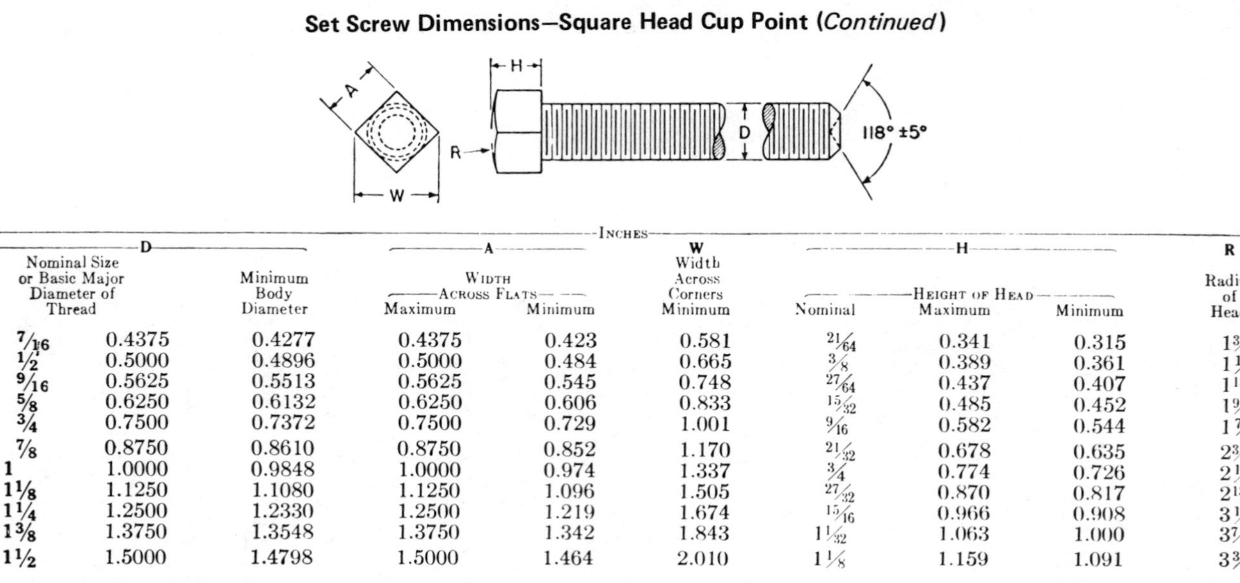

D			A		W	H			R
Nominal Size or Basic Major Diameter of Thread		Minimum Body Diameter	Width Across Flats		Width Across Corners	Height of Head			Radius of Head
			Maximum	Minimum	Minimum	Nominal	Maximum	Minimum	
7/16	0.4375	0.4277	0.4375	0.423	0.581	21/64	0.341	0.315	1 3/32
1/2	0.5000	0.4896	0.5000	0.484	0.665	3/8	0.389	0.361	1 1/4
9/16	0.5625	0.5513	0.5625	0.545	0.748	27/64	0.437	0.407	1 13/32
5/8	0.6250	0.6132	0.6250	0.606	0.833	15/32	0.485	0.452	1 9/16
3/4	0.7500	0.7372	0.7500	0.729	1.001	9/16	0.582	0.544	1 7/8
7/8	0.8750	0.8610	0.8750	0.852	1.170	21/32	0.678	0.635	2 3/16
1	1.0000	0.9848	1.0000	0.974	1.337	3/4	0.774	0.726	2 1/2
1 1/8	1.1250	1.1080	1.1250	1.096	1.505	27/32	0.870	0.817	2 13/16
1 1/4	1.2500	1.2330	1.2500	1.219	1.674	15/16	0.966	0.908	3 1/8
1 3/8	1.3750	1.3548	1.3750	1.342	1.843	1 1/32	1.063	1.000	3 7/16
1 1/2	1.5000	1.4798	1.5000	1.464	2.010	1 1/8	1.159	1.091	3 3/4

All dimensions in inches.

Table 17-8. Hexagon and Square Machine Screw and Stove Bolt Nuts.

Hexagon and Square Machine Screw and Stove Bolt Nuts

Nominal Size or Basic Major Diameter of Thread		F Width Across Flats			G Width Across Corners				H Thickness		
					Square		Hex.				
		Max.	(Basic)	Min.	Max.	Min.	Max.	Min.	Nom.	Max.	Min.
0	.0600	5/32	.1562	.150	.221	.206	.180	.171	3/64	.050	.043
1	.0730	5/32	.1562	.150	.221	.206	.180	.171	3/64	.050	.043
2	.0860	3/16	.1875	.180	.265	.247	.217	.205	1/16	.066	.057
3	.0990	3/16	.1875	.180	.265	.247	.217	.205	1/16	.066	.057
4	.1120	1/4	.2500	.241	.354	.331	.289	.275	3/32	.098	.087
5	.1250	5/16	.3125	.302	.442	.415	.361	.344	7/64	.114	.102
6	.1380	5/16	.3125	.302	.442	.415	.361	.344	7/64	.114	.102
8	.1640	11/32	.3438	.332	.486	.456	.397	.378	1/8	.130	.117
10	.1900	3/8	.3750	.362	.530	.497	.433	.413	1/8	.130	.117
12	.2160	7/16	.4375	.423	.619	.581	.505	.482	5/32	.161	.148
1/4	.2500	7/16	.4375	.423	.619	.581	.505	.482	3/16	.193	.178
5/16	.3125	9/16	.5625	.545	.795	.748	.650	.621	7/32	.225	.208
3/8	.3750	5/8	.6250	.607	.884	.833	.722	.692	1/4	.257	.239

Finished Hexagon Castellated (Castle) Nuts

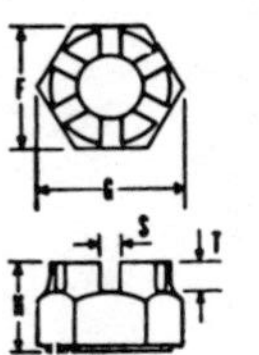

Nominal Size or Basic Major Diameter of Thread		F Width Across Flats			G Width Across Corners		H Thickness			Nominal Height of Flats	S Slot Width	T Slot Depth	Radius of Fillet (±.010)	Dia. of Cylindrical Part
		Max.	(Basic)	Min.	Max.	Min.	Nom.	Max.	Min.					
1/4	.2500	7/16	.4375	.428	.505	.488	9/32	.288	.274	3/16	.078	.094	3/32	.371
5/16	.3125	1/2	.5000	.489	.577	.557	21/64	.336	.320	15/64	.094	.094	3/32	.425
3/8	.3750	9/16	.5265	.551	.650	.628	13/32	.415	.398	9/32	.125	.125	3/32	.478
7/16	.4375	11/16	.6875	.675	.794	.768	29/64	.463	.444	19/64	.125	.156	3/32	.582
1/2	.5000	3/4	.7500	.736	.866	.840	9/16	.573	.552	13/32	.156	.156	1/8	.637
9/16	.5625	7/8	.8750	.861	1.010	.982	39/64	.621	.598	27/64	.156	.188	5/32	.744
5/8	.6250	15/16	.9375	.922	1.083	1.051	23/32	.731	.706	1/2	.188	.219	5/32	.797
3/4	.7500	1 1/8	1.1250	1.088	1.299	1.240	13/16	.827	.798	9/16	.188	.250	3/16	.941
7/8	.8750	1 5/16	1.3125	1.269	1.516	1.447	29/32	.922	.890	21/32	.188	.250	3/16	1.097
1	1.0000	1 1/2	1.5000	1.450	1.732	1.653	1	1.018	.982	23/32	.250	.281	3/16	1.254
1 1/8	1.1250	1 11/16	1.6875	1.631	1.949	1.859	1 5/32	1.176	1.136	13/16	.250	.344	1/4	1.411
1 1/4	1.2500	1 7/8	1.8750	1.812	2.165	2.066	1 1/4	1.272	1.228	7/8	.312	.375	1/4	1.570
1 3/8	1.3750	2 1/16	2.0625	1.994	2.382	2.273	1 3/8	1.399	1.351	1	.312	.375	1/4	1.726
1 1/2	1.5000	2 1/4	2.2500	2.175	2.598	2.480	1 1/2	1.526	1.474	1 1/16	.375	.438	1/4	1.881

Table 17-8. *(Continued)*

Finished Hexagon Nuts (Full and Jam)

Nominal Size or Basic Major Diameter of Thread		F Width Across Flats Max.	F (Basic)	F Min.	G Width Across Corners Max.	G Min.	H Thickness Full Nuts Nom.	H Max.	H Min.	h Thickness Jam Nuts Nom.	h Max.	h Min.
1/4	.2500	7/16	.4375	.428	.505	.488	7/32	.226	.212	5/32	.163	.150
5/16	.3125	1/2	.5000	.489	.577	.557	17/64	.273	.258	3/16	.195	.180
3/8	.3750	9/16	.5625	.551	.650	.628	21/64	.337	.320	7/32	.227	.210
7/16	.4375	11/16	.6875	.675	.794	.768	3/8	.385	.365	1/4	.260	.240
1/2	.5000	3/4	.7500	.736	.866	.840	7/16	.448	.427	5/16	.323	.302
9/16	.5625	7/8	.8750	.861	1.010	.982	31/64	.496	.473	5/16	.324	.301
5/8	.6250	15/16	.9375	.922	1.083	1.051	35/64	.559	.535	3/8	.387	.363
3/4	.7500	1 1/8	1.1250	1.088	1.299	1.240	41/64	.665	.617	27/64	.446	.398
7/8	.8750	1 5/16	1.3125	1.269	1.516	1.447	3/4	.776	.724	31/64	.510	.458
1	1.0000	1 1/2	1.5000	1.450	1.732	1.653	55/64	.887	.831	35/64	.575	.519
1 1/8	1.1250	1 11/16	1.6875	1.631	1.949	1.859	31/32	.999	.939	39/64	.639	.579
1 1/4	1.2500	1 7/8	1.8750	1.812	2.165	2.066	1 1/16	1.094	1.030	23/32	.751	.687
1 3/8	1.3750	2 1/16	2.0625	1.994	2.382	2.273	1 11/64	1.206	1.138	25/32	.815	.747
1 1/2	1.5000	2 1/4	2.2500	2.175	2.598	2.480	1 9/32	1.317	1.245	27/32	.880	.808

Heavy Semi-Finished Hexagon Nuts (Full and Jam)

Nominal Size or Basic Major Diameter of Thread		F Width Across Flats Max.	F (Basic)	F Min.	G Width Across Corners Max.	G Min.	H Thickness (Heavy Nuts) Nom.	H Max.	H Min.	h Thickness (Heavy Jam Nuts) Nom.	h Max.	h Min.
1/4	.2500	1/2	.5000	.488	.577	.556	15/64	.250	.218	11/64	.188	.156
5/16	.3125	9/16	.5625	.546	.650	.622	19/64	.314	.280	13/64	.220	.186
3/8	.3750	11/16	.6875	.669	.794	.763	23/64	.377	.341	15/64	.252	.216
7/16	.4375	3/4	.7500	.728	.866	.830	27/64	.441	.403	17/64	.285	.247
1/2	.5000	7/8	.8750	.850	1.010	.969	31/64	.504	.464	19/64	.317	.277
9/16	.5625	15/16	.9375	.909	1.083	1.037	35/64	.568	.526	21/64	.349	.307
5/8	.6250	1 1/16	1.0625	1.031	1.227	1.175	39/64	.631	.587	23/64	.381	.337
3/4	.7500	1 1/4	1.2500	1.212	1.443	1.382	47/64	.758	.710	27/64	.446	.398
7/8	.8750	1 7/16	1.4375	1.394	1.660	1.589	55/64	.885	.833	31/64	.510	.458
1	1.0000	1 5/8	1.6250	1.575	1.876	1.796	63/64	1.012	.956	35/64	.575	.519
1 1/8	1.1250	1 13/16	1.8125	1.756	2.093	2.002	1 7/64	1.139	1.079	39/64	.639	.579
1 1/4	1.2500	2	2.0000	1.938	2.309	2.209	1 7/32	1.251	1.187	23/32	.751	.687
1 3/8	1.3750	2 3/16	2.1875	2.119	2.526	2.416	1 11/32	1.378	1.310	25/32	.815	.747
1 1/2	1.5000	2 3/8	2.3750	2.300	2.742	2.622	1 15/32	1.505	1.433	27/32	.830	.808
1 5/8	1.6250	2 9/16	2.5625	2.481	2.959	2.828	1 19/32	1.632	1.556	29/32	.944	.868
1 3/4	1.7500	2 3/4	2.7500	2.662	3.175	3.035	1 23/32	1.759	1.679	31/32	1.009	.929
1 7/8	1.8750	2 15/16	2.9375	2.844	3.392	3.242	1 27/32	1.886	1.802	1 1/32	1.073	.989
2	2.0000	3 1/8	3.1250	3.025	3.608	3.449	1 31/32	2.013	1.925	1 3/32	1.138	1.050
2 1/4	2.2500	3 1/2	3.5000	3.388	4.041	3.862	2 13/64	2.251	2.155	1 13/64	1.251	1.155
2 1/2	2.5000	3 7/8	3.8750	3.750	4.474	4.275	2 29/64	2.505	2.401	1 29/64	1.505	1.401

Table 17-9. Round-Head Small Solid Rivets.

Round (Button) Head Small Solid Rivets

D Diameter of Body Nominal		D Max.	D Min.	A Diameter of Head Max.	A Diameter of Head Min.	H Height of Head Max.	H Height of Head Min.	r Radius of Head Approx.	Length Tolerance Plus	Length Tolerance Minus
3/32	.094	.096	.090	.182	.162	.077	.065	.084	.016	.016
1/8	.125	.127	.121	.235	.215	.100	.088	.111	.016	.016
5/32	.156	.158	.152	.290	.268	.124	.110	.138	.016	.016
3/16	.188	.191	.182	.348	.322	.147	.133	.166	.016	.016
7/32	.219	.222	.213	.405	.379	.172	.158	.195	.016	.016
1/4	.250	.253	.244	.460	.430	.196	.180	.221	.016	.016
9/32	.281	.285	.273	.518	.484	.220	.202	.249	.016	.016
5/16	.313	.317	.305	.572	.538	.243	.225	.276	.016	.016
11/32	.344	.348	.336	.630	.592	.267	.247	.304	.016	.016
3/8	.375	.380	.365	.684	.646	.291	.271	.332	.016	.016
7/16	.438	.443	.428	.798	.754	.339	.317	.387	.016	.016

Countersunk Head Small Solid Rivets

D Diameter of Body Nominal		D Max.	D Min.	A Diameter of Head Max.	A Diameter of Head Min.	H Height of Head	Length Tolerance Plus	Length Tolerance Minus
3/32	.094	.096	.090	.176	.171	.040	.016	.016
1/8	.125	.127	.121	.235	.227	.053	.016	.016
5/32	.156	.158	.152	.293	.284	.066	.016	.016
3/16	.188	.191	.182	.351	.340	.079	.016	.016
7/32	.219	.222	.213	.413	.400	.094	.016	.016
1/4	.250	.253	.244	.469	.455	.106	.016	.016
9/32	.281	.285	.273	.528	.511	.119	.016	.016
5/16	.313	.317	.305	.588	.569	.133	.016	.016
11/32	.344	.348	.336	.646	.626	.146	.016	.016
3/8	.375	.380	.365	.704	.682	.159	.016	.016
7/16	.438	.443	.428	.823	.797	.186	.016	.016

Table 17-9. *(Continued)*

Flat Head Small Solid Rivets

	D Diameter of Body			A Diameter of Head		H Height of Head		Length Tolerance	
	Nominal	Max.	Min.	Max.	Min.	Max.	Min.	Plus	Minus
3/32	.094	.096	.090	.200	.180	.038	.026	.016	.016
1/8	.125	.127	.121	.260	.240	.048	.036	.016	.016
5/32	.156	.158	.152	.323	.301	.059	.045	.016	.016
3/16	.188	.191	.182	.387	.361	.069	.055	.016	.016
7/32	.219	.222	.213	.453	.427	.080	.065	.016	.016
1/4	.250	.253	.244	.515	.485	.091	.075	.016	.016
9/32	.281	.285	.273	.579	.545	.103	.085	.016	.016
5/16	.313	.317	.305	.641	.607	.113	.095	.016	.016
11/32	.344	.348	.336	.705	.667	.124	.104	.016	.016
3/8	.375	.380	.365	.769	.731	.135	.115	.016	.016
7/16	.438	.443	.428	.896	.852	.157	.135	.016	.016

Table 17-10. Basic Thread Dimensions.

American & Unified Std. · Sharp V · Whitworth · Std. Pipe Thrd. · British Assoc. Std. · Brit. Std. Pipe · Löwenherz · Buttress · French & International · Square · Acme

$$P\ (\text{Pitch}) = \frac{1}{\text{No. threads per inch}}$$

$$D\ (\text{Depth}) = .649519\ P = \frac{.649519}{n}$$

$$H = .866025\ P$$

$$\frac{H}{8} = .108253\ P$$

$$F\ (\text{flat}) = .125\ P = \frac{P}{8}$$

n = No. of threads per in.

Size	Major Diam. Inches	Pitch Diam. Inches	Minor Diam. Inches
0-80	.0600	.0519	.0438
1-56	.0730	.0614	.0498
64	.0730	.0629	.0527
72	.0730	.0640	.0550
2-56	.0860	.0744	.0628
64	.0860	.0759	.0657
3-48	.0990	.0855	.0719
56	.0990	.0874	.0758
4-32	.1120	.0917	.0714
36	.1120	.0940	.0759
40	.1120	.0958	.0795
48	.1120	.0985	.0849
5-36	.1250	.1070	.0889
40	.1250	.1088	.0925

Nominal Size	*Basic Major Diam. Inches	*Basic Pitch Diam. Inches	Basic Minor Diam. Inches
3/16-24	.1875	.1604	.1334
32	.1875	.1672	.1469
7/32-24	.2188	.1917	.1646
32	.2188	.1985	.1782
1/4-20	.2500	.2175	.1850
24	.2500	.2229	.1959
28	.2500	.2268	.2036
32	.2500	.2297	.2094
5/16-18	.3125	.2764	.2403
20	.3125	.2800	.2476
24	.3125	.2854	.2584
32	.3125	.2922	.2719
3/8-16	.3750	.3344	.2938
20	.3750	.3425	.3100

Nominal Size	*Basic Major Diam. Inches	*Basic Pitch Diam. Inches	Basic Minor Diam. Inches
7/8-12	.8750	.8209	.7668
14	.8750	.8286	.7822
16	.8750	.8344	.7938
18	.8750	.8389	.8028
20	.8750	.8425	.8100
15/16-9	.9375	.8654	.7932
12	.9375	.8834	.8293
16	.9375	.8969	.8563
20	.9375	.9050	.8725
1-8	1.0000	.9188	.8376
12†	1.0000	.9459	.8918
14†	1.0000	.9536	.9072
16	1.0000	.9594	.9188
20	1.0000	.9675	.9350

Table 17.10. (Continued)

Nominal Size	*Basic Major Diam. Inches	*Basic Pitch Diam. Inches	Basic Minor Diam. Inches
44	.1250	.1102	.0955
6-32	.1380	.1177	.0974
36	.1380	.1200	.1019
40	.1380	.1218	.1055
8-30	.1640	.1423	.1207
32	.1640	.1437	.1234
36	.1640	.1460	.1279
40	.1640	.1478	.1315
10-24	.1900	.1629	.1359
28	.1900	.1668	.1436
30	.1900	.1684	.1467
32	.1900	.1697	.1494
12-24	.2160	.1889	.1619
28	.2160	.1928	.1696
32	.2160	.1957	.1754
14-20	.2420	.2095	.1770
24	.2420	.2149	.1879
1/16-64	.0625	.0524	.0422
3/32-48	.0938	.0803	.0667
1/8-40	.1250	.1088	.0925
5/32-32	.1563	.1360	.1157
36	.1563	.1382	.1202

Nominal Size	*Basic Major Diam. Inches	*Basic Pitch Diam. Inches	Basic Minor Diam. Inches
24	.3750	.3479	.3209
32	.3750	.3547	.3344
7/16-14	.4375	.3911	.3447
20	.4375	.4050	.3726
24	.4375	.4104	.3834
28	.4375	.4143	.3911
1/2-12	.5000	.4459	.3918
13	.5000	.4500	.4001
20	.5000	.4675	.4351
24	.5000	.4729	.4459
28	.5000	.4768	.4536
9/16-12	.5625	.5084	.4542
18	.5625	.5624	.4903
24	.5625	.5354	.5084
5/8-11	.6250	.5660	.5069
12	.6250	.5709	.5168
18	.6250	.5889	.5528
24	.6250	.5979	.5709
11/16-11	.6875	.6285	.5694
12	.6875	.6334	.5793
16	.6875	.6469	.6063
24	.6875	.6604	.6334
3/4-10	.7500	.6850	.6201
12	.7500	.6959	.6418
16	.7500	.7094	.6688
20	.7500	.7175	.6850
13/16-10	.8125	.7476	.6826
12	.8125	.7584	.7042
16	.8125	.7719	.7313
20	.8125	.7800	.7475
7/8- 9	.8750	.8028	.7307

Nominal Size	*Basic Major Diam. Inches	*Basic Pitch Diam. Inches	Basic Minor Diam. Inches
1 1/16 -12	1.0625	1.0084	.9543
16	1.0625	1.0219	.9813
18	1.0625	1.0264	.9903
1 1/8 - 7	1.1250	1.0322	.9394
8	1.1250	1.0438	.9626
12	1.1250	1.0709	1.0168
16	1.1250	1.0844	1.0438
18	1.1250	1.0889	1.0528
1 3/16 -12	1.1875	1.1334	1.0793
16	1.1875	1.1469	1.1063
18	1.1875	1.1514	1.1153
1 1/4 - 7	1.2500	1.1572	1.0644
8	1.2500	1.1688	1.0876
12	1.2500	1.1959	1.1418
16	1.2500	1.2094	1.1688
18	1.2500	1.2139	1.1778
1 5/16 -12	1.3125	1.2584	1.2043
16	1.3125	1.2719	1.2313
18	1.3125	1.2764	1.2403
1 3/8 - 6	1.3750	1.2667	1.1585
8	1.3750	1.2938	1.2126
12	1.3750	1.3209	1.2668
16	1.3750	1.3344	1.2938
18	1.3750	1.3389	1.3028
1 7/16 -12	1.4375	1.3834	1.3293
16	1.4375	1.3969	1.3563
18	1.4375	1.4014	1.3653
1 1/2 - 6	1.5000	1.3917	1.2835
8	1.5000	1.4188	1.3376
12	1.5000	1.4459	1.3918
16	1.5000	1.4594	1.4188
18	1.5000	1.4639	1.4278

†1″-12 is the Unified Thread for all classes of fit under American Standard ASA B1.1-1949, Second Edition. 1″-14 is the former American National Fine Thread for all classes of fit except 1A.

*These figures are correct for Unified Thread Form as computed in American Standard Publication ASA B1.1-1949.

Table 17-11. Ultimate Strength in Pounds of Various Sizes and Grades of Bolts.

Material Strength in Pounds per Square Inch		55,000	60,000	65,000	70,000	80,000	90,000	110,000	120,000	130,000	150,000	170,000
								Medium Carbon Steel Heat Treated		Alloy Steels Heat Treated		
Bolt Size	Stress* Area Square Inches	Low Carbon Steel Naval Bronze	Aluminum 24S-T4	Cold Rolled Low Carbon Steel	Yellow Brass	Medium Carbon Steel Low Silicon Bronze Monel	Medium Carbon Steel Type 304 Stainless					
10–24	.0175	960	1,050	1,140	1,220	1,400	1,570	1,920	2,100	2,270	2,620	2,970
10–32	.0200	1,100	1,200	1,300	1,400	1,600	1,800	2,200	2,400	2,600	3,000	3,400
1/4–20	.0318	1,750	1,910	2,070	2,230	2,540	2,860	3,500	3,820	4,130	4,770	5,410
1/4–28	.0364	2,000	2,180	2,370	2,550	2,910	3,280	4,000	4,370	4,730	5,460	6,190
5/16–18	.0524	2,880	3,140	3,410	3,670	4,190	4,720	5,760	6,290	6,810	7,860	8,910
5/16–24	.0581	3,190	3,490	3,780	4,070	4,650	5,230	6,390	6,970	7,550	8,710	9,880
3/8–16	.0775	4,260	4,650	5,040	5,420	6,200	6,970	8,520	9,300	10,100	11,600	13,200
3/8–24	.0878	4,830	5,270	5,710	6,150	7,020	7,900	9,660	10,500	11,400	13,200	14,900
7/16–14	.1063	5,850	6,380	6,910	7,440	8,500	9,570	11,700	12,800	13,800	15,900	18,100
7/16–20	.1187	6,530	7,120	7,710	8,310	9,500	10,700	13,100	14,200	15,400	17,800	20,200
1/2–13	.1419	7,800	8,510	9,220	9,930	11,300	12,800	15,600	17,000	18,400	21,300	24,100
1/2–20	.1599	8,790	9,590	10,400	11,200	12,800	14,400	17,600	19,200	20,800	24,000	27,200
9/16–12	.1819	10,000	10,900	11,800	12,700	14,500	16,400	20,000	21,800	23,600	27,300	30,900
9/16–18	.2029	11,200	12,200	13,200	14,200	16,200	18,300	22,300	24,300	26,400	30,400	34,500
5/8–11	.2260	12,400	13,600	14,700	15,800	18,100	20,300	24,900	27,100	29,400	33,900	38,400
5/8–18	.2559	14,100	15,300	16,600	17,900	20,500	23,000	28,100	30,700	33,300	38,400	43,500
3/4–10	.3344	18,400	20,100	21,700	23,400	26,700	30,100	36,800	40,100	43,500	50,200	56,800
3/4–16	.3729	20,500	22,400	24,200	26,100	29,800	33,600	41,000	44,700	48,500	55,900	63,400
7/8– 9	.4617	25,400	27,700	30,000	32,300	36,900	41,600	50,800	55,400	60,000	69,300	78,500
7/8–14	.5095	28,000	30,600	33,100	35,700	40,800	45,800	56,000	61,100	66,200	76,400	86,600
1– 8	.6057	33,300	36,300	39,400	42,400	48,500	54,500	66,600	72,700	78,700	90,800	103,000
1–14	.6799	37,400	40,800	44,200	47,600	54,400	61,200	74,800	81,600	88,400	102,000	115,600
1 1/8– 7	.7632	42,000	45,800	49,600	53,400	61,100	68,700	84,000	91,600	99,200	114,500	129,700
1 1/8–12	.8557	47,100	51,300	55,600	59,900	68,500	77,000	94,100	102,700	111,200	128,400	145,500
1 1/4– 7	.9691	53,300	58,100	63,000	67,800	77,500	87,200	106,600	116,300	126,000	145,400	164,700
1 1/4–12	1.0729	59,000	64,400	69,700	75,100	85,800	96,600	118,000	128,700	139,500	160,900	182,400

*Stress area is calculated as the area of the circle whose diameter is the mean between the root and pitch diameters. This closely approximates the actual stress condition.

Table 17-12. Suggested Torquing Values.[a]

Bolt Size	18-8 St. St.	Brass	Silicon Bronze	Aluminum 2024-T4	316 St. St.	Monel	Nylon*
	Values are stated in Inch Pounds						
2-56	2.5	2.0	2.3	1.4	2.6	2.5	.44
2-64	3.0	2.5	2.8	1.7	3.2	3.1	
3-48	3.9	3.2	3.6	2.1	4.0	4.0	
3-56	4.4	3.6	4.1	2.4	4.6	4.5	
4-40	5.2	4.3	4.8	2.9	5.5	5.3	1.19
4-48	6.6	5.4	6.1	3.6	6.9	6.7	
5-40	7.7	6.3	7.1	4.2	8.1	7.8	
5-44	9.4	7.7	8.7	5.1	9.8	9.6	
6-32	9.6	7.9	8.9	5.3	10.1	9.8	2.14
6-40	12.1	9.9	11.2	6.6	12.7	12.3	
8-32	19.8	16.2	18.4	10.8	20.7	20.2	4.3
8-36	22.0	18.0	20.4	12.0	23.0	22.4	
10-24	22.8	18.6	21.2	13.8	23.8	25.9	6.61
10-32	31.7	25.9	29.3	19.2	33.1	34.9	8.2
1/4"-20	75.2	61.5	68.8	45.6	78.8	85.3	16.0
1/4"-28	94.0	77.0	87.0	57.0	99.0	106.0	20.8
5/16"-18	132	107	123	80	138	149	34.9
5/16"-24	142	116	131	86	147	160	
3/8"-16	236	192	219	143	247	266	
3/8"-24	259	212	240	157	271	294	
7/16"-14	376	317	349	228	393	427	
7/16"-20	400	327	371	242	418	451	
1/2"-13	517	422	480	313	542	584	
1/2"-20	541	443	502	328	565	613	
9/16"-12	682	558	632	413	713	774	
9/16"-18	752	615	697	456	787	855	
5/8"-11	1110	907	1030	715	1160	1330	
5/8"-18	1244	1016	1154	798	1301	1482	
3/4"-10	1530	1249	1416	980	1582	1832	
3/4"-16	1490	1220	1382	958	1558	1790	
7/8"-9	2328	1905	2140	1495	2430	2775	
7/8"-14	2318	1895	2130	1490	2420	2755	
1"-8	3440	2815	3185	2205	3595	4130	
1"-14	3110	2545	2885	1995	3250	3730	
	Values are stated in Foot Pounds						
1-1/8"-7	413	337	383	265	432	499	
1-1/8"-12	390	318	361	251	408	470	
1-1/4"-7	523	428	485	336	546	627	
1-1/4"-12	480	394	447	308	504	575	
1-1/2"-6	888	727	822	570	930	1064	
1-1/2"-12	703	575	651	450	732	840	

[a]Values up to 1.00 in. in diameter are in inch-pounds. Values over 1.00 in. in diameter are in foot-pounds. Nylon values are breaking torque.

Table 17-13. Electrical Units.

The electrical units are as follows:

Volt–The unit of electrical motive force. Force required to send one ampere of current through one ohm of resistance.

Ohm–Unit of resistance. The resistance offered to the passage of one ampere, when impelled by one volt.

Ampere–Unit of current. The current which one volt can send through a resistance of one ohm.

Coulomb–Unit of quantity. Quantity of current which, propelled by one volt, would pass through one ohm in one second.

Farad–Unit of capacity. A conductor or condenser which will hold one coulomb under the pressure of one volt.

Joule–Unit of work. The work done by one watt in one second.

Watt–The unit of electrical energy, and the product of one ampere and volt. That is, one ampere of current flowing under a pressure of one volt, gives one watt of energy.

One electrical horsepower is equal to 746 watts.

One kilowatt is equal to 1000 watts.

Ohm's Law for Direct or Single Phase Noninductive Alternating Current

Let

I = ampere = unit of current strength or rate of flow.
E = volt = unit of electro-motive force or electric pressure.
R = ohm = unit of resistance to flow of current.
W = watt = unit of power.

Then

$$I = \frac{E}{R} \qquad E = IR \qquad R = \frac{E}{I}$$

$$W = IE = \frac{E^2}{R} = I^2R$$

1000W = 1 K. W. or kilowatt; this is the usual unit of measure of electric power. 1 Kilowatt hour or K. W. hour is the work done by one Kilowatt in one hour.

For alternating current circuits, the following rules are useful for finding the power of a polyphase circuit.

Let $P.\ F.$ = Power Factor.

3 Phase Alternating Current

$$\text{K. W.} = \frac{1.73 \times E \times I \times P.F.}{1000}$$

Table 17-13. (*Continued*)

2 Phase Alternating Current

$$\text{K. W.} = \frac{2 \times I \times P.F.}{1000}$$

1 Phase Alternating Current

$$\text{K. W.} = \frac{E \times I \times P.F.}{1000}$$

1 Horsepower, H. P.	746 watts 33000 foot-pounds per minute[a] 2545 heat units per hour, Btu[b] 2.64 lb water evaporated at 212°F.

[a] 1 foot-pound = raising one pound one foot.
[b] 1 Btu (British thermal unit) = heat required to raise the temperature of 1 lb of pure water 1° F.

BOLT AND CAP SCREW HEADS

Hexagon Head (Trimmed):

Most commonly specified finished machine bolt and cap screw head style. Six-sided, characterized by sharp, clean corners for easy wrenching, washer-faced to relieve the corners and protect work from defacement. Also supplied as "Heavy" Hexagon Heads.

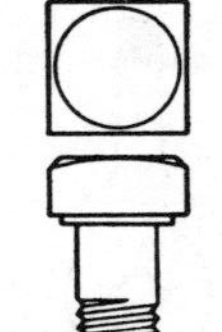
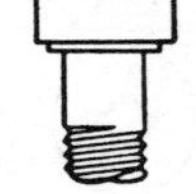

Square Head (Trimmed):

Applied principally to rough or unfinished bolts of low carbon steel and applications where special design make a square head desirable. Standard head design for Lag Bolts.

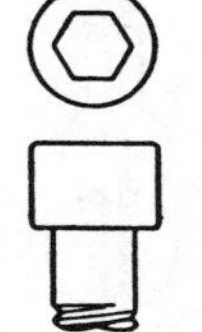

Socket Head:

A popular head design for cap screws characterized by a hexagonal recess for internal wrenching. A popular design for machinery and automotive applications where inherent strength is required, accessibility is limited or flush surfaces are required. This design allows application of high torque without damage or burring of head.

Round Head Square Neck (Carriage):

The most common carriage bolt head design with a low, convex head for clearance and a square neck which is either set up in wood or recessed in metal so the nut can be tightened without holding the head.

SET SCREW HEADS

Square Head Set Screw:

The most frequently specified set screw head for general applications where high wrenching torque is required. The square is trimmed with sharp, well defined corners. Threads extend to the head.

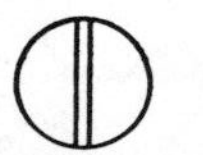

Headless Slotted Set Screw:

This type of set screw is frequently specified where flush surfaces are desired so that top of screw can be driven below the surface of the work. Also, where screw driver assembly is more practical.

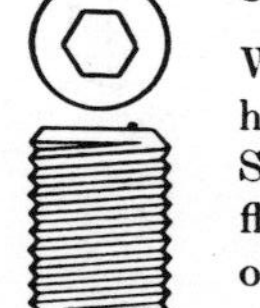

Socket Head Set Screw:

Where internal wrenching is desired this type of hexagonal recessed set screw head is specified. Socket heads also make it possible to provide flush surfaces. Widely used in the manufacture of machinery of all types where frequent disassembly or adjustment occurs.

†SCREW HEADS (MACHINE, WOOD, TAPPING, STOVE BOLTS)

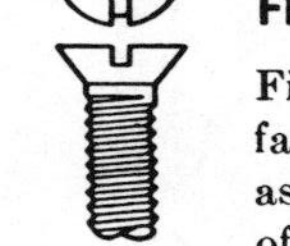

Flat Head:

Fits a pre-countersunk hole where finished surfaces are required. Countersink frequently speeds assembly by making it easy to locate hole. Angle of countersink approximately 80 degrees.

Figure 17-1. Fastener head styles.

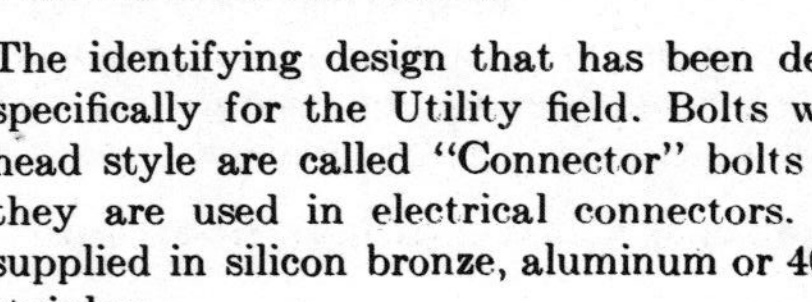

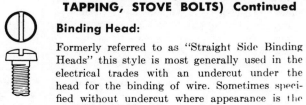

Pan Head Oval Neck:

The identifying design that has been developed specifically for the Utility field. Bolts with this head style are called "Connector" bolts because they are used in electrical connectors. Usually supplied in silicon bronze, aluminum or 400 series stainless.

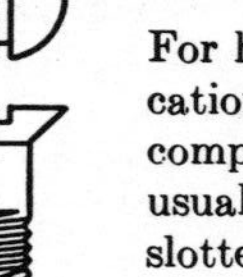

Countersunk Head:

For heavy duty machine bolt or cap screw applications where a flush surface is required in the completed assembly. Countersunk cap screws are usually standard slotted; bolts are usually unslotted. Angle of countersink is usually 82 degrees.

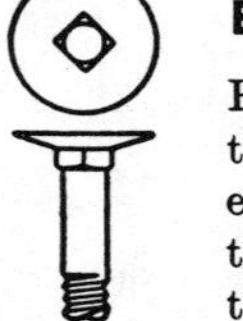

Elevator Bolt Head:

Fully specified as "Flat Countersunk Head Elevator Bolt." Characterized by a thin, circular head of extra-large diameter with a square shoulder under the head to prevent turning. Designed for application where a wide, thin bearing surface is desired.

†SCREW HEADS (MACHINE, WOOD, TAPPING, STOVE BOLTS) Continued

Binding Head:

Formerly referred to as "Straight Side Binding Heads" this style is most generally used in the electrical trades with an undercut under the head for the binding of wire. Sometimes specified without undercut where appearance is the principal factor.

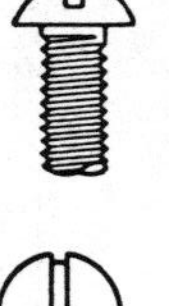

Round Head:

The most widely specified head style for general application. Ease of assembly, absence of sharp corners, rounded design, deep slots make it the most popular head style.

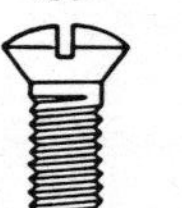

Oval Head:

A variation and relative of the flat (countersunk) head. Its characteristic rounded oval is frequently desirable in assemblies where a finished, rounded-off appearance is desired. Countersink: approximately 80 degrees.

Fillister Head:

Where the maximum amount of purchase must be applied in setting screws up tight or where minimum head diameter is desired in assembling next to channels or flanges the Fillister head is usually designated. Also frequently used in counterbored holes.

Knurled Head:

A high, circular head with ridges on the outer rim for thumb adjustment or manual fastening.

Figure 17-1. *(Continued)*

Truss Head:

Sometimes called "Oval Binding," "Oven Head" or "Stove Head" this large diameter, low clearance head is used both for appearance and low height and sometimes to span clearance holes or slots where wide tolerances are necessary.

Pan Head:

Principally a developed design for tapping screws where a low, attractive head style is desirable. The semi-squared outer periphery of the head offers full edges for high torque driving.

Jackson Head:

This style which is nothing more than an undersized oval (countersunk) head screw is used extensively in the architectural hardware trades for assembling moldings, etc., because of its neat, diminutive and finished style.

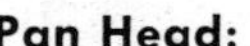

Welding Screw Head:

Characterized by multiple welding lugs on the top or underside of the head which are fused to the assembly by means of projection welding to provide a permanent threaded member.

RIVET HEADS

Round Head:

Like the typical screw head, this is the most common rivet head and is specified for most general applications.

Flat Head:

Unlike the same nomenclature in screw heads the rivet "flat" head is a low, round, flat type of head not countersunk. Specified where a low head will solve a clearance or other design problem.

Countersunk Head:

The sister head to the screw standard "Flat Head." For countersunk applications where flush surfaces are indicated. Angle of countersink on small diameter rivets is approximately 90 degrees; on rivets one-half inch and over, 78 degrees.

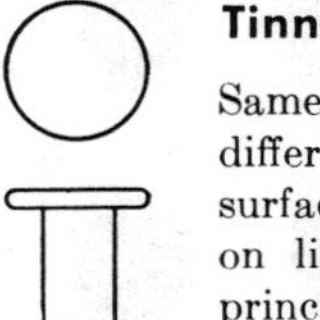

Tinners Rivet Head:

Same general contour as a "Flat Head" but to different standards. The head has a larger bearing surface to provide a greater distribution of load on light gauge metal. Tinners rivets are used principally in sheet metal work.

Figure 17-1. *(Continued)*

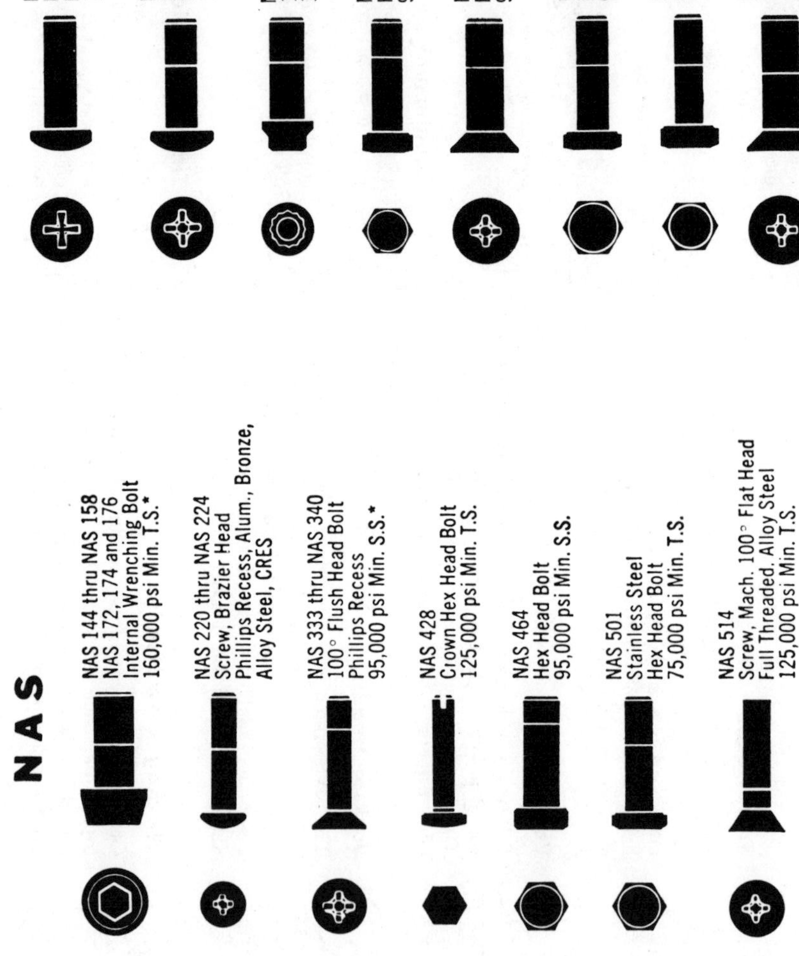
NAS
NAS 144 thru NAS 158
NAS 172, 174 and 176
Internal Wrenching Bolt
160,000 psi Min. T.S.*
NAS 220 thru NAS 224
Screw, Brazier Head
Phillips Recess, Alum., Bronze,
Alloy Steel, CRES
NAS 333 thru NAS 340
100° Flush Head Bolt
Phillips Recess
95,000 psi Min. S.S.*
NAS 428
Crown Hex Head Bolt
125,000 psi Min. T.S.
NAS 464
Hex Head Bolt
95,000 psi Min. S.S.
NAS 501
Stainless Steel
Hex Head Bolt
75,000 psi Min. T.S.
NAS 514
Screw, Mach. 100° Flat Head
Full Threaded, Alloy Steel
125,000 psi Min. T.S.
NAS 610 thru NAS 616
Pan Head Screw
Reed & Prince Recess
160,000 psi Min. T.S.
NAS 623
Pan Head Screw
Phillips Recess
160,000 psi Min. T.S.
NAS 624 thru NAS 644
12 Point Bolt
180,000 psi Min. T.S.
NAS 653 thru NAS 658
Hex Head Bolt, Titanium
95,000 psi Min. S.S.
NAS 663 thru NAS 668
Flush Head Bolt, Titanium
95,000 psi Min. S.S.
NAS 673 thru NAS 678
Hex Head Bolt, Titanium
95,000 psi Min. S.S.
NAS 1003 thru NAS 1020
Hex Head Bolt
140,000 psi Min. T.S.
NAS 1083 thru NAS 1088
100° Flush Head Bolt, Titanium
Phillips Recess
95,000 psi Min. S.S.

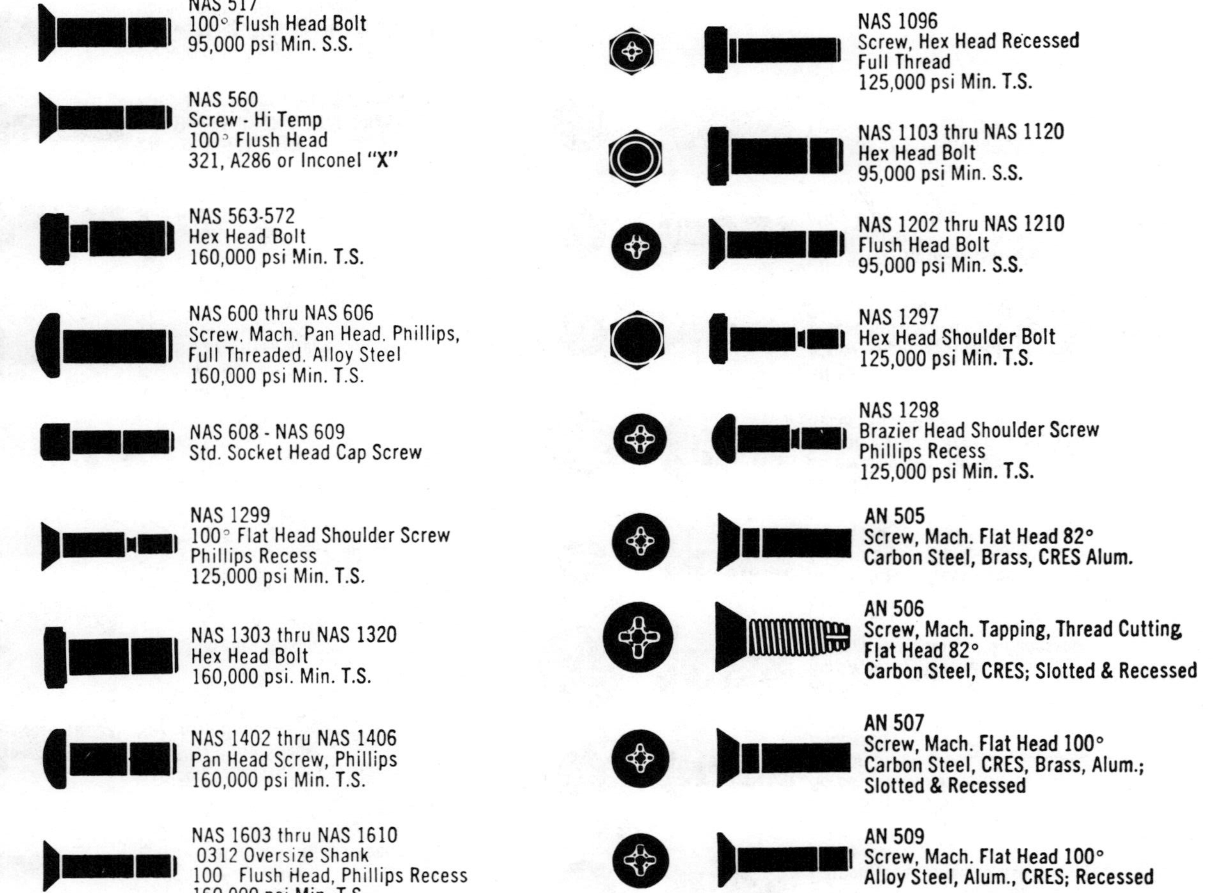
NAS 517
100° Flush Head Bolt
95,000 psi Min. S.S.
NAS 560
Screw - Hi Temp
100° Flush Head
321, A286 or Inconel "X"
NAS 563-572
Hex Head Bolt
160,000 psi Min. T.S.
NAS 600 thru NAS 606
Screw. Mach. Pan Head. Phillips,
Full Threaded. Alloy Steel
160,000 psi Min. T.S.
NAS 608 - NAS 609
Std. Socket Head Cap Screw
NAS 1299
100° Flat Head Shoulder Screw
Phillips Recess
125,000 psi Min. T.S.
NAS 1303 thru NAS 1320
Hex Head Bolt
160,000 psi. Min. T.S.
NAS 1402 thru NAS 1406
Pan Head Screw, Phillips
160,000 psi Min. T.S.
NAS 1603 thru NAS 1610
0312 Oversize Shank
100 Flush Head, Phillips Recess
160,000 psi Min. T.S.
NAS 1096
Screw, Hex Head Recessed
Full Thread
125,000 psi Min. T.S.
NAS 1103 thru NAS 1120
Hex Head Bolt
95,000 psi Min. S.S.
NAS 1202 thru NAS 1210
Flush Head Bolt
95,000 psi Min. S.S.
NAS 1297
Hex Head Shoulder Bolt
125,000 psi Min. T.S.
NAS 1298
Brazier Head Shoulder Screw
Phillips Recess
125,000 psi Min. T.S.
AN 505
Screw, Mach. Flat Head 82°
Carbon Steel, Brass, CRES Alum.
AN 506
Screw, Mach. Tapping, Thread Cutting
Flat Head 82°
Carbon Steel, CRES; Slotted & Recessed
AN 507
Screw, Mach. Flat Head 100°
Carbon Steel, CRES, Brass, Alum.;
Slotted & Recessed
AN 509
Screw, Mach. Flat Head 100°
Alloy Steel, Alum., CRES; Recessed

Figure 17-2. NAS, AN, and MS screws and bolts.

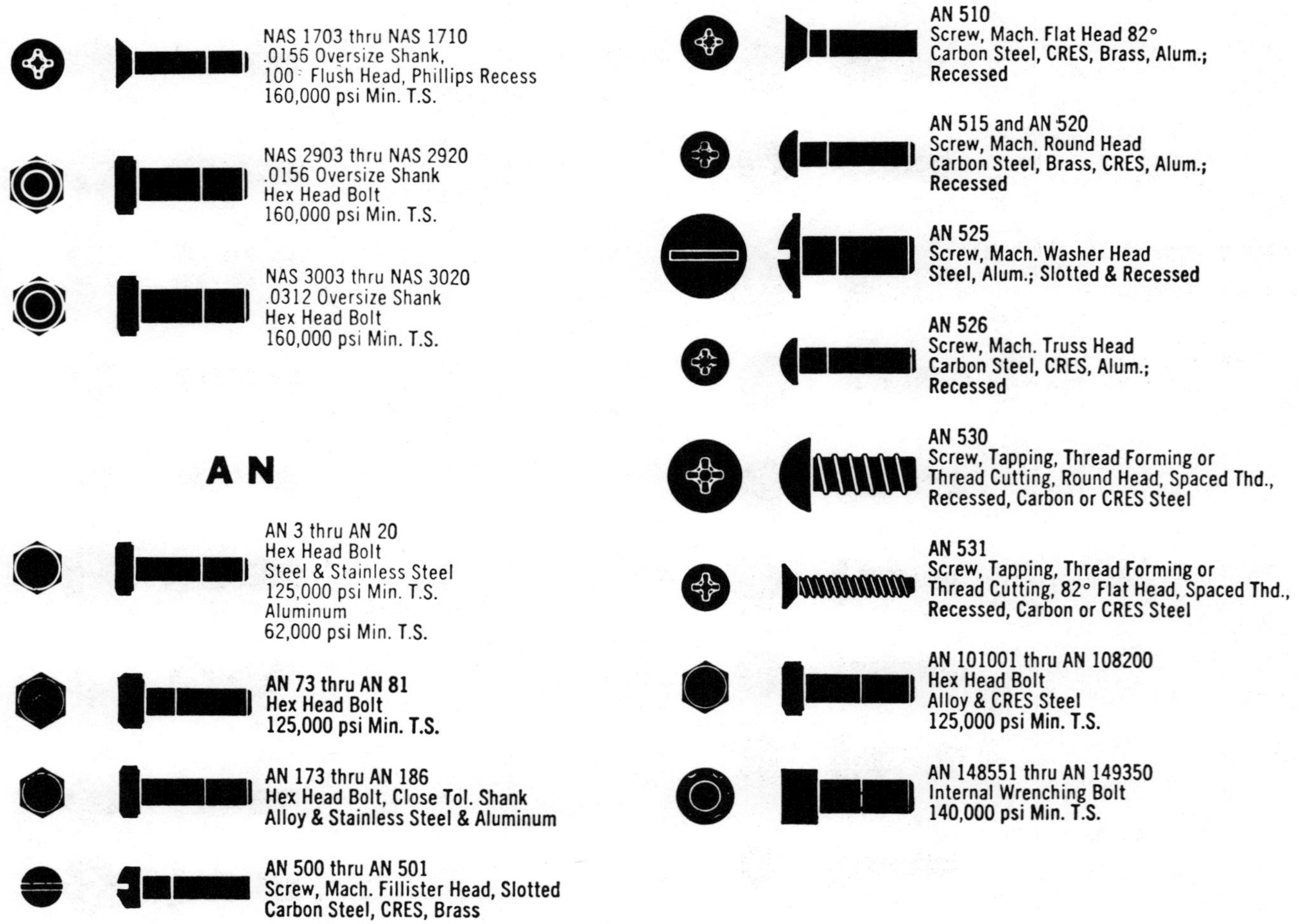

NAS 1703 thru NAS 1710
.0156 Oversize Shank,
100° Flush Head, Phillips Recess
160,000 psi Min. T.S.

NAS 2903 thru NAS 2920
.0156 Oversize Shank
Hex Head Bolt
160,000 psi Min. T.S.

NAS 3003 thru NAS 3020
.0312 Oversize Shank
Hex Head Bolt
160,000 psi Min. T.S.

A N

AN 3 thru AN 20
Hex Head Bolt
Steel & Stainless Steel
125,000 psi Min. T.S.
Aluminum
62,000 psi Min. T.S.

AN 73 thru AN 81
Hex Head Bolt
125,000 psi Min. T.S.

AN 173 thru AN 186
Hex Head Bolt, Close Tol. Shank
Alloy & Stainless Steel & Aluminum

AN 500 thru AN 501
Screw, Mach. Fillister Head, Slotted
Carbon Steel, CRES, Brass

AN 510
Screw, Mach. Flat Head 82°
Carbon Steel, CRES, Brass, Alum.;
Recessed

AN 515 and AN 520
Screw, Mach. Round Head
Carbon Steel, Brass, CRES, Alum.;
Recessed

AN 525
Screw, Mach. Washer Head
Steel, Alum.; Slotted & Recessed

AN 526
Screw, Mach. Truss Head
Carbon Steel, CRES, Alum.;
Recessed

AN 530
Screw, Tapping, Thread Forming or
Thread Cutting, Round Head, Spaced Thd.,
Recessed, Carbon or CRES Steel

AN 531
Screw, Tapping, Thread Forming or
Thread Cutting, 82° Flat Head, Spaced Thd.,
Recessed, Carbon or CRES Steel

AN 101001 thru AN 108200
Hex Head Bolt
Alloy & CRES Steel
125,000 psi Min. T.S.

AN 148551 thru AN 149350
Internal Wrenching Bolt
140,000 psi Min. T.S.

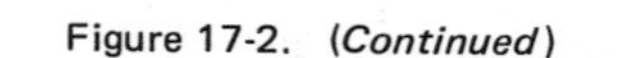

Figure 17-2. *(Continued)*

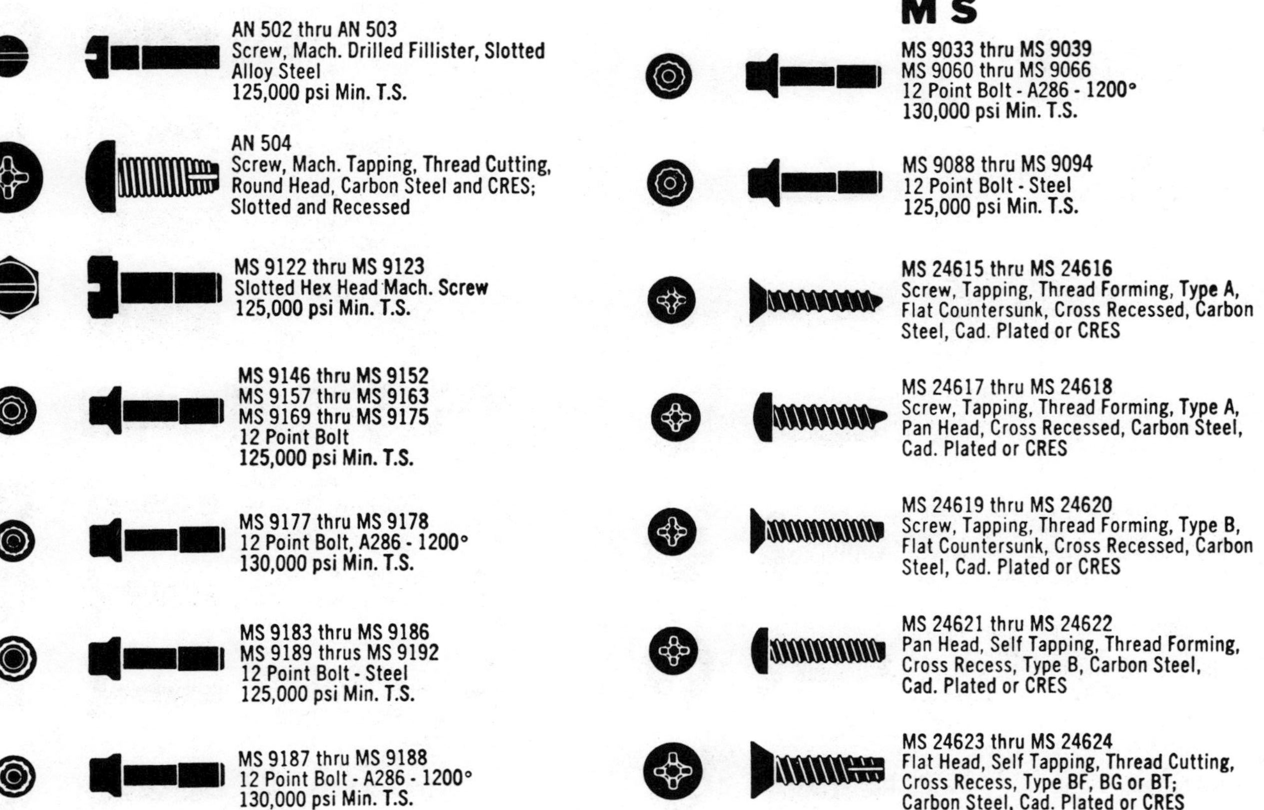
AN 502 thru AN 503
Screw, Mach. Drilled Fillister, Slotted
Alloy Steel
125,000 psi Min. T.S.
AN 504
Screw, Mach. Tapping, Thread Cutting,
Round Head, Carbon Steel and CRES;
Slotted and Recessed
MS 9122 thru MS 9123
Slotted Hex Head Mach. Screw
125,000 psi Min. T.S.
MS 9146 thru MS 9152
MS 9157 thru MS 9163
MS 9169 thru MS 9175
12 Point Bolt
125,000 psi Min. T.S.
MS 9177 thru MS 9178
12 Point Bolt, A286 - 1200°
130,000 psi Min. T.S.
MS 9183 thru MS 9186
MS 9189 thrus MS 9192
12 Point Bolt - Steel
125,000 psi Min. T.S.
MS 9187 thru MS 9188
12 Point Bolt - A286 - 1200°
130,000 psi Min. T.S.
M S
MS 9033 thru MS 9039
MS 9060 thru MS 9066
12 Point Bolt - A286 - 1200°
130,000 psi Min. T.S.
MS 9088 thru MS 9094
12 Point Bolt - Steel
125,000 psi Min. T.S.
MS 24615 thru MS 24616
Screw, Tapping, Thread Forming, Type A,
Flat Countersunk, Cross Recessed, Carbon
Steel, Cad. Plated or CRES
MS 24617 thru MS 24618
Screw, Tapping, Thread Forming, Type A,
Pan Head, Cross Recessed, Carbon Steel,
Cad. Plated or CRES
MS 24619 thru MS 24620
Screw, Tapping, Thread Forming, Type B,
Flat Countersunk, Cross Recessed, Carbon
Steel, Cad. Plated or CRES
MS 24621 thru MS 24622
Pan Head, Self Tapping, Thread Forming,
Cross Recess, Type B, Carbon Steel,
Cad. Plated or CRES
MS 24623 thru MS 24624
Flat Head, Self Tapping, Thread Cutting,
Cross Recess, Type BF, BG or BT;
Carbon Steel, Cad. Plated or CRES

Figure 17-2. *(Continued)*

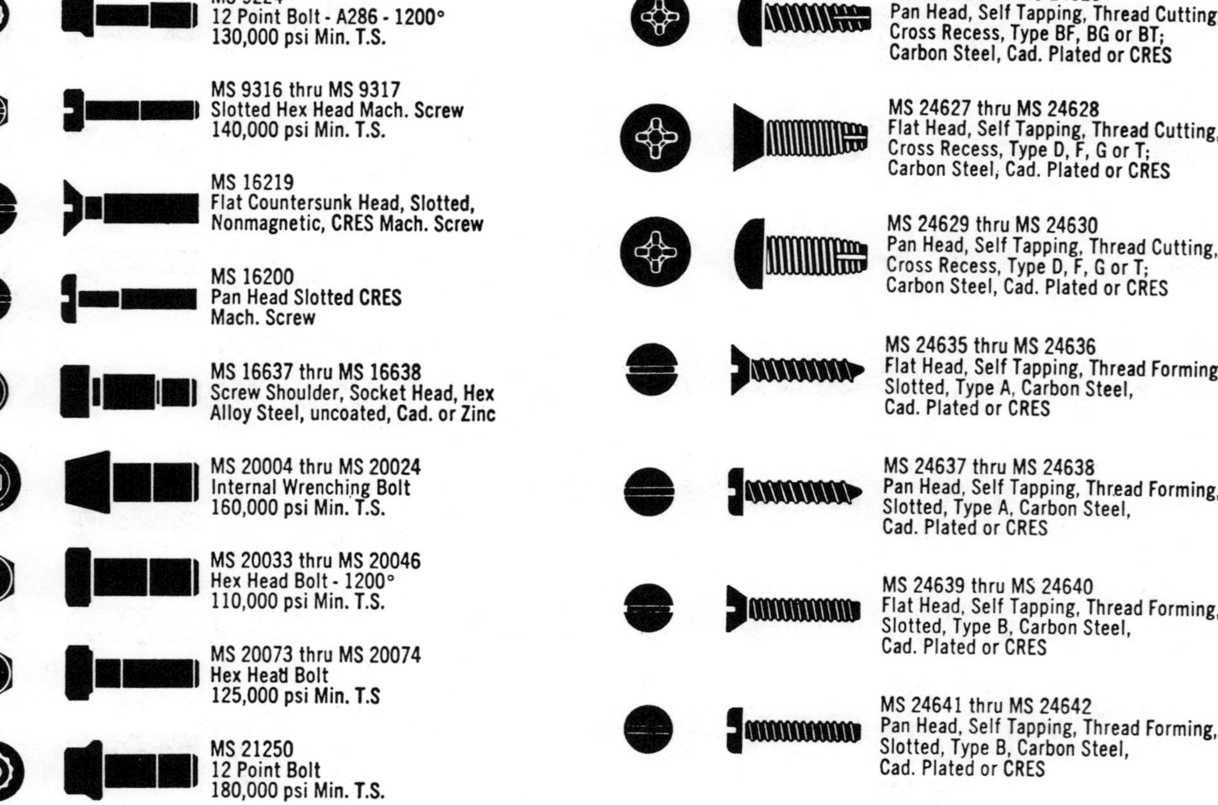
MS 9224
12 Point Bolt - A286 - 1200°
130,000 psi Min. T.S.
MS 9316 thru MS 9317
Slotted Hex Head Mach. Screw
140,000 psi Min. T.S.
MS 16219
Flat Countersunk Head, Slotted,
Nonmagnetic, CRES Mach. Screw
MS 16200
Pan Head Slotted CRES
Mach. Screw
MS 16637 thru MS 16638
Screw Shoulder, Socket Head, Hex
Alloy Steel, uncoated, Cad. or Zinc
MS 20004 thru MS 20024
Internal Wrenching Bolt
160,000 psi Min. T.S.
MS 20033 thru MS 20046
Hex Head Bolt - 1200°
110,000 psi Min. T.S.
MS 20073 thru MS 20074
Hex Head Bolt
125,000 psi Min. T.S
MS 21250
12 Point Bolt
180,000 psi Min. T.S.
MS 24625 thru MS 24626
Pan Head, Self Tapping, Thread Cutting,
Cross Recess, Type BF, BG or BT;
Carbon Steel, Cad. Plated or CRES
MS 24627 thru MS 24628
Flat Head, Self Tapping, Thread Cutting,
Cross Recess, Type D, F, G or T;
Carbon Steel, Cad. Plated or CRES
MS 24629 thru MS 24630
Pan Head, Self Tapping, Thread Cutting,
Cross Recess, Type D, F, G or T;
Carbon Steel, Cad. Plated or CRES
MS 24635 thru MS 24636
Flat Head, Self Tapping, Thread Forming,
Slotted, Type A, Carbon Steel,
Cad. Plated or CRES
MS 24637 thru MS 24638
Pan Head, Self Tapping, Thread Forming,
Slotted, Type A, Carbon Steel,
Cad. Plated or CRES
MS 24639 thru MS 24640
Flat Head, Self Tapping, Thread Forming,
Slotted, Type B, Carbon Steel,
Cad. Plated or CRES
MS 24641 thru MS 24642
Pan Head, Self Tapping, Thread Forming,
Slotted, Type B, Carbon Steel,
Cad. Plated or CRES

Figure 17-2. *(Continued)*

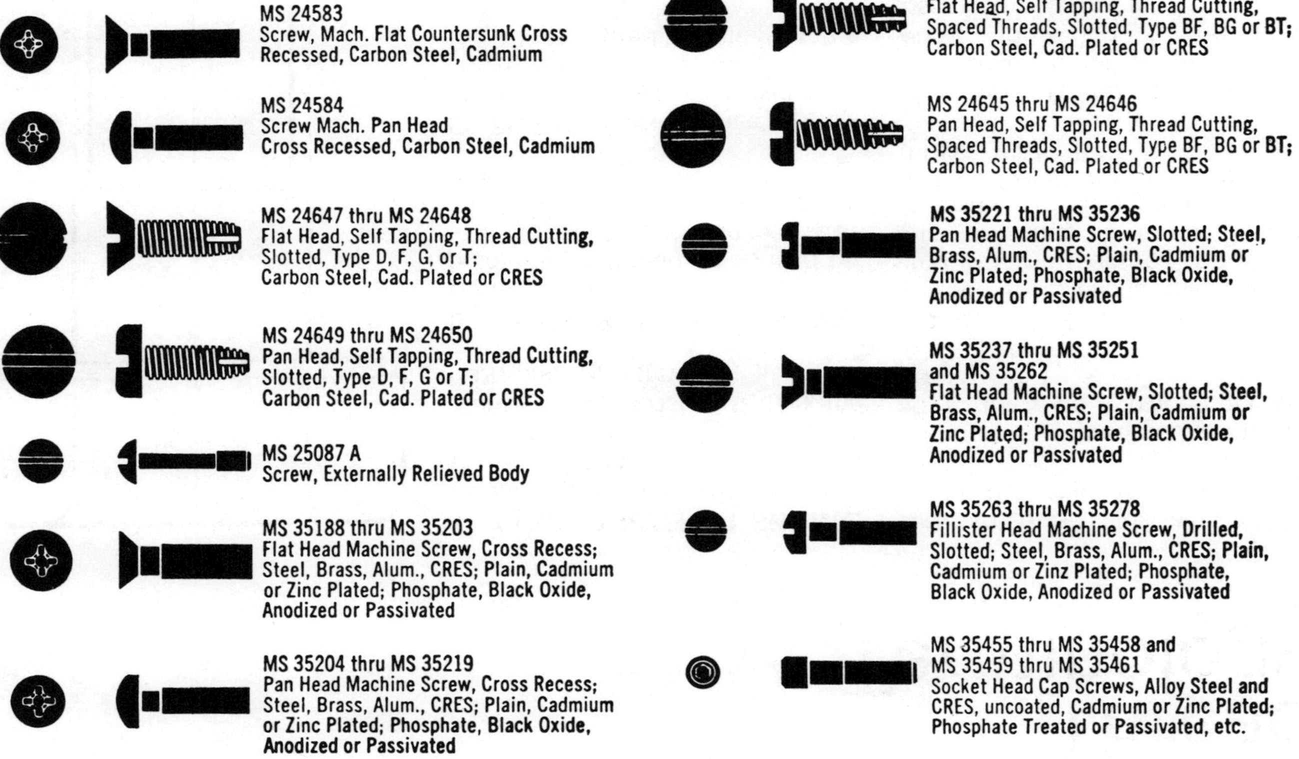
MS 24583
Screw, Mach. Flat Countersunk Cross
Recessed, Carbon Steel, Cadmium
MS 24584
Screw Mach. Pan Head
Cross Recessed, Carbon Steel, Cadmium
MS 24647 thru MS 24648
Flat Head, Self Tapping, Thread Cutting,
Slotted, Type D, F, G, or T;
Carbon Steel, Cad. Plated or CRES
MS 24649 thru MS 24650
Pan Head, Self Tapping, Thread Cutting,
Slotted, Type D, F, G or T;
Carbon Steel, Cad. Plated or CRES
MS 25087 A
Screw, Externally Relieved Body
MS 35188 thru MS 35203
Flat Head Machine Screw, Cross Recess;
Steel, Brass, Alum., CRES; Plain, Cadmium
or Zinc Plated; Phosphate, Black Oxide,
Anodized or Passivated
MS 35204 thru MS 35219
Pan Head Machine Screw, Cross Recess;
Steel, Brass, Alum., CRES; Plain, Cadmium
or Zinc Plated; Phosphate, Black Oxide,
Anodized or Passivated
MS 24643 thru MS 24644
Flat Head, Self Tapping, Thread Cutting,
Spaced Threads, Slotted, Type BF, BG or BT;
Carbon Steel, Cad. Plated or CRES
MS 24645 thru MS 24646
Pan Head, Self Tapping, Thread Cutting,
Spaced Threads, Slotted, Type BF, BG or BT;
Carbon Steel, Cad. Plated or CRES
MS 35221 thru MS 35236
Pan Head Machine Screw, Slotted; Steel,
Brass, Alum., CRES; Plain, Cadmium or
Zinc Plated; Phosphate, Black Oxide,
Anodized or Passivated
MS 35237 thru MS 35251
and MS 35262
Flat Head Machine Screw, Slotted; Steel,
Brass, Alum., CRES; Plain, Cadmium or
Zinc Plated; Phosphate, Black Oxide,
Anodized or Passivated
MS 35263 thru MS 35278
Fillister Head Machine Screw, Drilled,
Slotted; Steel, Brass, Alum., CRES; Plain,
Cadmium or Zinz Plated; Phosphate,
Black Oxide, Anodized or Passivated
MS 35455 thru MS 35458 and
MS 35459 thru MS 35461
Socket Head Cap Screws, Alloy Steel and
CRES, uncoated, Cadmium or Zinc Plated;
Phosphate Treated or Passivated, etc.

Figure 17-2. *(Continued)*

Thread Terminology

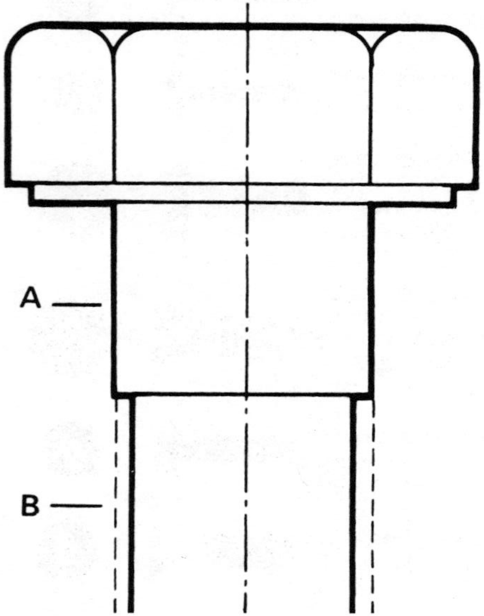

A. FULL DIAMETER SHANK: Equal to major diameter of thread. Produced by cut thread or by roll thread on extruded blank. Characteristic of machine bolts and cap screws.

B. UNDERSIZED SHANK: Equal approximately to pitch diameter of thread. Produced by roll threading a non-extruded blank. Characteristic of machine screws.

C. PITCH: The distance from a point on the screw thread to a corresponding point on the next thread measured parallel to the axis.

serrated dies. This acts to increase the major diameter of the thread over and above the diameter of unthreaded shank (if any), unless an extruded blank is used.

Classes of thread are distinguished from each other by the amounts of tolerance and allowance specified. External threads or bolts are designated with the suffix "A"; internal or nut threads with "B"

CLASSES 1A and 1B: For work of rough commercial quality where loose fit for spin-on-assembly is desirable.

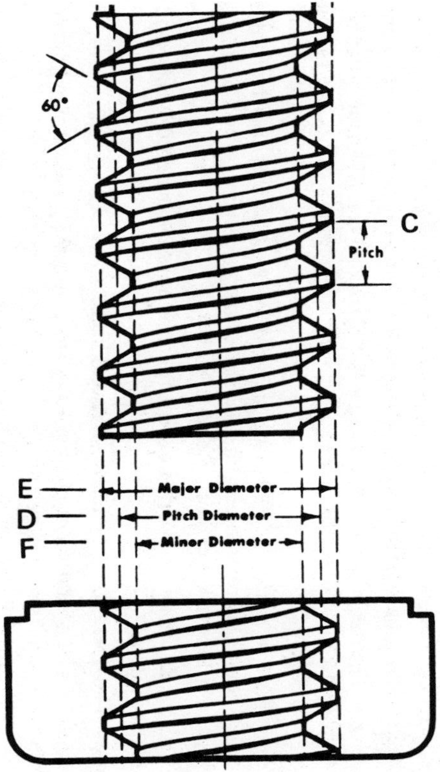

D. PITCH DIAMETER: The simple, effective diameter of screw thread. Approximately half way between the major and minor diameters.

E. MAJOR DIAMETER: The largest diameter of a screw thread.

F. MINOR DIAMETER: The smallest diameter of a screw thread.

LEAD: The distance a screw thread advances axially in one turn.

CUT THREAD: Threads are cut or chased; the unthreaded portion of shank will be equal to major diameter of thread.

ROLLED THREAD: Threads are cold formed by squeezing the blank between reciprocating

CLASSES 2A and 2B: The recognized standard for normal production of the great bulk of commercial bolts, nuts and screws.

CLASSES 3A and 3B: Used where a closed fit between mating parts for high quality work is required.

CLASS 4: A theoretical rather than practical class, now obsolete.

CLASS 5: For a wrench fit. Used principally for studs and their mating tapped holes. A force fit requiring the application of high torque for semi-permanent assembly.

Figure 17-3. Thread terminology.

18

TERMINOLOGY

Communications has been a major problem in the area of engineering design. Thus, the following glossary of engineering and manufacturing terminology is provided. The engineer and designer should use the terminology generally accepted within the industry in all forms of communication. All engineering documents should use only the standard accepted terminology.

The following glossary is presented by category, for example, "terminology related to manufacturing processes" and "terminology related to printed circuits."

TERMS RELATED TO MANUFACTURING PROCESSES FOR FLAT METALS

Beading. The operation of rolling over the edge of circular-shaped material, either by stamping or spinning.

Bending. The operation of forming flat materials, usually sheet metal or flat wire, into irregular shapes by the action of a punch forcing the material into a cavity or depression. Such forming is done in a punch press. When the operation involves long straight bends, the material is usually bent in a press brake.

Blanking. The operation of separating a flat piece of a certain shape a flat product preparatory to other processes or operations necessary to finalize the piece into a finished part.

Blasting. The operation of removing material by the action of abrasive material directed under air pressure on the surface of the piece part through controlled orifice openings at the end of a flexible hosing.

Blasting, sand. Same as Blasting except that the abrasive material is fine sand of some definite grit.

Boring. The operation of enlarging a round hole.

Broaching. The operation of cutting away the material by a succession of cutting teeth on a tool that is either pushed or pulled along the surface of the workpiece. Cutting may be applied to either holes or the outside edges of the piece.

Buffing. Essentially, the same as polishing except wheels and belts are made of softer material and finer abrasive is combined with a lubricant binder.

Bulging. The operation of exapnding the metal below the opening of a drawn cup or shell.

Burnishing. The operation of producing smooth-finished surfaces by compressing the outer layer of the metal, either by the application of highly polished tools or by the use of steel balls, in rolling contact with the surface of the piece part.

Coining. The operation of forming metallic material in a shaped-cavity die by impact. All the material is restricted to the cavity resulting in fine-line detail of the piece part.

Counterboring. The operation of enlarging some part of a cylindrical bore or hole.

Countersinking. The operation of producing a conical entrance to a bore or hole.

Curling. See Beading.

Cutting, flame. The operation of severing ferrous metals by rapid oxidation from a jet of pure oxygen directed at a point heated to the fusion point.

Dimpling. The operation of forming semispherical-shaped impressions in flat material.

Dinking. The operation of blanking a piece out of a soft flat material by the use of sharp beveled cutting edges arranged to produce a part of a specific configration. Dinking is similar to steel rule die.

Drawing. The operation of forming cylindrical and other shaped parts from flat stock by action of a punch pushing the metal into a cavity having the same shape as the punch.

Drawing, deep. Essentially the same as Drawing except that the final shape is the result of a series of redraws.

Drawing, shallow. Essentially the same as Drawing except that the operation is restricted to shallow pieces and operation is usually accomplished by a stamping single-action process.

Drilling. The operation of producing cylindrically shaped holes, usually a cone shape at the bottom.

Drilling, bottom. Same as Counterboring except performed with square-ended drill instead of the conical tip.

Embossing. The operation of production raised patterns on the surface of materials by use of dies or plates brought forcibly upon the material to be embossed.

Etching, acid. Same as Chemical etching except material used for metal removing is confined to action of acids.

Etching, chemical. The operation of removing material from the surface of a metal by the action of chemicals upon the surface.

Extruding, impact. The operation of producing cup-shaped parts by a single blow against a confined slug of cold metal until the desired length and wall thickness is reached.

Flanging. Essentially the same as Beading or Curling except the portion of metal rolled over is left in a flat form. The operation is only performed on parts with trimmed edges since the normal drawing operation will always begin with a flat piece forming the flange, which is later trimmed to size.

Forging, cold or hot. The operation of forming relatively thick sections of metallic material into a desired shape. Material may be worked cold or hot between shaped forms or shaped rolls. In the case of roll dies, the material is squeezed into shape. In the case of shaped dies, the material is formed by a succession of blows forcing the material into the shaped cavity, or the material may be squeezed into the shaped cavity by a steady pressure of the moving form on the metal.

Forming, contour. The operation of reforming material sections of all types, rolled formed sections, cold-drawn or rolled shapes, extrusions, etc., normally received in straight lengths, into various contours. (Section shape remains unchanged for full length of piece.)

Forming, roll. The operation of shaping flat-strip material into a desired form by passing it through a series of shaped rolls. Each roll brings the shape closer to its final form.

Forming, rubber die, Guerin process, Marform process. Essentially the same as Hydroforming except the rubber pad (or other suitable material) is not backed up with any pressure.

Forming, stretch. The operation of forming sheet or strip by gripping two sides or ends (depending on the shape of unformed blank or required contour) until the material takes a set.

Grinding. The operation of separating material from the surface of a material by means of abrasion. When restricted to flat materials, the term is more generally known as *surface grinding.* Abrasion may be by means of an abrasive wheel or belt. When performed by a belt, it is more commonly known as *abrasive belt grinding.*

Heading, cold. The operation of gathering sufficient stock at one end of a bar of metallic material to form a desired shape. Gathering the material is done by impact of the material against a shaped die.

Hobbing. The operation of producing teeth on the outside edge of the workpiece. Tool and workpiece move with respect to each other.

Hydroforming. The operation of drawing sheet metal using a shaped punch that is forced against the material and a rubber pad backing the material, while hydraulic pressure on the opposite side is applied.

Ironing. The operation of reducing the walls of drawn shells to assure uniform thickness.

Lancing. The operation of piercing the material in a shape that will leave the portion cut affixed to the part.

Lapping. The operation of refining a surface from undulations, roughness, and toolmarks left by previous operations.

Milling. Essentially the same as Sawing except the cutting operation leaves a trough of some specific shape depending on the form of the cutter used.

Necking. The operation of reducing the opening of a drawn cup or shell.

Nibbling. The operation of removing flat material by means of a rapidly reciprocating punch to form a contour of a desired shape.

Peening, shot. The operation of compressing a selective surface of a piece part by a rain of metal shot impelled from rotating blades of a wheel or air blast.

Perforating. The operation of piercing, in rows, large numbers of small (usually round) holes in a part that may have been previously formed.

Piercing. Essentially the same as Blanking except that the operation may be done on a previously formed piece part and that portion that is separated is usually scrap.

Planing. Essentially the same as Shaping except the operation is primarily intended to produce flat surfaces.

Polishing. The operation of smoothing the surface of a piece part by action of abrasive particles, which are adhered to the surface of moving belts or wheels, coming in contact with the piece part being polished.

Punching. The operation of separating or forming a flat piece of material in some particular shape.

Reaming. The operation of enlarging and finishing a hole to an accurate dimension.

Routing. The operation of cutting away material overhanging a templete by means of a rotating cylindrical cutter engaging the edge of the material to be profiled or cut. (The term *routing* is also applied to the operation of making shaped cavities and is more commonly known as *end milling.*)

Sawing. The operation of separating material by teeth performing successive cuts in material. Operation may follow a straight line or contour.

Serrating. The operation of producing teeth on the inside or outside edge of the workpiece by broaching. Workpiece is fixed, and tool moves.

Shaping. The operation of removing material from the surface of a material mass by a series of horizontally adjacent cuts. The tool is moved in forward strokes while the material moves in preset increments in a plane at right angles to the tool.

Shaving. The operation of finalizing a dimensional requirement on the shape of a part by cutting away a very small portion around its configuration.

Shearing. The operation of cutting the material by means of straight blades, one of which is fixed and the other moving progressively toward it until the material between is finally cleaved. Shearing may be in a straight line, circular, or other shape.

Sizing. The operation of finalizing a dimension on a part by direct pressure, impact, or a combination of both.

Slitting. Essentially the same as Milling except the trough produced is usually very narrow, as in the case of screw heads.

Slotting. Essentially the same as Shaping except the tool is moved vertically. Sometimes referred to as *vertical shaping.*

Spinning. The operation of forming sheet metal into circular shapes by means of a lathe, forms and hand tools, which press and shape the metal about the revolving form.

Spotfacing. The same as Counterboring except that the depth of the hole is restricted to just breaking through the surface of the material.

Squeezing. The operation of forming material by compressing it into a desired shape. The material may be confined in a shaped cavity or may be free-flowing.

Stamping. The process of stamping generally refers to that portion of the press working and forming fields, which is considered as including the following operations:

1. Punching
   a. Blanking
   b. Piercing
   c. Lancing
2. Bending
3. Forming
   a. Beading or curling
   b. Embossing
4. Shallow drawing
5. Extruding
6. Swedging
7. Coining
8. Shaving
9. Triming

Superfinishing. Essentially the same as Lapping except that the operation results in a higher degree of surface finish.

Swaging, cold. The operation of forming or reducing a metallic material by successive blows of a pair of dies or hammers. Material flows at right angles to the pressure.

Swaging, hot. Essentially the same as Cold swaging except that the metal is heated and the arrangement of the dies causes them to be carried in a slow rotary motion. This operation is also known as *rotary swaging.*

Swedging. The operation of forming or shaping by a single blow or squeeze of a die.

Tapping. The operation of cutting a screw thread into a round hole.

Tapping, bottom. The same as Tapping except extending the threads to the bottom of a hole.

Trimming. The operation of removing excess material on the ends or edges of articles resulting from some kind of forming operation.

Tumbling, barrel. The operation of removing burrs, fins, scale, and roughness by an abrasive material falling against finished piece parts in a revolving barrel.

Upsetting, hot. Essentially the same as Cold heading except that the material is heated in order to aid ductility in forming.

TERMS RELATING TO MATERIALS AND FABRICATION PROCESSES

Age hardening. Hardening by aging, usually after rapid cooling or cold working. *See* Aging.

Aging. In a metal or alloy, a change in properties that generally occurs slowly at room temperature and more rapidly at higher temperatures. *See* Age hardening, Artificial aging, Natural aging, Precipitation hardening, Precipitation heat treatment, Quench aging, and Strain aging.

Air-hardening steel. A steel containing sufficient carbon and other alloying elements to harden fully during cooling in air or other gaseous media, from a temperature above its transformation range. The term should be restricted to steels that are capable of being hardened by cooling in air in fairly large sections, about 2 inches or more in diameter. *See* Self-hardening steel.

Alclad. Composite sheet produced by bonding either corrosion-resistant aluminum alloy or aluminum of high purity to base metal of structurally stronger aluminum alloy.

Alkali metal. A metal in group IA of the periodic system (namely, lithium, sodium, potassium, rubidium, cesium, francium). These metals form strong alkaline hydroxides; hence, the name.

Alkaline earth metal. A metal in group IIA of the periodic system (including beryllium, magnesium, calcium, strontium, barium, and radium), so named because the oxides of calcium, strontium, and barium were found by early chemists to be alkaline in reaction.

Allotropy. The reversible phenomenon by which certain metals may exist in more than one crystal structure. If not reversible, the phenomenon is termed *polymorphism.*

Alloy. A substance having metallic properties and being composed of two or more chemical elements, of which at least one is an elemental metal.

Alpha ferrite. *See* Ferrite.

Alpha iron. The body-centered cubic form of pure iron, stable below 1670°F.

Alumel. A nickel-base alloy containing about 2.5% Mn, 2% Al, and 1% Si, used chiefly as a component of pyrometric thermocouples.

Aluminizing. Forming an aluminum or aluminum alloy coating on a metal by hot dipping, hot spraying, or diffusion.

Amorphous. Not having a crystal structure; noncrystalline.

Annealing. Heating to and holding at a suitable temperature and then cooling at a suitable rate, for such purposes as reducing hardness, improving machinability, facilitating cold working, producing a desired microstructure, or obtaining desired mechanical, physical, or other properties. When applicable, the following more specific terms should be used: black annealing, blue annealing, box annealing, bright annealing, flame annealing, full annealing, graphitizing, intermediate annealing, isothermal annealing, malleableizing, process annealing, quench annealing, recrystallization annealing, and spheroidizing.

When applied to ferrous alloys, the term *annealing*, without qualification, implies full annealing.

When applied to nonferrous alloys, the term implies a heat treatment designed to soften a cold-wroked structure by recrystallization or subsequent grain growth or th soften an age hardened alloy by causing a nearly complete precipitation of the second phase in relatively coarse form.

Any process of annealing will usually reduce stresses, but if the treatment is applied for the sole purpose of such relief, it should be designed *stress relieving.*

Anode. The electrode by which electrons leave (current enters) an operating system such as a battery, an electrolytic cell, an X-ray tube, or a vacuum tube. In the case of the battery, the anode is negative; in the other three, positive. In a battery or electrolytic cell, the anode is the electrode at which oxidation occurs. *Contrast with* Cathode.

Anode effect. The effect produced by polarization of the anode in the electrolysis of fused salts. It is characterized by a sudden increase in voltage and a corresponding decrease in current due to the virtual separation of the anode from the electrolyte by a gas film.

Anodizing. Froming a conversion coating on a metal surface by anodic oxidation; most frequently applied to aluminum.

Arc brazing. Brazing with an electric arc, usually with two nonconsumable electrodes.

Arc cutting. Metal cutting with an arc between an electrode and the metal itself. The terms *carbon-arc cutting* and *metal-arc cutting* refer, respectively, to the use of a carbon or metal electrode.

Arc welding. Welding with an electric arc-welding electrode. *See* Electrode.

Artificial aging. Aging above room temperature. *See* Aging; Precipitation heat treatment. *Compare with* Natural aging.

Atomic-hydrogen welding. Arc welding with heat from an arc between two tungsten or other suitable electrodes in a hydrogen atmosphere. The use of pressure and filler metal is optional.

Austempering. Quenching a ferrous alloy from a temperature above the transformation range, in a medium having a rate of heat abstraction high enough to prevent the formation of high-temperature transformation products, and then holding the alloy, until transformation is complete, at a temperature below that of pearlite formation and above that of martensite formation.

Austenite. A solid solution of one or more elements in face-centered cubic iron. Unless otherwise designated (such as nickel austenite), the solute is generally assumed to be carbon.

Austenitic steel. A alloy steel whose structure is normally austenitic at room temperature.

Austenitizing. Forming austenite by heating a ferrous alloy into the transformation range (partial austenitizing) or above the transformation range (complete austenitizing).

Backhand welding. Welding in which the back of the principal hand (torch or electrode hand) of the welder faces the direction of travel. It has special significance in gas welding because it provides postheating.

Backstep sequence. A longitudinal welding sequence in which the direction of general progress is opposite to that of welding the individual increments.

Bainite. A decomposition product of austenite consisting of an aggregate of ferrite and carbide. In general, it forms at temperatures lower than those where very fine pearlite forms and higher than that where martensite begins to form on cooling. Its appearance is feathery if formed in the upper part of the temperature range; acicular, resembling tempered martensite, if formed in the lower part.

Bead weld. A weld composed of one or more string or weave beads deposited on an unbroken surface.

Bearing load. A compressive load supported by a member, usually a tube or collar along a line where contact is made with a pin, rivet, axle, or shaft.

Bearing strength. The maximum bearing load at failure divided by the effective bearing area. In a pinned or riveted joint, the effective area is calculated as the product of the diameter of the hole and the thickness of the bearing member.

Bending moment. The algebraic sum of the couples or the moments of the external forces, or both, to the left or right of any section on a member subjected to bending by couples or transverse forces, or both.

Beveling. See Chamfering.

Biaxial stress. A state of stress in which only one of the principal stresses is zero, the other two usually being in tension.

Billet. A solid, semifinished round or square product that has been hot worked by forging, rolling, or extrusion. An iron or steel billet has a minimum width or thickness of $1\frac{1}{2}$ inches and the cross-sectional area varies from $2\frac{1}{4}$ to 36 square inches. For nonferrous metals, it may also be a casting suitable for finished or semifinished rolling or for extrusion. *See* Extrusion billet.

Black light. Electronmagnetic radiation not visible to the human eye. The portion of the spectrum generally used in fluorescent inspection falls in the ultraviolet region between 3300 to 4000 angstrom units, with the peak at 3650 angstrom units.

Blank nitriding. Simulating the nitriding operation without introducing nitrogen. This is usually accomplished by using an inert material in place of the nitriding agent or by applying a suitable protective coating to the ferrous alloy.

Block brazing. Brazing with heat from hot blocks.

Bottom drill. A flat-ended twist drill used to convert a cone at the bottom of a drilled hole into a cylinder.

Bottoming tap. A tap with a chamfer of 1 to $1\frac{1}{2}$ threads in length.

Brass. An alloy consisting mainly of copper (over 50%) and zinc, to which smaller amounts of other elements may be added.

Braze welding. Welding in which a groove, fillet, plug, or slot weld is made, using a nonferrous filler metal having a melting point lower than that of the base metal but higher than 800°F. The filler metal is not distributed by capillarity.

Brazing. Joining metals by following a thin layer, capillary thickness, of nonferrous filler metal into the space between them. Bonding results from the intimate contact produced by the dissolution of a small amount of base metal in the molten filler metal, without fusion of the base metal. Sometimes the filler metal is put in place as a thin solid sheet or as a clad layer, and the composite is heated as in furnace brazing. The term *brazing* is used where the temperature exceeds some arbitrary value, such as 800°F; the term *soldering* is used for temperatures lower than the arbitrary value.

Brazing alloy. See Brazing filler metal.

Brazing filler metal. A nonferrous filler metal used in brazing and braze welding.

Brazing sheet. Brazing filler metal in sheet form a flat-rolled metal clad with brazing filler metal on one or both sides.

Breaking stress. See Fracture stress (1).

Brinell hardness test. A test for determing the hardness of a material by forcing a hard steel or carbide ball of specified diameter into it under a specified load. The result is expressed as the Brinell hardness number, which is the value obtained by dividing the applied load in kilograms by the surface area of the resulting impression in square millimeters.

Brittle crack propagation. A very sudden propagation of a crack with the absorption of no energy except that stored elastically in the body. Microscopic examination may reveal some deformation even though it is not noticeable to the unaided eye.

Brittle fracture. Fracture with little or no plastic deformation.

Brittleness. The quality of a material that leads to crack propagation without appreciable plastic deformation.

Broach. A bar-shaped cutting tool provided with a series of cutting edges or teeth that increase in size or change in shape from the starting to finishing end.

The tool cuts in the axial direction when pushed or pulled and is used to shape either holes or outside surfaces.

Bronze. A cooper-rich copper-tin alloy with or without small proportions of elements such as zinc and phosphorus. By extension, certain copper-base alloys containing considerably less tin than other alloying elements, such as manganese bronze (copper-zinc plus manganese, tin, and iron) and leaded tin bronze (copper-lead plus tin and sometimes zinc). Also certain other essentially binary copper-base alloys containing no tin, such as aluminum bronze (copper-aluminum), silicon bronze (copper-silicon), and beryllium bronze (copper-beryllium). Also, trade designations for certain specific copper-base alloys that are actually brasses, such as architectural bronze (57% Cu, 40% Zn, 3% Ph) and commercial bronze (90% Cu, 10% Zn).

Brush plating. Plating with a concentrated solution or gel held in or fed to an absorbing medium, pad, or brush carrying the anode (usually insoluble). The brush is moved back and forth over the area of the cathode to be plated.

Buckling. Producing a bulge, bend, bow, kink, or other wavy condition in sheets or plates by compressive stresses.

Butt seam welding. See Seam welding.

Butt welding. Welding a butt joint.

Carat. A unit weight of diamond (abbreviated *c*); the international metric carat (abbreviated *mc*) is 200 mg.

Carbide. A compound of carbon with one or more metallic elements.

Carbide tools. Cutting tools, made of tungsten carbide, titanium carbide, tantalum carbide, or combinations of these, in a matrix of cobalt or nickel, having sufficient wear resistance and heat resistance to permit high maching speeds.

Carbon-arc cutting. Metal cutting by melting with the heat of an arc between a carbon electrode and the base metal.

Carbon-arc welding. Welding in which an arc is maintained between a non-consumable carbon electrode and the work.

Carbonitriding. Introducing carbon and nitrogen into a solid ferrous alloy by holding above Ac_1 in an atmosphere that contains suitable gases such as hydrocarbons, carbon monoxide, and ammonia. The carbonitrided alloy is usually quench-hardened.

Carbonization. Conversion of a substance into elemental carbon. (Should not be confused with *carburization.*)

Carbon steel. Steel containing carbon up to about 2% and only residual quantities of other elements except those added for deoxidation, with silicon usually limited to 0.60% and manganese to about 1.65%. Also termed *plain carbon steel*, *ordinary steel*, and *straight carbon steel.*

Carburizing. Introducing carbon into a solid ferrous alloy by holding above Ac_1 in contact with a suitable carbonaceous material, which may be a solid, liquid, or gas. The carburized alloy is usually quench-hardened.

Case hardening. Hardening a ferrous alloy so that the outer portion, or case, is made substantially harder than the inner portion, or core. Typical processes used for case hardening are carburizing, cyaniding, carbonitriding, nitriding, induction hardening, and flame hardening.

Cast-alloy tool. A cutting tool made by casting a cobalt-base alloy and used at machining speeds between those for high-speed steels and sintered carbides.

Casting. (1) An object at or near finished shape, obtained by solidification of a substance in a mold. (2) Pouring molten metal into a mold to produce an object of desired shape.

Casting copper. Fire-refined, tough pitch copper usually cast from melted secondary metal into ingot and ingot bars only, and used for making foundry castings but not wrought products.

Casting shrinkage. (a) *Liquid shirnkage*: the reduction in volume of liquid metal as it cools to the liquidus. (2) *Solidification shrinkage*: the reduction in volume of metal from the beginning to ending of solidification. (3) *Solid shrinkage*: the reduction in volume of metal from the solidus to room temperature. (4) *Total shrinkage*: the sum of the shrinkage in parts (1), (2), and (3).

Casting strains. Strains in a casting caused by casting stresses that develop as the casting cools.

Casting stresses. Stresses set up in a casting because of geometry and casting shrinkage.

Cast iron. An iron containing carbon in excess of the solubility in the austenite that exists in the alloy at the eutectic temperature. For the various forms (gray cast iron, white cast iron, malleable cast iron, and nodular cast iron), the word *cast* is often left out, resulting in "gray iron," "white rion," "malleable iron," and "nodular iron," respectively.

Cast steel. Steel in the form of castings.

Cast structure. The internal physical structure of a casing evidence by shape and orientation of crystals and segregation of impurities.

Catalyst. A substance capable of changing the rate of a reaction without itself undergoing any net change.

Cathode. The electrode where electrons enter (current leaves) an operating system such as a battery, an electrolytic cell, an X-ray tube, or a vacuum tube. In the first of these, it is positive; in the other three, negative. In a battery or electrolytic cell, it is the electrode where reduction occurs. *Contrast with* Anode.

Cathode compartment. In an electrolytic cell, the enclosure formed by a diaphragm around the cathode.

Cathode copper. Copper deposited at the cathode in electrolytic refining.

Cathode-ray tube. A special form of vacuum tube in which a focused beam of electrons is caused to strike a surface coated with a phosphor. This beam is deflected so that it traces an orthogonal presentation of two separate signals; a third independent signal may be presented as a variation of the intensity of the electron beam, and in turn, the fluorescent intensity.

Caustic dip. A strongly alkaline solution into which metal is immersed for etching, neutralizing acid, or removing organic materials such as grease or paints.

Cavitation. The formation and instantaneous collapse of innumerable tiny voids or cavities within a liquid subjected to rapid and intense pressure changes. Cavitation produced by ultrasonic radiation is sometimes used to give violent localized agitation. That caused by severe turbulent flow often leads to cavitation damage.

Center drilling. Drilling a conical hole (pit) in one end of a workpiece.

Centerless grinding. Grinding the outside or inside of a workpiece mounted on rollers rather than on centers. The workpiece may be in the form of a cylinder or the frustum of a cone.

Centrifugal casting. A casting made by pouring metal into a mold that is rotated or revolved.

Cermet (ceramal). A body consisting of ceramic particles bonded with a metal.

Chafting fatigue. Fatigue initiated in a surface damaged by rubbing against another body. *See* Fretting.

Chamfer. (1) A beveled surface to eliminate an otherwise sharp corner. (2) A relieved angular cutting edge at a tooth corner.

Chamfer angle. (1) The angle between a referenced surface and the bevel. (2) On a milling cutter, the angle between a beveled surface and the axis of the cutter.

Chamfering. Making a sloping surface on the edge of a member. Also called *beveling.*

Charpy test. A pendulum-type single-blow impact test in which the specimen, usually notched, is supported at both ends as a simple beam and broken by a falling pendulum. The energy absorbed, as determined by the subsequent rise of the pendulum, is a measure of impact strength or notch toughness.

Chromate treatment. A treatment of metal in a solution of a hexavalent chromium compound to produce a conversion coating consisting of trivalent and hexavalent chromium compounds.

Chromating. Performing a chromate treatment.

Chromel. (1) A 90 Ni-10 Cr alloy used in thermocouples. (2) A series of nickel-chromium alloys, some with iron, used for heat-resistant applications.

Clay. An earthy or stony mineral aggregate consisting essentially of hydrous silicates of alumina, plastic when sufficiently pulverized and wetted, rigid when dry, and vitreous when fired at a sufficiently high temperature. Clay minerals most commonly used in the foundry are montmorillonites and kaolinites.

Cleavage. The splitting (fracture) of a crystal on a crystallographic plane of low index.

Cleavage fracture. A fracture, usually of a polycrystalline metal, in which most of the grains have failed by cleavage, resulting in bright reflecting facets. It is one type of crystalline fracture. *Contrast with* Shear fracture.

Coalescence. The union of particles of a dispersed phase into larger units, usually effected at temperatures below the fusion point.

Coarsening. See Grain growth.

Coating abrasive. An abrasive product. Sandpaper is an example in which a layer of abrasive particles is firmly attached to a paper, cloth, or fiber backing by means of glue or synthetic-resin adhesive.

Cobalt-60. A radioisotope with a half-life of 5.2 years and dominant-characteristic gamma-radiation energies of 1.17 and 1.33 MeV. It is used as a gamma-radiation source in industrial radiography and in therapy.

Coefficient of elasticity. See Modulus of elasticity.

Cohesion. Force of attraction between the molecules (or atoms) within a single phase. *Contrast with* Adhesion.

Cohesive strength. (1) The hypothetical stress in an unnotched bar causing tensile fracture without plastic deformation. (2) The stress corresponding to the forces between atoms. (3) Same as technical cohesive strength or disruptive strength.

Coil weld. A butt weld joining the ends of two metal sheets to make a continuous strip for coiling.

Coin silver. An alloy containing 90% silver, with copper being the usual alloying element.

Cold treatment. Cooling to a low temperature, often near $-100°F$, for the purpose of obtaining desired conditions or properties, such as dimensional or structural stability.

Cold welding. Solid-phase welding in which pressure, without pressure, without added heat, is used to cause interface movements that bring the atoms of the faying surfaces close enough together that a weld ensues.

Cold work. Permanent strain produced by an external force in a metal below its recrystallization temperature.

Cold working. Deforming metal plastically at a temperature lower than the recrystallization temperature.

Compound die. Any die so designed that it performs more than one operation on a part with one stroke of the press, such as blanking and piercing where all functions are performed simultaneously within the confines of the particular blank size being worked.

Compressive strength. The maximum compressive stress that a material is capable of developing, based on original area of cross section. In the case of a material that fails in compression by a shattering fracture, the compressive strength has a very definite value. In the case of materials that do not fail in compression by a shattering fracture, the value obtained for compressive strength is an arbitrary value depending upon the degree of distortion that is regarded as indicating complete failure of the material.

Continuous casting. A casting technique in which an ingot, billet, tube, or other shape is continuously solidified while it is being poured, so that its length is not determined by mold dimensions.

Continuous precipitation. Precipitation from a supersaturated solid solution accompanied by a gradual change of lattice parameter of the matrix with aging time. It is characteristic of the alloys that produce uniform precipitate throughout the grains. *See* Discontinuous precipitation.

Continuous weld. A weld extending continuously from one end of a joint to the other; where the joint is essentially circular, completely around the joint. *Contrast with* Intermittent weld.

Contour forming. See Stretch forming, Tangent bending, Wiper forming.

Contour machining. Machining of irregular surfaces, such as those generated in tracer turning, tracer boring, and tracer milling.

Contour milling. Milling of irregular surfaces. *See* Tracer milling.

Conversion coating. A coating consisting of a compound of the surface metal, produced by chemical or electromechanical treaments of the metal. (Examples are chromate coatings on zinc, cadmium, magnesium, and aluminum; oxides or phosphate coatings on steel.)

Copper brazing. Brazing with copper as the filler metal.

Corona. In spot welding, an area sometimes surrounding the nugget at the faying surfaces, contributing slightly to overall bond strength.

Corrosion. The deterioration of a metal by chemical or electromechemical reaction with its environment.

Corrosion embrittlement. The severe loss of ductility of a metal resulting from corrosive attack, usually intergranular and often not visually apparent.

Corrosion fatigue. Effect of the application of repeated or fluctuating stresses in a corrosive environment characterized by shorter life than would be encountered as a result of either the repeated or fluctuating stresses alone or the corrosive environment alone.

Corundum. Natural abrasive of the aluminum oxide type that has higher purity than emery.

Coupon. A piece of metal from which a test specimen is to be prepared—often an extra piece as on a casting or forging.

Covalent bond. A bond between two or more atoms resulting from the completion of shells by the sharing of electrons.

Creep. Time-dependent strain occurring under stress. The creep strain occurring at a diminishing rate is called *primary creep*; that occurring at a minimum and almost constant rate, *secondary creep*; that occurring at an accelerating rate, *tertiary creep.*

Creep limit. (1) The maximum stress that will cause less than a specified quantity of creep in a given time. (2) The maximum nominal stress under which the creep strain rage decreases continuously with time under constant load and at constant temperature. Sometimes used synonymously with Creep strength.

Creep recovery. Time-dependent strain after release of load in a creep test.

Creep strength. (1) The constant nominal stress that will cause a specified quantity of creep in a given time at constant temperature. (2) The constant nominal stress that will cause a specified creep rate at constant temperature.

Crevice corrosion. A type of concentration-cell corrosion; corrosion of a metal that is caused by the concentration of dissolved salts, metal ions, oxygen or other gases, and such, in crevices or pockets remote from the principal fluid stream, with a resultant building-up of differential cells, which ultimately causes deep pitting.

Critical point. (1) The temperature or pressure at which a change in crystal structure, phase, or physical properties occurs. *See* Transformation temperature. (2) In an equilibrium diagram, that specific value of composition, temperature, and pressure, or combinations thereof, at which the phases of a heterogeneous system are in equilibrium.

Critical temperature. (1) Synonymous with critical point if the pressure is constant. (2) The temperature above which the vapor phase cannot be condensed to liquid by an increase in pressure.

Crystal. A solid composed of atoms, ions, or molecules arranged in a pattern which is repetitive in three dimensions.

Crystalline fracture. A fracture of a polycrystalline metal characterized by a grainy appearance.

Crystallization. The separation, usually from a liquid phase on cooling, of a solid crystalline phase.

Cyanide copper. Copper electrodeposited from an alkali-cyanide solution containing a complex ion made up of univalent copper and the cyanide radical; also, the solution itself.

Cyaniding. Introducing carbon and nitrogen into a solid ferrous alloy by holding above Ac_1 in contact with molten cyanide of suitable composition. The cyanided alloy is usually quench-hardened.

dc casting. *See* Direct-chill casting.

Decalescence. A phenomenon, associated with the transformation of alpha iron to gamma iron on the heating (superheating) of iron or steel, revealed by the darkening of the metal surface owing to the sudden decrease in temperature caused by the fast absorption of the latent heat of transformation.

Decarburization. The loss of carbon from the surface of a ferrous alloy as a result of heating in a medium that reacts with the carbon at the surface.

Decay curve. A graphic presentation of the manner in which a quantity decays with time or, rarely, with distance through matter; usually refers to radioactive decay or decay of electrical and acoustical signals.

Degreasing. Removing oil or grease from a surface. *See* Vapor degreasing.

Deionization. Removal of ions from solution by chemical means.

Dendrite. A crystal that has a treelike branching pattern, being most evident in cast metals slowly cooled through the solidification range.

Density ratio. Powdered metal. The ratio of the determined density of a compact to the absolute density of metal of the same composition, usually expressed as a percentage.

Deoxidized copper. Copper from which cuprous oxide has been removed by adding a deoxider, such as phosphorus, to the molten bath.

Deoxidizer. A substance that can be added to molten metal to remove either free or combined oxygen.

Deoxidizing. (1) The removal of oxygen from molten metals by use of suitable deoxidizers. (2) Sometimes refers to the removal of undesirable elements other than oxygen by the introduction of elements or compounds that readily react with them. (3) In metal finishing, the removal of oxide films from metal surfaces by chemical or electronchemical reaction.

Depolarization. Reduction of polarization by changing the electrode film.

Depolarizer. A substance that produces depolarization.

Dezincification. Corrosion of some copper-zinc alloys involving loss of zinc and the formation of a spongy porous copper.

Diamond pyramid hardness test. An indentation hardness test employing a 136° diamond pyramid indenter and variable loads enabling the use of one hardness scale for all ranges of hardness from very soft lead to tungsten carbide.

Diamond tool. A diamond, shaped or formed to the contour of a single-pointed cutting tool, for use in the precision machining of nonferrous or nonmetallic materials.

Dichromate treatment. A chromate conversion coating produced on magnesium alloys in a boiling solution of sodium dichromate.

Die. (1) Various tools used to impart shape to material primarily because of the shape of the tool itself. Examples are blanking dies, cutting dies, drawing dies, forging dies, punching dies, and threading dies. (2) Powdered metal. The part or parts making up the confining form into which a powder is pressed. The parts of the die may include some or all of the following: die body, punches, core rods. Synonym: "mold."

Die block. The tool steel block into which the desired impressions are machined and from which forgings are produced.

Die body. Powdered metal. The stationary or fixed part of a die.

Die casting. (1) A casting made in a die. (2) A casting process where molten metal is forced under high pressure into the cavity of a metal mold.

Die welding. Forge welding between dies.

Dilatometer. An instrument for measuring the expansion or contraction in a metal resulting from changes in such factors as temperature or allotropy.

Dip brazing. Brazing by immersion in a molten salt or metal bath. Where a metal bath is employed, it may provide the filler metal.

Direct-chill (dc) casting. A continuous method of making ingots or billets for sheet or extrusion by pouring the metal into a short mold. The base of the mold is a platform that is gradually lowered while the metal solidifies, the frozen shell or metal acting as a retainer for the liquid metal below the wall of the mold. The ingot is usually cooled by the impingement of water directly on the mold or on the walls of the solid metal as it is lowered. The length of the ingot is limited by the depth to which the platform can be lowered; therefore, it is often called *semicontinuous casting.*

Directional solidification. The solidification of molten metal in a casting in such a manner that feed metal is always available for that portion that is just solidifying.

Direct quenching. Quenching carburized parts directly from the carburizing operation.

Dow process. A process for the production of magnesium electrolysis of molten magnesium chloride.

Draft. (1) The angle or taper on the surface of a punch or die, or the parts made with them, which facilitates the removal of the work. (2) The change in cross section in rolling or wiredrawing. (3) Taper put on the surfaces of a pattern so that it can be withdrawn successfully from the mold.

Drift. (1) A flat piece or steel of tapering width used to remove taper shank drills and other tools from their holders. (2) A tapered rod used to force mismated holes in line for riveting or bolting. Sometimes called a *drift pin.*

Dross. The scum that forms on the surface of molten metals largely because of oxidation but sometimes because of the rising of impurities to the surface.

Dry cyaniding. See Carbonitriding.

Dry method. In magnetic-particle inspection, a method in which a dry powder is used to detect magnetic-leakage fields.

Dry sand mold. A mold of sand and then dried.

Ductile crack propagation. Slow crack propagation that is accompanied by noticeable plastic deformation and requires energy to be supplied from outside the body.

Dye penetrant. Penetrant with dye added to render it more readily visible under normal conditions.

Eddy-current testing. Nondestructive testing method in which eddy-current flow is induced in the test object. Changes in the flow caused by variations in the object are reflected into a nearby coil or coils for subsequent analysis by suitable instrumentation and techniques.

Elastic deformation. Change of dimensions accompanying stress in the elastic range, original dimensions being restored upon release of stress.

Elastic hysteresis. Erroneously used for mechanical hysteresis. The effect is inelastic.

Elasticity. That property of a material by virtue of which it tends to recover its original size and shape after deformation.

Elastic limit. The maximum stress to which a material may be subjected without any permanent strain remaining upon complete release of stress.

Elastic modulus. See Modulus of elasticity.

Electrochemical corrosion. Corrosion that occurs when current flows between cathodic and anodic areas metallic surfaces.

Electrochemical series. See Electromotive series.

Electrode. (1) In arc welding, a current-carrying rod which supports the arc between the rod and work, or between two rods as in twin carbon-arc welding. It may or may not furnish filler metal. (2) In resistance welding, a part of a resistance welding machine through which current and, in most cases, pressure are applied directly to the work. The electrode may be in the form of a rotating wheel, rotating roll, bar, cylinder, plate, clamp, chuck, or modification thereof. (3) An electrical conductor for leading current into or out of a medium.

Electroforming. Making parts by electrodeposition on a removable form.

Electrogalvanizing. The electroplating of zinc upon iron or steel.

Electroless plating. Immersion plating where a chemical reducing agent changes metal ions to metal.

Electrolysis. Chemical change resulting from the passage of an electric current through an electrolyte.

Electrolyte. (1) An ionic conductor. (2) A liquid, most often a solution, that will conduct an electric current.

Electrolytic cell. An assembly, consisting of a vessel, electrodes, and an electrolyte, in which electrolysis can be carried out.

Electrolytic cleaning. Removing soil from work by electrolysis, the work being one of the electrodes. The electrolyte is usually alkaline.

Electrolytic copper. Copper that has been refined by electrolytic deposition, including cathodes that are the direct product of the refining operation, refinery shapes cast from melted cathodes, and, by extension, fabricators' product made therefrom. Usually when this term is used alone, it refers to electrolytic tough-pitch copper without elements other than oxygen being present in significant amounts.

Electromachining. (1) Electrical discharge machining. (2) Electrolytic machining.

Electromotive series. A list of elements arranged according to their standard electrode potentials. In corrosion studies, the analogous but more practical galvanic series of metals is generally used. The relative position of a given metal is not necessarily the same in the two series.

Electroplating. Electrodepositing metal (may be an alloy) in an adherent form upon an object serving as a cathode.

Elongation. In tensile testing, the increase in the gauge length, measured after fracture of the specimen within the gauge length, usually expressed as a percentage of the original gauge length.

Elutriation. Powdered metal. Classification of powder particles by means of a rising stream of gas or liquid.

Embrittlement. Reduction in the normal ductility of a metal due to a physical or chemical change.

Emery. An impure mineral of the corundum or aluminum oxide type used extensively as an abrasive before the development of electric-furnace products.

Emulsifying agent. A material that increases the stability of a dispers of one liquid in another.

Emulsion. A suspension of one liquid phase in another.

Emulsion cleaner. A cleaner consisting of organic solvents dispersed in an aqueous medium with the aid of an emulsifying agent.

Erosion. Destruction of metals or other materials by the abrasive action of moving fluids, usually accelerated by the presence of solid particules or matter in suspension. When corrosion occurs simultaneously, the term *erosion-corrosion* is often used.

Eutectic. (1) An isothermal reversible reaction in which a liquid solution is converted into two or more intimately mixed solids on cooling, the number of solids formed being the same as the number of components in the system. (2) An alloy having the composition indicated by the eutectic point on an equilibrium diagram. (3) An alloy structure of intermixed solid constituents formed by an eutectic reaction.

Eutectic melting. Melting of localized microareas whose composition corresponds to that of the eutectic in the system.

Eutectoid. (1) An isothermal reversible reaction in which a solid solution is converted into two or more intimately mixed solids on cooling, the number of solids formed being the same as the number of components in the system. (2) An alloy having the composition indicated by the eutectoid point on an equilibrium diagram. (3) An alloy structure of intermixed solid constituents formed by an eutectoid reaction.

Extensometer. An instrument for measuring changes in a linear dimension of a body caused by stress.

Extrusion. Conversion of a billet into lengths of uniform cross section by forcing the plastic metal through a die orifice of the desired cross-sectional outline. In "direct extrusion," the die and ram are at opposite ends of the billet, and the product and ram travel in the same direction. In "indirect extrusion" (rare), the die is at the ram end of the billet, and the product travels through and in the opposite direction to the hollow ram. A "stepped extrusion" is a single product with one or more abrupt cross-section changes and is obtained by interrupting the extrusion by die changes. "Impact extrusion" (cold extrusion) is the process or resultant product of a punch striking an unheated slug in a confining die. The metal flow may be either between the punch and die or through another opening. (*See* Hooker process, which uses a pierced slug.) "Hot extrusion" is similar to cold extrusion except that a preheated slug is used and the pressure application is slower.

Extrusion billet. A cast or wrought metal slug used for extrusion.

Extrusion defect. A defect of flow in extruded products caused by the oxidized outer surface of the billet flowing into the center of the extrusion. It normally occurs in the last 10% to 20% of the extruded bar. Also called *pipe* or *core.*

Extrusion ingot. A solid or hollow cylindrical casting used for extruding into rods, bars, shapes, or tubes.

Fatigue. The phenomenon leading to fracture under repeated or fluctuating stresses having a maximum value less than the tensile strength of the material. Fatigue fractures are progressive, beginning as minute cracks that grow under the action of the fluctuating stress.

Fatigue life. The number of cycles of stress that can be sustained prior to failure for a stated test condition.

Fatigue limit. The maximum stress below which a material can presumably endure an infinite number of stress cycles. If the stress is not completely reversed, the value of the mean stress, the minimum stress, or the stress ratio should be stated.

Fatigue ratio. The ratio of the fatigue limit for cycles of reversed flexural stress to the tensile strength.

Fatigue strength. The maximum stress that can be sustained for a specified number of cycles without failure, the stress completely reversed within each cycle unless otherwise stated.

Fatigue-strength reduction factor (K_f). The ratio of the fatigue strength of a member or specimen with no stress concentration to the fatigue strength with stress concentration. K_f has no meaning unless the geometry, size, and material of the member or specimen and stress range are stated.

Faying surface. The surface of a piece of metal (or a member) in contact with another to which it is to be joined.

Ferrite. (1) A solid solution of one or more elements in body-centered cubic iron. Unless otherwise designated (for instance, as chromium ferrite), the solute is generally assumed to be carbon. On some equilibrium diagrams, there are two ferrite regions separated by an austenite area. The lower area is alpha ferrite; the upper delta ferrite. If there is no designation, alpha ferrite is assumed. (2) In the field or magnetics, substances having the general formula:

$$M^{++}O \,.\, M_2^{+++}O_3,$$

the trivalent metal often being iron.

File hardness. Hardness as determined by the use of a file of standardized hardness on the assumption that a material that cannot be cut with the file is as hard as, or harder than, the file. Files covering a range of hardnesses may be employed.

Fillet. (1) A radius (curvature) imparted to inside meeting surfaces. (2) A concave cornerpiece used on foundry patterns.

Fillet weld. A weld, approximately triangular in cross section, joining two surfaces essentially at right angles to each other in a lap, tee or corner joint.

Fine silver. Silver with a fineness of 999; equivalent to a minimum content of 99.9% silver with the remaining content not restricted.

Flame annealing. Annealing in which the heat is applied directly by a flame.

Flame straightening. Correcting distortion in metal structure by locatized heating with a gas flame.

Flash welding. A resistance butt-welding process in which the weld is produced over the entire abutting surface by pressure and heat, the heat being produced by electric arcs between the numbers.

Flow brazing. Brazing by pouring molten filler metal over a joint.

Fluorescence. The emission of characteristics electromagnetic radiation by a substance as a result of the absorption of electromagnetic or corpuscular radiation having a greater unit energy than that of the fluorescent radiation. It occurs only as long as the stimulus responsible for it is maintained.

Fluorescent magnetic-particle inspection. Inspection with either dry magnetic particles or those in a liquid suspension, the particles being coated with a fluorescent substance to increase the visibility of the indications.

Fluoroscopy. An inspection procedure in which the radiographic image of the subject is viewed on a fluorescent screen, normally limited to low-density materials or thin section of metals because of the low light output of the fluorescent screen at safe levels of radiation.

Flute. (1) As applied to drills, reamers, and taps, the channels or grooves formed in the body of the tool to provide cutting edges and to permit passage of cutting fluid and chips. (2) As applied to milling cutters and hobs, the chip space between the back of one tooth and face of the following tooth.

Flux. (1) In metal refining, a material used to remove undesirable substances, such as sand, ash, or dirt, as a molten mixture. It is also used as a protective covering for certain molten-metal baths. Lime or limestone is generally used to remove sand, as in iron smelting; sand, to remove iron oxide in copper refining. (2) In brazing, cutting, soldering, or welding, material used to prevent the formation of, or to dissolve and facilitate removal of, oxides and other undesirable substances.

Flux density. (1) In magnetism, the number of flux lines per unit area passing through a cross section at right angles. It is given by $B = \mu H$, where μ and H are permeability and magnetic field intensity, respectively. (2) In neutron radiation, the neutron flux, total.

Forge welding. Welding hot metal by pressure or blows only.

Forging. Plastically deforming metal, usually hot, into desired shapes with compressive force, with or without dies.

Forming. Making a change, with the exception of shearing or blanking, in the shape or contour of a metal part without intentionally altering the thickness.

Foundry. A commercial establishment or building where metal castings are produced.

Fracture stress. (1) The maximum principal true stress at fracture. Usually refers to unnotched tensile specimens. (2) The (hypothetical) true stress, which will cause fracture without further deformation at any given strain.

Free carbon. The part of the total carbon in steel or cast iron that is present in the elemental form as graphite or temper carbon.

Free cyanide. The cyanide not combined in complex ions.

Free ferrite. Ferrite that is structurally separate and distinct, as may be formed without the simultaneous formation of carbide when cooling hypoeutectoid austenite into the critical temperature range. Also proeutectoid ferrite.

Free fit. Various clearance fits for assembly by hand and free rotation of parts. *See* Running fit.

Free machining. Pertains to the machining characteristics of an alloy to which an ingredient has been introduced to give small broken chips, lower power consumption, better surface finish, and longer tool life; among such additions are sulfur or lead to steel, lead to brass, lead and bismuth to aluminum, sulfur or selenium to stainless steel.

Fretting (*fretting corrosion*). Action that results in surface damage, especially in a corrosive environment, when there is relative motion between solid surfaces in contact under pressure. *See* Chafing fatigue.

Full annealing. Annealing a ferrous alloy by austenitizing and then cooling slowly through the transformation range. The austenitizing temperature for hypoeutectoid steel is usually above Ac_1; and for hypereutectoid steel, usually between Ac_1 and Ac_{cm}.

Fusion welding. Welding, without pressure, in which a portion of the base metal is melted.

Galling. Developing a condition on the rubbing surface of one or both mating parts where excessive friction between high spots results in localized welding with subsequent spalling and a further roughening of the surface.

Galvanic cell. A cell in which chemical change is the source of electrical energy. It usually consists of two dissimilar conductors in contact with each other and with an electrolyte or of two similar conductors in contact with each other and with dissimilar electrolytes.

Galvanic corrosion. Corrosion associated with the current of a galvanic cell consisting of two dissimilar conductors in an electrolyte or two similar conductors in dissimilar electrolytes. Where the two dissimilar metals are in contact, the resulting reaction is referred to as "couple action."

Galvanic series. A series of metals and alloys arranged according to their relative electrode potentials in a specified enviornment.

Gamma. (1) In photography, the slope of the straight-line portion of a film's characteristic curve. (2) Also used inexactly in photography to refer to film contrast, gradient, or average gradient.

Gammagraphs. A radiograph produced by gamma rays.

Gamma iron. The face-centered cubic form of pure iron, stable from 1670 to 2550°F.

Gamma ray. Short-wavelength electromagnetic radiation of nuclear origin with a range of wavelengths from about 10^{12} to 10^{9} cm.

Gas-shielded arc welding. Arc welding in which the arc and molten metal are shielded from the atmosphere by a stream of gas, such as argon, helium, argon-hydrogen mixtures, or carbon dioxide.

Gold-filled. Covered on one or more surfaces with a layer of gold alloy to form a clad metal. By commercial agreement, a quality mark showing the quantity and fineness of gold alloy may be affixed, which shows the actual proportional weight and karat finess of the gold-alloy cladding. For example, "$\frac{1}{10}$ 12K Gold Filled" means that the article consists of base metal covered on one or more surfaces with a gold alloy of 12-karat fineness comprising one-tenth part by weight of the entire metal in the article. No article having a gold alloy coating of less than 10-karat fineness may have any quality mark affixed. No article having a gold-alloy portion of less than one-twentieth by weight may by marked "Gold Filler," but may be marked "Rolled Gold Plate" provided the proportional fraction and fineness designation precede. These standards do not necessarily apply to watchcases.

Grain. An individual crystal in a polycrystalline metal or alloy.

Grain-fineness number. A weighted-average grain size of a granular material. The AFS grain-fineness number is calculated with prescribed weighting factors from the standard screen analysis.

Grain growth (coarsening). An increase in the size of grains in polycrystalline metal, usually effected during heating at elevated temperatures. The increase may be gradual or abrupt, resulting in either uniform or nonuniform grains after growth has ceased. A mixture of nonuniform grains is sometimes termed *duplexed.* Abnormal grain growth (exaggerated grain growth) implies the formation of excessively large grains, uniform or nonuniform. The abrupt form of abnormal grain growth is also termed *germinative grain growth* when a critical amount of strain or other nuclei are present to promote the growth. Secondary recrystallization is the selective grain growth of a few grains only, as distinct from uniform coarsening, when the new set of grains resulting from primary recrystallization is subjected to further annealing.

Grain size. (1) For metals, a measure of the areas or volumes of grains in a polycrystalline material, usually expressed as an average when the individual sizes are fairly uniform. Grain sizes are reported in terms of number of grains per unit area or volume, average diameter, or as a grain-size number derived from area measurements. (2) For grinding wheels, *see* the preferred term, Grit size.

Granular fracture. A type of irregular surface produced when metal is broken, which is characterized by a rough, grainlike appearance as differentiated from a smooth silky, or fibrous, type. It can be subclassified into transgranular and intergranular forms. This type of fracture is frequently called *crystalline fracture*, but the inference that the metal has crystallized is not justified.

Granulation. The production of coarse metal particles by pouring the molten metal through a screen into water or by agitating the molten metal violently during its solidification.

Grit size. Nominal size of abrasive particles in a grinding wheel corresponding to the number of openings per linear inch in a screen through which the particles can just pass. Sometimes called *grain size.*

Hard chromium. Chromium deposited for engineering purposes, such as increasing the wear resistance of sliding metal surfaces, rather than as a decorative coating. It is usually applied directly to basis metal and is customarily thicker than a decorative deposit.

Hardener. An alloy, rich in one or more alloying elements, added to a melt to permit closer composition control than possible by addition of pure metals or to introduce refractory elements not readily alloyed with the base metal.

Hardness. (1) Resistance of metal to plastic deformation usually by indentation. However, the term may also refer to stiffness or temper, or to resistance to scratching, abrasion, or cutting. Indentation hardness may be measured by various hardness tests, such as Brinell, Rockwell, and Vickers. (2) For grinding wheels, same as grade.

Heat treatment. Heating and cooling a solid metal or alloy in such a way as to obtain desired conditions or properties. Heating for the sole purpose of hot working is excluded from the meaning of this definition.

Hooke's law. Stress is proportional to strain. The law holds only up to the proportional limit.

Hydrogen brazing. Brazing in a hydrogen atmosphere, usually in a furnace.

Hydrogen embrittlement. A condition of low ductility in metals resulting from the absorption of hydrogen.

Impedance. (1) Acoustical impedance is the complex ratio of the sound pressure to the product of the product of the sound velocity and the area at a given surface. It is frequently approximated by only the product of the density and velocity. (2) Electrical impedance is the complex property of an electrical circuit, or the components of a circuit, that opposes the flow of an alternating current. The real part represents the resistance, and the imaginary part represents the reactance of the circuit.

Impregnation. (1) The treatment of porous castings with a sealing medium to stop pressure leaks. (2) The process of filling the pores of a sintered compact, usually with a liquid such as a lubricant. (3) The process of mixing particles of a nonmetallic substance in a matrix of metal powder, as in diamond-impregnated tools.

Indirect extrusion. See Extrusion.

Induction brazing. Brazing with induction heat.

Inert-gas shielded-arc cutting. Metal cutting with the heat of an arc in an inert gas such as argon or helium.

Inert-gas shielded-arc welding. Arc welding in an inert gas such as argon or helium.

Intergranular corrosion. Corrosion occurring preferentially at grain boundaries.

Intermediate annealing. Annealing wrought metals at one or more stages during manufacture and before final treatment.

Intermittent weld. A weld in which the continuity is broken by recurring unwelded spaces.

Investment casting. (1) Casting metal into a mold produced by surrounding (investing) an expendable pattern with a refractory slurry that sets at room temperature after which the wax, plastic, or frozen mercury pattern is removed through the use of heat. Also called *precision casting* or *lost-wax* process. (2) A casting made by the process.

Investment compound. A mixture of a graded refractory filler, a binder, and a liquid vehicle, used to make molds for investment casting.

Ion. An atom, or group of atoms, that has gained or lost one or more outer electrons and thus carries an electric charge. Positive ions, or cations, are deficient in outer electrons. Negative ions, or anions, have an excess of outer electrons.

Iridium-192. A radioisotope with a half-life of 74 days and 12 dominant, characteristic, gamma-radiation energies ranging from 0.14 to 0.65 MeV. It is suitable as a gamma radiation source, mostly in radiography.

Iron. (1) Element No. 26 of the periodic system, the average atomic weight of the naturally occurring isotopes being 55.85. (2) Iron-base materials not falling into the steel classifications. *See* Gray cast iron, Ingot iron, Malleable cast iron, Nodular cast iron, White cast iron, and Wrought iron.

Isothermal annealing. Austenitizing a ferrous alloy and then cooling to and holding at a temperature at which austenite transforms to a relatively soft ferrite carbide aggregate.

Isothermal transformation. A change in phase at any constant temperature.

Isotropy. Quality of having identical properties in all directions.

Izod test. A pendulum type of single-blow impact test in which the specimen, usually notched, is fixed at one end and broken by a falling pendulum. The energy absorbed, as measured by the subsequent rise of the pendulum, is a measure of impact strength or notch toughness.

Jig boring. Boring with a single-point tool where the work is positioned upon a table that can be located so as to bring any desired part of the work under the tool. Thus, holes can be accurately spaced. This type of boring can be done on milling machines or "jig borers."

Joggle. An offset in a flat plane consisting of two parallel bends in opposite directions by the same angle.

Karat. A 24th part, used to designate the fineness of gold. Abbreviated *K*, *Kt.* Sometimes spelled "carat."

Knoop hardness. Microhardness determined from the resistance of metal to indentation by a pyramidal diamond indenter, having edge angles of 172°, 30′, and 130°, making a rhombohedral impression with one long and one short diagonal.

Knurling. Impressing a design into a metallic surface, usually by means of small, hard rollers that carry the corresponding design on their surfaces.

Latent heat. Thermal energy absorbed or released when a substance undergoes a phase change.

Lay. Direction of predominant surface pattern remaining after cutting, grinding, lapping, or other processing.

Liquid honing. Polishing metal by bombardment with an air-ejected liquid containing fine solid particles in suspension. If an impeller wheel is used to propel the suspension, the process is called *wet blasting.*

Liquidus. In a constitution or equilibrium diagram, the locus of points representing the temperatures at which the various compositions in the system begin to freeze on cooling or to finish melting on heating.

Lost-wax process. An investment casting process in which a wax pattern is used.

Lubricant. Any substance used to reduce friction between two surfaces in contact.

Machining. Removing material, in the form of chips, from work, usually through the use of a machine.

Macro-etch. Etching of a metal surface for accentuation of gross structural details and defects for observation by the unaided eye or at magnifications not exceeding 10 diameters.

Macrograph. A graphic reproduction of the surface of a prepared specimen at a magnification not exceeding 10 diameters. When photographed, the reproduction is known as a *photomacrograph.*

Macroscopic. Visible at magnifications from 1 to 10 diameters.

Macrostructure. The structure of metals as revealed by examination of the etched surface of a polished specimen at a magnification not exceeding 10 diameters.

Magnesite wheel. A grinding wheel bonded with magnesium oxychloride.

Magnetic-particle inspection. A nondestructive method of inspection for determining the existence and extent of possible defects in ferromagnetic materials. Finely divided magnetic particles, applied to the magnetized part, are attracted to and outline the pattern of any magnetic-leakage fields created by discontinuities.

Malleability. The characteristic of metals that permits plastic deformation in compression without rupture.

Martempering. Quenching an austenitized ferrous alloy in a medium at a temperature in the upper part of the martensite range, or slightly above that range, and holding it in the medium until the temperature throughout the alloy is substantially uniform. The alloy is then allowed to cool in air through the martensite range.

Martensite. (1) In an alloy, a metastable transitional structure intermediate between two allotropic modifications whose abilities to dissolve solute differ considerably, the high-temperature phase having the greater solubility. The amount of the high-temperature phase transformed to martensite depends to a large extent upon the temperature attained in cooling, there being a rather distinct beginning temperature. (2) A metastable phase of steel, formed by a transformation of austenite below the M (or Ar") temperature. It is an interstitial supersaturated solid solution of carbon in iron having a body-centered tetragonal lattice. Its microstructure is characterized by an acicular, or needle-like, pattern.

Martensitic transformation. A reaction that takes place in some metals on cooling, with the formation of an acicular structure called *martensite.*

Metal. (1) An opaque lustrous elemental chemical substance that is a good conductor of heat and electricity and, when polished, a good reflector of light. Most elemental metals are malleable and ductile and are, in general, heavier than other elemental substances. (2) As to structure, metals may be distinguished from nonmetals by their atomic binding and electron availablity. Metallic atoms tend to lose electrons from the outer shells, the positive ions thus formed arc held together by the electron gas produced by the separation. The ability of these free electrons to carry an electric current, and the fact that the conducting power decreases as temperature increases, establish one of the prime distinctions of a metallic solid. (3) From the chemical viewpoint, an elemental substance whose hydroxide is alkaline. (4) An alloy.

Metal-arc welding. Arc welding with metal electrodes. Commonly refers to shielded metal-arc welding using covered electrodes.

Metallurgy. The science and technology of metals. Process (chemical) metallurgy is concerned with the extraction of metals from their ores and with the refining of metals; physical metallurgy, with the physical and mechanical properties of metals as affected by composition, mechanical working, and heat treatment.

Micrograph. A graphic reproduction of the surface of a prepared specimen, usually etched, at a magnification greater than 10 diameters. If produced by photographic means, it is called a *photomicrograph* (not a microphotograph).

Microshrinkage. A casting defect, not detectable at magnifications lower than 10 diameters, consisting of interdendritic voids. This defect results from contraction during solidification where there is not an adequate opportunity to supply filler material to compensate for shrinkage. Alloys with a wide range in solidification temperature are particularly susceptible.

Misrun. A casting not fully formed, resulting from the metal solidifying before the mold is filled.

Modulus of elasticity. A measure of the rigidity of metal. Ratio of stress, within proportional limit, to corresponding strain. Specifically, the modulus obtained in tension or compression is Young's modulus, stretch modulus, or modulus of extensibility; the modulus obtained in torsion or shear is modulus of rigidity, shear modulus, or modulus of torsion; the modulus covering the ratio of the mean normal stress to the change in volume per unit volume is the bulk modulus. The tangent modulus and secant modulus are not restricted within the proportional limit; the former is the slope of the stress-strain curve at a specified point; the latter is the slope of a line from the origin to a specified point on the stress-strain curve. Also called *elastic modulus* and *coefficient of elasticity.*

Modulus of rigidity. See Modulus of elasticity.

Modulus of rupture. Nominal stress at fracture in a bend test or torsion test. In bending, modulus of rupture is the bending moment at fracture divided by the section modulus. In torsion, modulus of rupture is the torque at fracture divided by the polar section modulus.

Monotectic. An isothermal reversible reaction in a binary system, in which a liquid on cooling decomposes into a second liquid of a different composition and a solid. It differs from a eutectic in that only one of the two products of the reaction is below its freezing range.

Natural aging. Spontaneous aging of a supersaturated solid solution at room temperature. *See* Aging. *Compare with* Artificial aging.

Nitriding. Introducing nitrogen into a solid ferrous alloy by holding at a suitable temperature (below Ac_1 for ferritic steels) in contact with a nitrogenous material, usually ammonia or molten cyanide of appropriate composition. Quenching is not required to produce a hard case.

Noble metal. (1) A metal whose potential is highly positive relative to the hydrogen electrode. (2) A metal with marked resistance to chemical reaction, particularly to oxidation and to solution by inorganic acids. The term as often used is synonymous with *precious metal.*

Nondestructive inspection. Inspection by methods that do not destroy the part to determine its suitability for use.

Nondestructive testing. See Nondestructive inspection.

Normalizing. Heating a ferrous alloy to a suitable temperature above the transformation range and then cooling in air to a temperature substantially below the transformation range.

Optical pyrometer. An instrument for measuring the temperature of heated material by comparing the intensity of light emitted with a known intensity of an incandescent lamp filament.

Overaging. Aging under conditions of time and temperature greater than those required to obtain maximum change in a certain property, so that the property is altered in the direction of the initial value. *See* Aging.

Particle size. Powdered metal. The controlling lineal dimension of an individual particle, as determined by analysis with screens or other suitable instruments.

Parting. (1) In the recovery of precious metals, the separation of silver from gold. (2) The zone of separation between cope and drag portions of mold or flask in sand casting. (3) A composition sometimes used in sand molding to facilitate the removal of the pattern. (4) Cutting simultaneously along two parallel lines or along two lines that balance each other in the matter of side thrust. (5) A shearing operation used to produce two or more parts from a stamping.

Parting line. A plane on a pattern or a line on a casting corresponding to the separation between the cope and drag portions of a mold.

Passivation. The changing of the chemically active surface of a metal to a much less reactive state.

Passivity. A condition in which a piece of metal, because of an impervious covering of oxide or other compound, has a potential much more positive than where the metal is in the active state.

Pearlite. A lamellar aggregate of ferrite and cementite, often occurring in steel and cast iron.

Penetrant. A liquid with low surface tension used in penetrant inspection to flow into surface openings of parts being inspected.

Penetrant inspection. A method of nondestructive testing for determining the existence and extent of discontinuities that are open to the surface in the part being inspected. The indications are made visible through the use of a dye or fluorescent chemical in the liquid employed as the inspection medium.

Penetration. (1) In founding, a defect on a casting surface caused by metal running into voids between sand grains. (2) In welding, the distance from the original surface of the base metal to that point at which fusion ceased.

Percussion welding. Resistance welding simultaneously over the entire area of abutting surfaces with arc heat, the pressure being applied by a hammerlike blow during or immediately following the electrical discharge.

Peritectoid. An isothermal reversible reaction in which a solid phase reacts with a second solid phase to produce yet a third solid phase on cooling.

Permanent mold. A metal mold (other than an ingot mold) of two or more parts that is used repeatedly for the production of many castings of the same form. Liquid metal is poured in by gravity.

Permanent set. Plastic deformation that remains upon releasing the stress that produces the deformation.

Permeability. (1) *Founding.* The characteristics of molding materials that permit gases to pass through them. *Permeability number* is determined by a standard test. (2) *Powdered metal.* A property measured as the rate of passage under specified conditions of a liquid or gas through a compact. (3) *Magnetism.* A general term used to express various relationships between magnetic induction and magnetizing force. These relationships are either "absolute permeability," which is the quotient of a change in magnetic induction divided by the corresponding change in magnetizing force, or "specific (relative) permeability," the ratio of the absolute permeability to the permeability of free space.

pH. The negative logarithm of the hydrogen ion activity; it denotes the degree of acidity or basicity of a solution. At 25°C, seven is the neutral value. Acidity increases with decreasing values below seven; basicity increases with increasing values above seven.

Plaster molding. Molding wherein a gypsum-bonded aggregate flour in the form of a water slurry is poured over a pattern, permitted to harden, and, after removal of the pattern, thoroughly dried. The technique is used to make smooth nonferrous castings of accurate size.

Plastic deformation. Deformation that does or will remain permanent after removal of the load that caused it.

Platen. (1) The sliding member of slide of a hydraulic press. *See* Slide and ram. (2) A part of a resistance welding, mechanical testing, or other machine with a flat surface, to which dies, fixtures, backups, or electrode holders are attached, and which transmits pressure.

Plating. Forming an adherent layer of metal upon an object.

Plug. (1) A rod or mandrel over which a pierced tube is forced. (2) A rod or mandrel that fills a tube as it is drawn through a die. (3) A punch or mandrel over which a cup is drawn. (4) A protruding portion of a die impression for forming a corresponding recess in the forging. (5) A false bottom in die. Also called a *peg.*

Plug tap. A tap with chamfer extending from three to five threads.

Poisson's ratio. The absolute value of the ratio of the transverse strain to the corresponding axial strain, in a body subjected to uniaxial stress; usually applied to elastic conditions.

Polarization. In electrolysis, the formation of a film on an electrode such that the potential necessary to get a desired reaction is increased beyond the reversible electrode potential.

Precious metal. One of the relatively scarce and valuable metals: gold, silver, and the platinum-group metals.

Precipitation hardening. Hardening caused by the precipitation of a constituent from a supersaturated solid solution. *See* Age hardening, Aging.

Precipitation heat treatment. Artificial aging in which a constituent precipitates from a supersaturated solid solution.

Proof stress. (1) The stress that will cause a specified small permanent set in a material. (2) A specified stress to be applied to a member or structure to indicate its ability to withstand service loads.

Proportional limit. The maximum stress at which strain remains directly proportional to stress.

Pyrometallurgy. Metallurgy involved in winning and refining metals where heat is used, as in roasting and smelting.

Pyrometer. A device for measuring temperatures above the range of liquid thermometers.

Quarter hard. See Temper of copper, Copper alloys.

Quench aging. Aging induced by rapid cooling after solution heat treatment.

Quench annealing. Annealing an austenitic ferrous alloy by solution-heat treatment.

Quench hardening. Hardening a ferrous alloy by austenitizing and then cooling rapidly enough so that some or all of the austenite transforms to martensite. The austenitizing temperature for hypoeutectoid steels is usually above Ac_3; for hypereutectoid steels, it usually ranges from Ac_1 and Ac_{cm}.

Quenching. Rapid cooling. When applicable, the following more specific terms should be used: *direct quenching*, *fog quenching*, *hot quenching*, *interrupted quenching*, *selective quenching*, *spray quenching*, and *time quenching.*

Quench time. In resistance welding, the time from the finish of the weld to the beginning of temper. Also called *chill time.*

Radiation energy. The energy of a given photon or particle in a beam of radiation, usually expressed in electron volts.

Radioactive element. An element that has at least one isotope that undergoes spontaneous nuclear disintegration to emit positive alpha particles, negative beta particles, or gamma rays.

Radioactivity. The spontaneous nuclear disintegration with emission of corpuscular or electromagnetic radiation.

Radiograph. A photographic shadow image resulting from uneven absorption of radiation in the object being subjected to penetrating radiation.

Radiographic contrast. The difference in density between an image, or part of an image, and its immediate surroundings on a radiograph. Radiographic contrast depends upon both subject contrast and film contrast.

Radiographic equivalence factor. The reciprocal of the thickness of a given material taken as a standard. It not only depends on the standard, but also on the radiation quality.

Radiographic sensitivity. A measure of quality of radiographs whereby the minimum discontinuity that may be detected on the film is expressed as a percentage of the base thickness. It depends on subject and film contrast and on geometrical and film graininess factors.

Radiography. A nondestructive method of internal examination in which metal or other objects are exposed to a beam of X-ray or gamma radiation. Differences in thickness, density, or absorption, caused by internal discontinuities, are apparent in the shadow image either on a fluorescent screen or on photographic film place behind the object.

Radioisotope. An isotope that emits ionizing radiation during its spontaneous decay. *See* Cesium-137, Cobalt-60, Iridium-192, Radium, Radon, Strontium-90, and Thulium-170, which are used commercially.

Radium. A radioactive element. It is found in nature as radium-226, which has a half-life of 1620 years. In equilibrium with its daughter products, it emits 11 principal gamma rays, ranging from 0.24 to 2.20 MeV. It is used as a gamma-radiation source, especially in radiography and therapy.

Rare earth metal. One of the group of 15 similar metals with atomic numbers 57 through 71.

Recrystallization. (1) The change from one crystal structure to another, as occurs on heating or cooling through a critical temperature. (2) The formation of a new, strain-free grain structure from that existing in cold-worked metal, usually accomplished by heating.

Recrystallization annealing. Annealing cold-worked metal to produce a new grain structure without phase change.

Recrystallization temperature. The approximate minimum temperature at which complete recrystallization of a cold-worked metal occurs within a specified time.

Reduction of area. (1) Commonly, the difference, expressed as a percentage of original area, between the original cross-sectional area of a tensile test specimen and the minimum cross-sectional area measured after complete separation. (2) The difference, expressed as a percentage of original area, between original cross-sectional area and that after straining the specimen.

Refractory. (1) The material of very high melting point with properties that make it suitable for such uses as furnace linings and kiln construction. (2) The quality of resisting heat.

Refractory alloy. (1) A heat-resistant alloy. (2) An alloy having an extremely high melting point. *See* Refractory metal. (3) An alloy difficult to work at elevated temperatures.

Refractory metal. A metal having an extremely high melting point. In the broad sense, it refers to metals having melting points above the ranges of iron, cobalt, and nickel.

Resistance brazing. Brazing by resistance heating, the joint being part of the electrical circuit.

Resistance welding. Welding with resistance heating and pressure, the work being part of the electrical circuit. Example: resistance spot welding, resistance seam welding, projection welding, and flash butt welding.

Rockwell hardness test. A test for determining the hardness of a material based upon the depth of penetration of a specified penetrator into the specimen under certain arbitrarily fixed conditions of test.

Rough machining. Machining without regard to finish, usually to be followed by a subsequent operation.

Roughness. Relatively finely spaced, surface irregularities, of which the height, width, and direction establish the predominant surface pattern.

Running fit. Any clearance fit in the range used for parts that rotate relative to each other. Actual values of clearance resulting from stated shaft and hole tolerances are given for nine classes of running and sliding fits for 21 nominal shaft sizes in ASA B4.1-1955.

Runout. (1) The unintentional escape of molten metal from a mold, crucible, or furnace. (2) The defect in a casting caused by the escape of metal from the mold. (3) *See* Axial runout, Radial runout.

Rust. A corrosion product consisting of hydrated oxides of iron. Applied only ferrous alloys.

Salt fog test. An accelerated corrosion test in which specimens are exposed to a fine mist of a solution usually containing sodium chloride but sometimes modified with other chemicals. For testing details, see ASTM B117 and B287.

Sand. A granular material resulting from the disintegration of rock. Foundry sands are mainly silica. *Bank sand* is found in sedimentary deposits and contains less than 5% clay. *Dune sand* occurs in windblown deposits near large bodies of water and is very high in silica content. *Molding sand* contains more than 5% clay, usually between 10 and 20%. *Silica sand* is a granular material containing at least 95% silica and often more than 99%. *Core sand* is nearly pure silica. *Miscellaneous sands* include zircon, olivine, calcium carbonate, lava, and titanium minerals.

Scleroscope test. A hardness test when the loss in kinetic energy of a falling metal "tup," absorbed by indentation upon impact of the tup on the metal being tested, is indicated by the height of rebound.

Seam welding. (1) Arc or resistance welding in which a series of overlapping spot welds is produced with rotating electrodes, rotating work, or both. (2) Making a longitudinal weld in sheet metal or tubing.

Shear. (1) That type of force that causes, or tends to cause, two contiguous parts of the same body to slide relative to one another in a direction parallel to their plane of contact.

Shear fracture. A fracture in which a crystal (or a polycrystalline mass) has separated by sliding or tearing under the action of shear stresses.

Shearing strain (shear strain). See Strain.

Shear modulus. See Modulus of elasticity.

Shear plane. A confined zone along which shear takes place in metal cutting. It extends from the cutting edge to the work surface.

Shear strength. The stress required to produce fracture in the plane of cross section, the conditions of loading being such that the directions of force and of resistance are parallel and opposite although their paths are offset a specified minimum amount.

Shielded-arc welding. Arc welding in which the arc and the weld metal are protected by a gaseous atmosphere, the products of decomposition of the electrode covering, or a blanket of fusible flux.

Shielded metal-arc welding. Arc welding in which the arc and the weld metal are protected by the decomposition products of the covering on a consumable metal electrode.

Shore hardness test. See Scleroscope test.

Shrink fit. A fit that allows the outside member, when heated to a practical temperature, to assemble easily with the inside member.

Sigma phase. A hard brittle nonmagnetic intermediate phase with a tetragonal crystal structure, containing 30 atoms per unit cell, space group $P4_2/mnm$, occurring in many binary and ternary alloys of the transition elements. The composition of this phase in the various systems is not the same, and the phase usually exhibits a wide range in homogeneity. Alloying with a third transition element usually enlarges the field of homogeneity and extends it deep into the ternary section.

Silver brazing. Brazing with silver-base alloys as the filler metal.

Silver brazing alloy. Filler metal used in silver brazing.

Silver solder. See Silver brazing alloy.

Sliding fit. A series of nine classes of running and sliding fits of 21 nominal shaft sizes defined in terms of clerance and tolerance of shaft and hole in ASA B4.1-1955.

Smelting. Thermal processing wherein chemical reactions take place to produce liquid metal from a beneficiated ore.

Snug fit. A loosely defined fit implying the closest clearances that can be assembled manually for firm connection between parts and comparable to one or more of the 11 classes of clearance locational fits given in ASA B4.1-1955.

Soldering. Similar to brazing, with the filler metal having a melting temperature range below an arbitrary value, generally 800°F. Soft solders are usually lead-tin alloys.

Solidus. In a constitution or equilibrium diagram, the locus of points representing the temperatures at which various compositions finish freezing or cooling or begin to melt on heating.

Spalling. The cracking and flaking of particles out of a surface.

Spot welding. Welding of lapped parts in which fusion is confined to a relatively small circular area. It is generally resistance welding, but may also be gas-shielded tungsten-arc, gas-shielded metal-arc, or submerged-arc welding.

Steel. An iron-base alloy, malleable in some temperature range as initially cast, containing manganese, usually carbon, and often alloying elements. In carbon steel and low-alloy steel, the maximum carbon is about 2.0%; in high-alloy steel, about 2.5%. The dividing line between low-alloy and high-alloy steel is generally regarded as being at about 5% metallic alloying elements.

Steel is to be differentiated from two general classes of "irons": the cast irons on the high-carbon side, and relatively pure irons such as ingot iron, carbonyl iron, and electrolytic iron on the low-carbon side. In some steels containing extremely low carbon, the maganese content is the principal differentiating factor, steel usually containing at least 0.25%; ingot iron contains considerably less.

Strain. A measure of the change in the size or shape of a body, referred to its original size or shape. *Linear strain* is the change per unit length of a linear dimension. *True strain* (or *natural strain*) is the natural logarithm of the ratio of the length at the moment of observation to the original gauge length. *Conventional strain* is the linear strain referred to the original gauge length. *Shearing strain* (or *shear strain*) is the change in angle (expressed in radians) between two lines originally at right angles. When the term *strain* is used alone, it usually refers to the linear strain in the direction of the applied stress.

Strain aging. Aging induced by cold working. *See* Aging.

Strain hardening. An increase in hardness and strength caused by plastic deformation at temperature below the recrystallization range.

Stress. Force per unit area, often thought of as force acting through a small area within a plane. It can be divided into components, normal and parallel to the plane, called *normal stress* and *shear stress*, respectively. *True stress* denotes the stress where force and area are measured at the same time. *Conventional stress*, as applied to tension and compression tests, is force divided by the original area. *Nominal stress* is the stress computed by simple elasticity formula, ignoring stress raisers and disregarding plastic flow. In a notch-bend test, for example, it is bending moment divided by minimum section modulus.

Stress-corrosion cracking. Failure by cracking under combined action of corrosion and stress, either external (applied) or internal (residual). Cracking may be either intergranular or transgranular, depending on metal and corrosive medium.

Stress relieving. Heating to a suitable temperature, holding long enough to reduce residual stresses and then cooling slowly enough to minimize the development of new residual stresses.

Stud welding. Welding a metal stud or similar part to another piece of metal, the heat being furnished by an arc between the two pieces just before pressure is applied.

Superalloy. An alloy developed for very-high-temperature service where relatively high stresses (tensile, thermal, vibrator, and shock) are encountered and where oxidation resistance is frequently required.

Supercooling. Cooling below the temperature at which an equilibrium phase transformation can take without actually obtaining the transformation.

Superficial Rockwell hardness test. Form of Rockwell hardness test using relatively light loads, which produce minimum penetration. Used for determining surface hardness or hardness of thin sections or small parts, or where large hardness impression might be harmful.

Superfinishing. A form of honing in which the abrasive stones are spring-supported.

Superheating. (1) Heating a phase above a temperature at which an equilibrium can exist between it and another phase having more internal energy, without obtaining the high-energy phase. (2) Heating molten metal above the normal casting temperature so as to obtain more complete refining or greater fluidity.

Taper tap. A tap with a chamber of seven to nine threads in length.

Tarnish. Surface discoloration of a metal caused by formation of a thin of corrosion product.

Temper. (1) In heat treatment, reheating hardened steel or hardened cast iron to some temperature below the eutectoid temperature for the purpose of decreasing the hardness and increasing the toughness. The process is also sometimes applied to normalized steel. (2) In tool steels, "temper" is sometimes used, but inadvisedly, to denote the carbon content. (3) In nonferrous alloys and in some ferrous alloys (steels that cannot be hardened by heat treatment), the hardness and strength produced by mechanical or thermal treatment, or both, and characterized by a certain structure, mechanical properties, or reduction in area during cold working. Refer to section G-IX.

Temper brittleness. Brittleness that results when certain steels are held within, or are cooled slowly through, a certain range of temperature below the transformation range. The brittleness is revealed by notched-bar impact tests at or below room temperature.

Tempering. Reheating a quench-hardened or normalized ferrous alloy to a temperature below the transformation range and then cooling at any rate desired.

Tensile strength. In tensile testing, the ratio of maximum load to original cross-sectional area. Also called *ultimate strength.*

Thermal fatigue. Fracture resulting from the presence of temperature gradients, which vary with time in a manner to produce cyclic stresses in a structure.

Thermal shock. The development of a steep temperature gradient and accompanying high stresses within a structure.

Thermit welding. Welding with heat produced by the reaction of aluminum with a metal oxide. Filler metal, if used, is obtained from the reduction of the appropriate oxide.

Thermocouple. A device for measuring temperatures, consisting of two dissimilar metals that produce an electromotive force roughly proportional to the temperature difference between their hot and cold junction ends.

Toughness. Ability of a metal to absorb energy and deform plastically before fracturing. It is usually measured by the energy absorbed in a notch-impact test, but the area under the stress-strain curve in tensile testing is a also a measure of toughness.

Transducer. A device actuated by one transmission system and supplying related waves to another transmission system; the input and output energies may be of different forms. (Ultrasonic transducers accept electrical waves and deliver ultrasonic waves, the reverse also being true.)

Transformation ranges (transformation temperature ranges). Those ranges of temperature within which austenite forms during heating and transforms during cooling. The two ranges are distinct, sometimes overlapping, but never coinciding. The limiting temperatures of the ranges depend on the composition of the alloy and on the rate of change of temperature, particularly during cooling. *See* Transformation temperature.

Transformation temperature. The temperature at which a change in phase occurs. The term is sometimes used to denote the limiting temperature of a transformation range. The following symbols are used for iron and steel:

Ac_{cm}. In hypereutectoid steel, the temperature at which the solution of cementite in austenite is completed during heating.

Ac_1. The temperature at which austenite begins to form during heating.

Ac_3. The temperature at which transformation of ferrite to austenite is completed during heating.

Ac_4. The temperature at which austenite transforms to delta ferrite during heating.

Ac_{cm}, Ae_1, Ae_3, Ae_4. The temperatures of phase changes at equilibrium.

Ac_{cm}. In Hypereutectoid steel, the temperature at which precipitation of cementite starts during cooling.

Ar_1. The temperature at which transformation of austenite to ferrite, or to ferrite, or to ferrite plus cementite is completed during cooling.

Ar_3. The temperature at which austenite begins to transform to ferrite during cooling.

Ar_4. The temperature at which delta ferrite transforms to austenite during cooling.

M_s (or Ar$''$). The temperature at which transformation of austenite to martensite starts during cooling.

M_f. The temperature at which martensite formation finishes during cooling.

Note: All these changes except the formation of martensite occur at lower temperatures during cooling than heating, and depend on the rate of change of temperature.

Transitional fit. A fit, which may have clearance of interference resulting from specified tolerances on hole and shaft as given by six classes of transition locational fits of 13 nominal shaft sizes in ASA B4.1-1955.

Tungsten-arc welding. Inert-gas, shielded-arc welding using a tungsten electrode.

Ultimate strength. The maximum conventional stress (tensile, compressive, or shear) that a material can withstand.

Ultrasonic beam. A beam of acoustical radiation with a frequency higher than the frequency range for audible sound.

Ultrasonic cleaning. Immersion cleaning aided by ultrasonic waves which cause microagitation.

Ultrasonic frequency. A frequency, associated with elastic waves, that is greater than the highest audible frequency, generally regarded as being higher than 15 kilocycles per second.

Ultrasonics. The acoustic field involving ultrasonic frequencies.

Undercut. A groove melted into the base metal adjacent to the toe of a weld and left unfilled.

Vacuum deposition. Condensation of thin metal coatings on the cool surface of work in a vacuum.

Vacuum melting. Melting in a vacuum to prevent contamination from air, as well as to remove gases already dissolved in the metal; the solidification may also be carried out in a vacuum or at low pressure.

Vacuum refining. *See* Vacuum melting.

Vapor blasting. *See* Liquid honing.

Vapor degreasing. Degreasing work in vapor over a boiling liquid solvent, the vapor being considerably heavier than air. At least one constituent of the soil must be soluble in the solvent.

Vapor plating. Deposition of a metal or compound upon a heater surface by reduction or decomposition of a volatile compound at a temperature below the melting points of the deposit and the basis material. The reduction is usually accomplished by a gaseous reducing agent such as hydrogen. The decomposition process may involve thermal dissociation or reaction with the basis material. Occasionally used to designate deposition on cold surfaces by vacuum evaporation. *See* Vacuum deposition.

Vickers hardness test. *See* Diamond pyramid hardness test.

Welding. (1) Joining two or more pieces of material by applying heat, pressure, or both, with or without filler material, to produce a localized union through fusion or recrystallization across the interface. The thickness of the filler material is much greater than the capillary dimensions encountered in brazing. (2) May also be extended to include brazing.

Welding current. The current flowing through a welding circuit during the making of a weld. In resistance welding, the current used during preweld or postweld intervals is excluded.

Welding cycle. The complete series of events involved in making a resistance weld. Also applies to semiautomatic mechanized fusion welds.

Welding force. See Electrode force.

Welding generator. A generator used for supplying current for welding.

Welding ground. See Work lead.

Welding lead (welding cable). A work lead or an electrode lead.

Welding machine. Equipment used to perform the welding operation; for example, spot-welding machine, arc-welding machine, seam-welding machine.

Welding procedure. The detailed methods and practices, including joint-welding procedures, involved in the production of a weldment.

Welding rod. Filler metal in rod or wire form used in welding.

Welding schedule. A record of all welding machine settings plus identification of the machine for a given material, size, and finish.

Welding sequence. The order of welding the various component parts of a weldment or structure.

Welding stress. Residual stress caused by localized heating and cooling during welding.

Welding technique. The details of a welding operation that, within the limitations of a welding procedure, are performed by the welder.

Welding tip. (1) A replaceable nozzle for a gas torch that is especially adapted for welding. (2) A spot-welding or projection-welding electrode.

Weld line. The junction of the weld metal and the base metal, or the junction of the base-metal parts when filler metal is not used.

Weldment. An assembly whose component parts are joined by welding.

Weld metal. That portion of a weld that has been melted during welding.

Wetting. A phenomenon involving a solid and a liquid in such intimate contact that the adhesive force between the two phases is greater than the cohesive force within the liquid. Thus, a solid that is wetted, on being removed from the liquid bath, will have a thin continuous layer of liquid adhering to it. Foreign substances such as grease may prevent wetting. Addition agents, such as detergents, may wetting by lowering the surface tension of the liquid.

Wetting agent. A surface-active agent that produces wetting by decreasing the cohesion within the liquid.

Work hardening. See Strain hardening.

Wringing fit. A fit of zero to negative allowance comparable to fits assigned to the first six nominal shaft sizes listed under class LN2 of interference locational fits in ASA B4.1-1955.

Wrought iron. A commercial iron consisting of slag (iron silicate) fibers entrained in a ferrite matrix.

Xeroradiography. A process utilizing a layer of photoconductive material on an aluminum sheet upon which an electrical charge is placed. After X-ray exposure, the electrical potential remaining on the plate in the form of a latent electrical pattern is developed by contact with a cloud of finely dispersed powder.

X-ray. Electromagnetic radiation, of wavelength less than about 500 angstrom units, emitted as the result of deceleration of fast-moving electrons (*bremsstrahlung,* continuous spectrum) or decay of atomic electrons from excited orbital states (characteristic radiation). Specifically, the radiation produced when an electron beam of sufficient energy impinges upon a target of suitable material.

Yield point. The first stress in a material, usually less than the maximum attainable stress, at which an increase in strain occurs without an increase in stress. Only certain metals exhibit a yield point. If there is a decrease in stress after yielding, a distinction may be made between upper and lower yield points.

Yield strength. The stress at which a material exhibits a specified deviation from proportionality of stress and strain. An offset of 0.2% is used for many metals.

Young's modulus. *See* Modulus elasticity.

Ziron sand. A very refractory mineral, composed chiefly of zironium silicate of extreme fineness, having low thermal expansion and high thermal conductivity.

TERMS RELATING TO PHYSICAL PROPERTIES OF MATERIALS

Analysis, modulation. Instrumentation method used in electromagnetic testing which separates responses due to factors influencing the total magnetic field by separating and interpreting individually, frequencies or frequency bands in the modulation envelope of the (carrier frequency) signal. (ASTM E-268)

Acoustic impedance. Ratio of sound pressure to product of sound velocity and area at a given surface. Acoustic impedance is frequently approximated as the product of only the density and velocity.

Acoustic impedance, specific normal. Ratio of sound pressure to component of particle velocity normal to the surface (ASTM C-384).

Acoustic reactance, resistance. Components of acoustic impedance of a material.

Anisotropy. If physical properties of a material differ along different directions, material is said to be anisotropic. Some crystals are easier to magnetize along one axis than along another and are therefore anisotropic.

Arc resistance. A measure of resistance of the surface of an electrical insulting material to breakdown under electrical stress. The time in seconds during which an arc of increasing severity is applied intermittently to the surface until failure occurs. Failure may be one of four general types: (1) material becomes incandescent and hence capable of conducting current, regaining its insulting qualities upon cooling; (2) material bursts into flame, although no visible conducting forms; (3) a thin wiry line ("tracking") forms between electrodes; or (4) surface carbonizes until there is sufficient carbon to carry current (ASTM D-495).

Autoradiography. Inspection technique in which radiation spontaneously emitted by a material is recorded photographically. The radiation is emitted by radioisotopes that are produced in or added to a material. The technique serves to locate the position of the radioactive element or compound.

Bearing strength. The maximum bearing load at failure divided by the effective bearing area. In a pinned or riveted joint, the effective area is calculated as the product of the diameter of the hole and the thickness of the bearing member. (ASTM D-953, ASTME-E-238)

Betatron. A device for accelerating electrons by means of magnetic induction.

Black light. Electromagnetic radiation not visible to the human eye. The portion of the spectrum generally used in fluorescent inspection falls in the ultraviolet region between 3300 to 4000 angstrom units, with the peak at 3650 angstrom units.

Bond strength. Stress, i.e., tensile load divided by area of bond, required to rupture a bond formed by an adhesive between two metal blocks. (ASTM D-952)

Breakdown ratio, surface. Ratio of arc resistance after tracking to arc resistance before tracking. Also called *surface breakdown voltage ratio.* Can be determined by allowing specimen to cool after an arc-resistance test, then repeating the test. (ASTM D-495)

Breakdown voltage. Voltage at which a material fails in a dielectric strength test.

Breaking load. The load that causes fracture in a tension, compression, flexure, or torsion test.

Brittleness temperature. Temperature at which plastics and elastomers exhibit brittle failure under impact conditions specified in ASTM D-746. A test for determining brittleness temperature in plastic film is presented in ASTM D-1790.

Brightness. Relative amount of light reflected by a material. Measured by the reflectance of a material.

Bulk factor. Ratio of volume of powdered material to volume of solid piece. Also, ratio of density of solid material to apparent density of loose powder.

Bursting strength. A measure of the ability of materials in various forms to withstand hydrostatic pressure. (ASTM D-1180, ASTM D-751)

Capacitance. Capacity of material for storing electrical energy. Is often used in calculating dielectric constant of a material. (ASTM D-150)

Capacity, specific. An alternate term for *dielectric constant.*

Charpy test. A pendulum-type single-blow impact test in which the specimen, usually notched, is supported at both ends as a simple beam and broken by a falling pendulum. The energy absorbed, as determined by the subsequent rise of the pendulum, is a measure of impact strength or notch toughness. (ASTM E-23, ASTM A-327, ASTM D-256, ASTM D-758)

Chroma. Color intensity or purity of tone, being a degree of freedom from gray.

Chromaticity coordinates. Two of three parameters commonly used in specifying color and describing color difference.

Chromaticity diagram. Plot of chromaticity coordinates useful in comparing color of materials.

Coefficient of expansion. Fractional change in length (or sometimes volume) of a material for a unit change in temperature. Values for plastics range from 0.01 to 0.2 mil/in/°C.

Coercive force. A magnetizing force (HC) required to bring induction of a material to zero, when the material is in a symmetrically, cyclically magnetized condition. (ASTM E-269)

Coercive force, intrinsic. A magnetizing force (H_{ci}) rquired to bring intrinsic induction of a magnetic material to zero, when the material is in a symmetrically, cyclically magnetized condition.

Coercive force, relaxation. Reversed magnetizing force (H_{cr}) of such value that when it is reduced to zero, induction becomes zero.

Coercivity. Maximum coercive force (H_{cs}) for a material.

Cold cracks. Straight or jagged lines, usually continuous throughout their length. Cold cracks generally appear singly and start at the surface. (ASTM E-192)

Color. Property of light by which an observer may distinguish between two structure-free patches of light of the same size and shape. Neutral color qualities such as black, white, and gray which have a zero saturation or chroma are called *achromatic colors.* Colors that have a finite saturation or chroma are said to be *chromatic* or *colored.* Color has three attributes: hue, lightness, and saturation. (ASTM D-307 and D-791)

Color difference. Difference in color between two materials. Both quantitative and qualitative tests are used. (ASTM D-1365)

Color, mass. Color, when viewed by reflected light, of a pigmented coating of such thickness that it completely obscures the background. Sometimes called *overtone* or *masstone.*

Compressive strength. The maximum compressive stress that a material is capable of developing, based on original area of cross section. In the case of a material that fails in compression by a shattering fracture, the compressive strength has a very definite value. In the case of materials that do not fail is compression by a shattering fracture, the value obtained for compressive strength is an arbitrary value depending upon the degree of distortion that is regarded as indicating complete failure of the material.

Conductivity, electrical. Ability of a material to conduct an electric current.

The reciprocal of resistivity electrical conductivity is often expressed in percent, based on a value of 100 for the International Annealed Copper Standard (IACS), which has a resistivity of 10.371 ohm-cir mil/foot or 1.7241 microhm-cm at 68°F.

Contact scanning. In ultrasonics, a planned systematic movement of the beam relative to the object being inspected, the search unit being in contact with and coupled to this object by a thin film of coupling material.

Contrast ratio. Measure of hiding power or opacity. Ratio of the reflectance of a material having a black backing to its reflectance with a white backing.

Continuous method. A method of magnetic particle testing in which the indicating medium is applied while the magnetizing force is present.

Core loss. Active power (watts) expended in a magnetic circuit in which there is a cyclically alternative induction. Measurements are usually made with sinusoidally alternating induction. (ASTM A-346)

Core loss, specific apparent. Product of rms-induced voltage and rms-exciting current for a ferromagnetic core, where induced voltage is approximately sinusoidal. Specific apparent core loss in apparent W/lb or kg is apparent core loss divided by core weight.

Core loss, standard. Specific core loss at an induction of 10 kilogausses and a frequency of 60 Hz, designated P/10/60.

Covering power. Alternative term for *opacity* or *hiding power of paints.*

Cracks, base metal. Discontinuity resulting from very narrow separations of metal. (ASTM E-99)

Cracks, welds, transverse, longitudinal. Discontinuity resulting from very narrow separation of metal. (ASTM E-99)

Creep. Deformation that occurs over a period of time when a material is subjected to constant stress at constant temperature. In metals, creep usually occurs only at elevated temperatures. Creep at room temperature is more common in plastic materials.

Current-flow method. A method of magnetizing by passing current through a component via prods or contact heads. The current may be alternating, rectified alternating, or direct. (ASTM E-269)

Currie point. Temperature at which ferromagnetic materials can no longer be magnetized by outside forces, and at which they lose their residual magnetism (approximately 1200° to 1600°F for many metals).

Demagnetization. Reduction in degree of residual magnetism in ferromagnetic materials to an acceptable level.

Demagnetizing coefficient. Ratio (D_D) of extent to which an applied magnetizing force, as measured in a vacuum, exceeds the magnetizing force in a material, to the extent to which the induction in a material exceeds the induction in a vacuum for the same magnetizing force (i.e., intrinsic induction).

Demagnetizing curve. Part of normal induction curve or hysteresis loop that lies between the residual induction point B_r, and the coercive force point, H_c.

Density. Fundamental property of matter that is a measure of the compactness of its particles. Density is expressed as the ratio of mass to volume and depends on the composition of the specimen, its homogeneity, temperature, and, especially in the case of gases, on pressure. It is equal to specific gravity multiplied by the difference in weight between a unit volume of air and a unit volume of water at the same temperature.

Density, apparent. Most commonly measured for powders. It is a measure of their fluffiness or bulk, and is useful in calculating bulk factor for determining proper charges to the mold. (ASTM D-212 for metal powder, D-1182 for thermoplastics, D-1457 for TFE powders, and B-329 for refractory metal powders.)

Density, bulk. Mass per unit volume of a powdered material as determined in a reasonably large volume. Recommended test for plastic molding powders is ASTM D-1182. Term is also used for refractory brick (ASTM D-134), fired whiteware ceramics (ASTM C-373), and granular refractories (ASTM C-357).

Density, green or pressed. Density of an unsintered compact.

Density, tap. Apparent density of a powder obtained by measuring volume in a receptacle that is tapped or vibrated during loading in a specified manner.

Density, true. Alternate term for density, as opposed to apparent density.

Depth of penetration. Depth at which magnetic field strength or intensity of induced eddy currents has decreased to 37 percent of its surface value. Also known as standard depth of penetration or skin depth. (ASTM E-268)

Dielectric constant. Normally the relative dielectric constant. For practical purposes, the ratio of capacitance of two electrodes separated solely by an insulating material to its capacitance when electrodes are separated by air (ASTM D-150). Also called relative are permittivity and specific capacity. Low values are desirable when material is used as an insulator, high values used in a capacitor. Dielectric constant generally increases with temperature, humidity, exposure to weather, and deterioration. For most materials, dielectric constant varies considerably with frequency, and to a lesser extent with voltage as a result of polarization.

Dielectric proof voltage test. An acceptance test, nondestructive to acceptable material (ASTM D-1389). An insulating material passes between two electrodes at a uniform rate. Proof voltage across electrodes is generally selected as percentage of dielectric strength (short-term) of material, or as a multiple of dielectric breakdown of an air gap of equivalent thickness. Results are reported in terms of frequency of breakdown occurrence.

Dielectric strength. (1) That property of an insulating material that enables it to withstand electric stresses successfully. (2) The highest electric stress that an insulating material can withstand for a specified time without occurrence of electrical breakdown by any path through its bulk. Its value is given by potential difference in volts divided by thickness of test specimen in thousandths of an inch, i.e., in volts per mil (ASTM D-149). (3) The highest potential difference that a specimen of insulating material of given thickness can withstand for a specified time without occurrence of electrical breakdown through its bulk. Four basic tests are used: short-time, step-by-step, slow rate of rise (generally an alternate to step-by-step), and long-time.

Diffuse light transmission factor. Ratio of transmitted to incident light for translucent reinforced plastic building panels (ASTM D-1494). Property is an arbitrary index of comparison and is not related directly to luminous transmittance.

Diffusion value. Measure of reflective and transmissive diffusion characteristics of plastics (ASTM D-636). Index of scattering or diffusion of light by a material compared to a theoretically perfect light-scattering material (rated as 1.0).

Dimensional stability. Ability of a material to retain precise shape in which it was molded, fabricated, or cast.

Discontinuity. Any interruption in the normal physical structure or configuration of a part such as cracks, laps, seams, inclusions, or porosity.

Dry method. Magnetic-particle inspection in which particles employed are in dry-powder form. (ASTM E-269)

Ductility. The extent to which a material can sustain plastic deformation without rupture. Elongation and reduction of area are common indexes or ductility.

Eddy currents. Currents caused to flow in an electrical conductor by time or space variation, or both, of an applied magnetic field. (ASTM E-268)

Eddy current loss The part of core loss that is due to current circulation in magnetic materials as a result of electromotive forces induced by varying induction.

Eddy-current testing. Nondestructive testing method in which eddy-current flow is induced in the object being tested. Changes in flow caused by variations in the specimen are reflected into coils for subsequent analysis.

Elasticity. The ability of a material to return to its original configuration when the load causing deformation is removed.

Electromagnetic testing. A nondestructive test method for engineering materials, including magnetic materials, which uses electromagnetic energy having frequencies less than those of visible light to yield information regarding the quality of the tested material. (ASTM E-268)

Elongation. In tensile testing, the increase in gauge length, measured after fracture of the specimen within the gauge length, usually expressed as a percentage of the original gauge length.

Excitation rms. The rms alternating current required to produce a specified induction in a material. For induction of less than 10 kilogausses, rms excitation (H_z) is usually expressed in oersteds; for high inductions, in amp-turns/inch.

False indication. In nondestructive inspection, an indication that may be interpreted erroneously as a discontinuity.

Fatigue. Permanent structural change that occurs in a material subjected to fluctuating stress and strain. In general, fatigue failure can occur with stress levels below the elastic limit.

Fatigue strength. The magnitude of fluctuating stress required to cause failure in a fatigue test specimen after a specified number of loading cycles.

Ferromagnetic. A term applied to materials that can be magnetized or strongly attracted by a magnetic field. (ASTM E-269)

Film contrast. A qualitative expression of the slope of the characteristic curve of a film; that property of a photographic material that is related to magnitude of density difference resulting from a given exposure difference. (ASTM E-94)

Flash point. Temperature at which vapor of a material will ignite when exposed to a flame in a specially designed testing apparatus.

Flash radiography. High-speed radiography in which exposure times are sufficiently short to give an ublurred photograph of moving objects such as fired projectiles or high-speed machinery.

Flexivity. Index of change in flexure with temperature of thermostat bimetals (ASTM D-106). Change of curvature of longitudinal centerline of specimen per unit temperature change per unit thickness.

Fluorescence. Property of emitting visible light as a result of, and only during absorption of, radiation from other energy sources. (ASTM E-270)

Fluorescent liquid penetrant. Highly penetrating liquid used in performance of liquid penetrant testing and characterized by its ability to fluoresce under black light. (ASTM E-270)

Fluorescent magnetic particle inspection. A magnetic particle inspection process employing a powdered ferromagnetic inspection medium coated with material that fluoresces when activated by light of suitable wavelength. (ASTM E-269)

Flux density. The strength of a magnetic field. (ASTM E-269)

Flux leakage. Magnetic lines of force that leave and enter the surface of a part due to discontinuity that forms poles at the surface of the part. (ASTM E-268)

Freezing point. Temperature at which a material solidifies on cooling from molten state under equilibrium conditions.

Gas holes. Round or elongated, smooth-edged dark spots, which may occur individually, in clusters, or distributed throughout a casting section. Gas holes are usually caused by trapped air or mold gases. (ASTM E-310, ASTM E-272)

Gel point. Stage at which a liquid begins to exhibit pseudoelastic properties. Gel-point stage may be conveniently observed from inflection point on a viscosity-time plot.

Glassy transition. Change in an amorphous polymer or amorphous regions of partially crystalline polymer from (or to) viscous or rubbery condition to (or from) a hard and relatively brittle one.

Gloss. Ratio of light reflected from surface of a material to light incident on surface when angles of incidence and reflectance are equal numerically but opposite in sign.

Gravity, absolute specific. Ratio of weight in a vacuum of a given volume of material to weight in a vacuum of an equal volume of gas-free distilled water. Sometimes called simply *specific gravity*, but it is not the same as specific gravity, which is based on weight measurements in air. For practical purposes, *specific gravity* is more commonly used.

Hall effect. Deflection by a magnetic field of an electric current traveling through a thin film. Force experienced by the current is perpendicular to both magnetic field and direction of current flow.

Hardness. A measure of the resistance of a material to localized plastic deformation. Hardness is tested and expressed in numerous ways (Brinell hardness, Rockwell hardness, scleroscope, durometer, etc.). A table relating various hardness values appears in ASTM E-140.

Haze. A measure of the extent to which light is diffused in passing through a transparent material. Percentage of transmitted light that, in passing through the material, deviates from the incident beam by forward scattering. (ASTM D-1003)

Heat distortion point. Temperature at which a standard test bar (ASTM D-048) defects 0.010 inches under a stated load of either 86 or 264 psi.

Heat of fusion. Amount of heat per unit weight absorbed by a material in melting.

Heat, specific. Component of thermal diffusivity of a material. Quantity of heat required to change temperature of unit mass of material 1°, commonly expressed in Btu/lb/°F. Mean specific heat is average specific heat over specified heat over specified temperature range. (ASTM C-351)

Hiding power. Ability of paint to oscure surface. Also called *opacity*. Usually measured by contrast ratio.

Hole size. Diameter of the hole in a reference block, which determines the area of the hole bottom. (ASTM E-127)

Hot tears. Ragged dark lines of variable width and numerous branches. Hot tears have no denitie lines of continuity and may exist in groups. They may originate internally or at the surface. (ASTM E-192)

Hue. Attribute of color that determines its position in wavelengths spectrum; i.e., whether it is red, yellow, green, etc.

Hysteresis. Retardation of magnetic effect when magnetizing forces acting on a ferromagnetic body are changed. (ASTM E-269)

Hystersis loop, magnetic. Curve showing relationship between magnetizing force and magnetic induction in a material in a cyclically magnetized condition. For each value and direction of magnetizing force, there are two values of induction; when the magnetizing force is increasing, and when the magnetizing force is decreasing. Result is actually two smooth curves joined at ends to form a loop. Area within loop represents energy expended in material as heat. Where alternating current is used, the loop indicates amount of energy transformed into heat during each cycle. Magnetic hysteresis loop is commonly determined by normal induction measurements giving values of residual induction (B_r) and coercive force (H_c).

Hysteresis loss, magnetic. Power (P_a) expended in magnetic material as result of magnetic hysteresis when induction is cyclic. Enclosed area of magnetic hysteresis loop. It is one component of specific core loss and is a function of coercivity and retentivity.

Impact strength. The energy required to fracture a specimen subjected to shock loading as in an impact test.

Impedance analysis. A type of signal processing sometimes used in eddy-current analysis. A signal that represents a change, in both amplitude and phase, of an impedance vector is resolved into any pair of its components separated 90° in phase.

Inclusions. Isolated, irregular, or elongated variations of magnetic particles occurring singly, in a linear distribution, or scattered randomly in feathery streaks. (ASTM E-125)

Incomplete fusion. Fusion that is less than complete. Failure of weld metal to fuse completely with base metal or preceding leads. (ASTM E-99)

Indication. That which marks or denotes presence of a discontinuity. In liquid-penetrant inspection, the presence of detectable bleedout of liquid penetrant from material discontinuities (ASTM E-270). In magnetic-particle inspection, the detectable magnetic particle buildup resulting from interruption of the magnetic field. (ASTM E-269)

Indications. Eddy-current signals caused by any change in the uniformity of a tube. (ASTM E-215)

Induction, intrinsic. Extent to which induction in magnetic material exceeds induction in vacuum for a particular magnetizing force (B_1). The measured induction minus the product of magnetizing force and permeability and a vacuum. In the cgs electromagnetic system (gausses, oersteds), permeability of a vacuum is arbitrarily taken as unity, so that intrinsic induction is numerically, equal to normal induction minus magnetizing force. Intrinsic induction is sometimes plotted against magnetizing force.

Induction, normal. Limiting induction (in either directional) or material in a symmetrically, cyclically magnetized condition. Induction is measured for both increasing and decreasing magnetizing force, and for magnetizing force in both directions, to develop magnetizing and demagnetizing curves. Complete magnetizing and demagnetizing curves form a magnetic hysteresis loop. Other important properties that can be determined from the normal induction curve include residual induction, coercive force, and normal permeability. (ASTM A-341)

Induction, residual. Measure of permanence of magnetization. Magnetic induction (B_r) corresponding to zero magnetizing force in a magnetic material in a symmetrically, cyclically magnetized condition. Induction at either of two points where magnetic hysteresis loop intersects the B axis. Determined from normal induction data.

Induction, saturation. Maximum intrinsic induction possible in a material (B_s).

Iron loss. Alternate term for *specific core loss.*

Lamination factor. Measure of effective volume of laminated structure, which is composed of strips of magnetic material. Ratio of volume of structure as calculated from weight and density of strips, to measured solid volume of structure under pressure. (ASTM A-344)

Latent heat. Heat that must be applied to material to effect change in state without change in temperature. For example, latent heat of fusion of ice water is 80 cal/gm.

Leak testing. Technique of liquid penetrant testing wherein penetrant is applied to one side of a material and observation is made on the opposite side to ascertain the presence of voids extending throughout the material. (ASTM E-270)

Lightness. Attribute of color that permits any color to be classified as equivalent to one of a series of grays ranging between black and white. The term *shade* is often used to describe differences in lightness. Lightness difference is measured as part of the color-difference test.

Linear discontinuities. Ragged lines of variable width that may appear as a single jagged line or exist in groups. They may or may not have a definite line of continuity, often originate at the casting surface, and usually become smaller as a function of depth. (ASTM E-215)

Liquid temperature. Temperature at which an alloy finishes melting during heating or starts freezing during cooling, under equilibrium conditions. For pure metals, same as solidus temperature and known simply as *melting point*. Effective liquids temperature is raised by fast heating and lowered by fast cooling.

Loss angle. Measure of electrical power losses in insulating material subjected to alternating current. The arc-tangent of dissipating factor and, thus, the angle between a material's parallel resistance and its total parallel impedance in a vector diagram when the material is used as dielectric in a capacitor. Complement of phase angle. Sometimes called *phase defect angle*. Conventionally designated as δ.

Loss factor. Measure of electrical-power loss in insulating material subjected to alternating current. Product of dissipation factor and dielectric constant. Loss factor is expressed in the same units as dissipation factor. Low loss factor is generally desirable. Loss factor generally increases with humidity, weathering, deterioration, and exponentially with temperature.

Magnetic energy product, maximum. Maximum external energy that contributes to magnetization of material. Corresponds to maximum value of the abscissa for magnetic-energy-product curve.

Magnetic-energy-product curve. Curve obtained by plotting product of induction and demagnetizing force aginst corresponding values of induction; i.e., product of coordinates of demagnetizing curve (B_d, H_d) as abscissa versus induction (B_d) as ordinate.

Magnetic field strength. The measured intensity of a magnetic field at any given point, usually expressed in oersteds. (ASTM E-269)

Magnetic hysteresis. In a magnetic material, such as iron, a lagging in the values of the resulting magnetization due to a changing magnetic force. (ASTM E-269)

Magnetic-inspection flaw indications. Accumulation of ferromagnetic particles along areas of flaws or discontinuities due to distortion of magnetic lines of force in those areas. (ASTM E-269)

Magnetic-particle inspection. A nondestructive method for detecting cracks or other discontinuities at or near the surface of ferromagnetic materials. (ASTM Method E-138, ASTM E-269)

Melt index. Amount, in grams, of a material (usually a thermoplastic) that can be forced through a 0.0825-inch orifice when subjected to a 2160-gram force for 10 minutes at 190°C.

Melting point. For a pure metal, the temperature at which liquefaction occurs on heating or solidification occurs on cooling under equilibrium conditions. Alloys and other materials have a melting range, and melting point is often desired as a temperature near the bottom of this range at which observable change caused by melting occurs. In general, melting point is determined by heating a small specimen to a temperature not too far below the melting point, then raising temperature slowly (a few degrees a minute) and watching closely for first indication of liquefaction.

Melting range. Range of temperature between solidus temperature and allow liquidus temperature.

Moisture-vapor transmission. Rate at which water vapor penetrates film over given time at specified temperature and relative humidity (e.g., gm-mil/24 hr/100 in.^{2}).

Molecular weight, average. Molecular weight of polymeric materials determined by viscosity of the polymer in solution at a specific temperature. Gives average molecular weight of molecular chains in a polymer independent of specific chain length.

Neutron absorption cross section. Measure of probability that a single nucleus of material subjected to nuclear bombardment will intercept and interact with an incoming neutron. When expressed in square centimeters, it is the fraction of neutrons contained in a beam of 1-cm^2 cross section that can be expected to be intercepted by a single nucleus. Common units are barns/atom, where 1 barn = 10^{-24} cm^2. Low cross section is desirable for nuclear-reactor core materials. High value for reactor shielding. Effective neutron-absorption cross section depends on level of incident energy; for test neutron scattering, it approximates theoretical area of nucleus.

Opacity. Degree to which material or coating obstructs transmittance of visible light. Term is used primarily with nearly opaque materials, for which opacity is generally reported by contrast ratio. Opacity of light-transmitting materials is usually expressed by its opposite or complementary property; i.e., total transmittance of diffuse light transmission factor.

Optical density. Term indicating absorption of light. Numerically equal to the logarithm to base 10 of reciprocal of transmittance of material.

Optical distance. Product of actual length of light path and refractive index of material.

Optical distortion. Any apparent alteration of geometric pattern of an object when seen through a plastic or as reflection from plastic surface.

Peel strength. A measure of the strength of an adhesive bond. The average load per unit width of bond line required to part bonding materials where the angle of separation is 180° and the separation rate is 6 inch/minute. (ASTM D-903)

Penetrant. A liquid that has unique properties that render it highly capable of entering small openings. This characteristic makes the liquid especially suitable for detecting discontinuities in material. (ASTM E-270)

Penetrant inspection. A method of nondestructive testing for determining the existence and extent of discontinuities that are open to the surface of the material being inspected. The indications are made visible through the use of a dye or fluorescent chemical in the liquid employed as the inspection medium.

Permeability. (1) Passage or diffusion of gas, vapor, liquid, or solid through a barrier without affecting it. (2) Rate of such passage.

Permeability, ac. Ratio of maximum value of induction to maximum value of magnetizing force for material in a symmetrically, cyclically magnetized condition.

Permeability, apparent impedance. Index of ease of magnetization. Ratio of induction to rms excitation in material that is symmetrically, cyclically, magnetized.

Permeability, effective ac. Denotes capacitance needed to balance effective induction of a material when magnetized as a core.

Permeability, incremental ac. When compared with normal ac permeability, an index of increase in mean induction that can be expected when ac magnetization is superimposed upon dc magnetization. (ASTM A-343)

Permeability, initial. Slope of normal induction curve at zero magnetizing force.

Permeability, intrinsic. Ratio of intrinsic induction to corresponding magnetizing force.

Permeability, magnetic. A factor, characteristic of material, that is proportional to magnetic flux density (magnetic induction) B, produced in material by the magnetic field divided by the intensity of field H. Different names are given to permeabilities measured under different circumstances: *initial*, *incremental*, *reversible*.

Permeability, maximum. Maximum value of normal permeability for a magnetic material.

Permeability, normal. Ratio of normal induction to corresponding magnetizing force. Slope of magnetizing or normal induction curve at specified value of magnetizing force.

pH value. Numerical expression to describe hydrogen ion concentration. Therefore, pH is simply a number denoting degree of acidity or alkalinity.

Phase angle. Angle by which voltage leads current in material subjected to ac current. Angle between the parallel reactance of the material and its total parallel impedance in a vector diagram when the material is used as dielectric in capacitor. Complement of loss angle.

Phase transition. Abrupt change in physical properties as temperature is changed continuously. Freezing of water at $0^\circ C$ is an example.

Porosity. Relative extent of volume of open pores in a material. Ratio of pore volume to overall volume of the material in percent. (ASTM D-328, D-116, E-125)

Power factor. Ratio of power expended in circuit to the product of emf acting in circuit and current in it; i.e., ratio of watts to volt-amperes. Because energy loss is directly proportional to frequency, low power factor is essential in materials used at high frequencies.

Quality factor. Reciprocal of dissipation factor or loss tangent. Also called *storage factor*. When materials have approximately the same dielectric constant, a higher quality factor indicates less power loss.

Radiograph. A permanent visible image on a recording medium produced by penetrating radiation passing through the material being tested. (ASTM E-142)

Radiographic inspection. The use of X-rays or nuclear radiation, or both, to detect discontinuities in a material, and to present their images on a recording medium. (ASTM E-142)

Reactive power, specific. For a specified normal induction in a material, the component of applied ac power, which is "reactive"; i.e, returned to the source, when polarity is reversed. Product of induced voltage and reactive current per unit weight of material; i.e., product of voltage, current, and sine of phase angle. (ASTM A-346)

Reactive power, specific incremental. Specific reactive power in a magnetic material when subjected simultaneously to a unidirectional ("biasing") and an alternating magnetizing force. This can be determined by the test used for incremental ac permeability. (ASTM A-343)

Reflectance, infrared. Measure of infrared brightness. Two direct optical methods are used. In each, a standard beam of light is thrown on a specimen normal to the surface and reflectance is measured from the same position, using filters to select specified wavelengths for viewing in the infrared range. (Fed 141-6341, 141-6242)

Reflectance, luminous. Measure of brightness. Ratio of light reflected to light incident. Luminous reflectance of a material of such thickness that any increase in thickness would fail to change the value, is called *luminous reflectivity*. Calculated from spectral reflectance (including spectral component) and spectral luminosity, it is a function of spectral and angular distribution of incident light energy. Also called *total luminous reflectance* to distinguish it more readily from luminous directional reflectance.

Reflectance, luminous directional. Ratio of brightness to brightness that an ideally diffusing, completely reflecting light surface would have when illuminated and viewed in the same manner. Luminous directional reflectance is commonly determined for an opaque material illuminated at 45° and viewed normal to the surface.

Reflectance, spectral. Measure of brightness. Ratio of light reflected by a specimen to homogeneous light energy incident on it. It depends on the angular distribution of the incident energy. The term *spectral reflectivity* is used to indicate the inherent property of a material of such thickness that any increase in thickness would fail to change the spectral reflectance. It may be expressed in the form of a cure of luminous directional reflectance versus wavelength.

Reflectance, spectral directional. Measure of brightness. The spectral reflectance which an ideal diffusing surface would need to appear the same as the specimen under test when illuminated and viewed the same way. It depends on the angular distribution of incident light and on the direction of viewing. The term *spectral directional reflectivity* is used to denote an inherent property of the material as opposed to that of a particular object.

Reflectivity, apparent. Term used for *luminous directional reflectance* (reflectivity) of paper. (TAPPI T442m)

Refraction, index of. Also called *refractive index*. Ratio of velocity of light in vacuum to its velocity in the material. Ratio of the sine of the angle of incidence of light to the sine of the angle of refraction. (ASTM D-542)

Reluctivity. Reciprocal of magnetic normal permeability of a material.

Remanence. Magnetization remaining when ferromagnetic material is removed from a magnetic field. Conversely, a material that exhibits remanence is ferromagnetic.

Residual method. A method of magnetic particle testing in which the indicating medium is applied after the magnetizing force has been discontinued. (ASTM E-138)

Reisistance, insulation. A measure of the ability of a nonconducting material to resist flow of electric current. Ratio, in ohms or megohms, of direct voltage applied between two electrodes in contact with material, to total current between electrodes. It is a property of form, not material, and this useful only in comparing materials having the same form. (ASTM D-257)

Resistance, inherlamination. Measure of resistance of a laminated magnetic structure to stray, intersheet power losses. Generally, a measure of the effectiveness of sheet surface oxides or coatings in reducing intersheet losses. (ASTM A-344)

Resistivity. Measure of electrical resistance properties of a conducting material, given by the resistance of a unit length of unit cross-section area. Reciprocal of conductivity. ASTM tests include D-193 for electrical conductor materials; D-63 for materials used in resistors, heating elements, and electrical contacts; and A-344 for magnetic materials. The term *resistivity* and *volume resistivity* are used interchangeably for conductors.

Resistivity, electrical. The electrical resistance of a material: (1) per unit length and unit cross-section area or (2) per unit length and unit weight.

Resistivity, insulation. A measure of the ability of a material to resist flow of electric current. Insulation resistance per unit volume. This property makes no distinction between volume resistivity and surface resistivity, even where no such distortion is made in the test. However, it is more common to use the term *volume resistivity.*

Resistivity, surface. A measure of the ability of surface of a dielectrical material to resist flow of electric current. Ratio of potential gradient parallel to current along surface of material, to current per unit width of surface. Together with volume resistivity, surface resistivity is often used to check purity of an insulating material during development and its uniformity during processing. (ASTM D-257)

Resistivity, volume. A measure of the ability of material to resist flow of an electric current. The ratio, of electrical potential gradient parallel to current in a material to current density. Although volume resistivity is of interest, knowledge of its changes with temperature and humidity are even more important, and curves plotted against temperature are sometimes particularly desirable. (ASTM D-257)

Resistivity, weight. A measure of the ability of a material to resist flow of electric current. Also called *mass resistivity*. Electrical resistance of a material per unit length and unit weight. A term generally reserved for metallic materials, i.e., conductors. Weight resistivity in ohm-lb/mile^2 or ohm-gm/m^2 is equal to WR/L_1L_2 where W is the weight of the specimen in pounds or grams, R is the measured resistance in ohms, L_1 is the gauge length in miles or meters used to determine R, and L_2 is the length of the specimen.

Retentivity. The maximum value of residual induction in magnetic material (B_{rs}). A measure of the permanence of magnetization. Low values are desirable for electromagnetic devices; high values, for permanent magnets, relays, and magnetos.

Rupture strength. The nominal stress developed in a material at rupture. This is not necessarily equal to ultimate strength. Since necking is not taken into account in determining rupture strength, it seldom indicates true stress at rupture.

Saturation. Attribute of a color that determines the extent to which it differs from a gray of the same lightness.

Saturation, magnetic. If a ferromagnetic material is magnetized by an external field, magnetization cannot increase indefinitely with increasing external field. The maximum value of magnetization is saturation magnetization.

Shade. Term used to characterize difference in lightness between two surface colors, the other attributes of color being essentially constant. A lighter shade of a color has higher lightness but about the same hue and saturation.

Shear strength. The maximum shear stress that can be sustained by a material before rupture. The ultimate strength of the material when subjected to shear loading. For shear strength in plastics, see ASTM-D-732; for structural adhesives, ASTM E-229; for timber, ASTM D-143 and ASTM D-198.

Sheen. The 85° specular gloss of nonmetallic materials.

Softening point. Indication of maximum temperature to which a nonmetallic material can be heated without loss of its normal "body." Minimum temperature at which a specified deformation occurs under a specified load. Softening point is often determined for materials that have no definite melting point, i.e., materials that gradually change from brittle or very thick and slow-flowing materials, to softer materials or loss viscous liquids. Vicat softening point is commonly used for some plastics. (ASTM D-1525)

Softening range. Range of temperature in which a plastic changes from rigid to soft state. Values depend on the method of test. Sometimes referred to as *softening point.*

Solidus temperature. Temperature at which an alloy begins to melt during heating or finishes freezing during cooling. For pure metals, same as liquids temperature and known simply as *melting point.*

Sound-absorption coefficient, normal incidence. A measure of the effectiveness of material in absorbing sound energy. Fraction of normally incident sound energy absorbed by a material assumed to have an infinite surface. (ASTM C-384)

Sound-absorption coefficient, statistical. A measure of the effectiveness of a material in absorbing sound energy. Fraction of incident sound energy absorbed by a material under conditions where it is subject to equal sound energy from all directions over a hemisphere, i.e., under reverberant sound conditions. Measurement is expensive and time-consuming, requiring a specially constructed reverberation chamber. Thus, property is sometimes estimated from specific normal sound-absorption coefficient or specific normal acoustic impedance. The normal coefficient is about half statistical coefficient for very low values and approaches equality with it at very high values. Maximum numerical difference occurs at intermediate values and is about 0.25 to 0.35.

Specific gravity. Ratio of the mass of a unit volume of a material at a stated temperature to the mass of same volume of gass-free distilled water at a stated temperature. If material is a solid, volume shall be that of the impermeable portion. Two basic methods are used for solids: liquid displacement (ASTM D-792, D-176, C-135, and C-329), and change in weight (D-792, D-311, and D-328). Three methods are commonly used for liquids: plummet displacement (D-176), volumetric (D-115), and hydrometer (D-901).

Sponge shrinkage. Found in heavier sections (generally more than 2 inches thick). Sponge shrinkages appear on radiographs as dark areas, lacy in texture, usually with a diffuse outline. (ASTM E-310, ASTM E-272)

Storage factor. Reciprocal of dissipation factor or loss tangent. Also called *quality factor.*

Storage life. Period of time during which resin or adhesive can be stored under specified temperature conditions and remain suitable for use. Storage life is sometimes called *shelf life.*

Stress corrosion cracking. Failure of a material due to the combined effects of corrosion and stress. Generally, stress corrosion cracking refers to the phenomenon by which stress increases the corrosion rate.

Subsurface disscontinuity. Any defect that does not open onto the surface of the part in which it exists. (ASTM E-269)

Surge method. Inspection by first using a high surge of magnetizing force, followed by a reduced magnetic field during application of a powdered ferromagnetic inspection medium. (ASTM E-269)

Tear resistance. A measure of the ability of sheet or film materails to resist tearing. For paper, it is the force required to tear a single ply or paper after the tear has been started (ASTM D-698). Three standard methods are available for determining tear resistance of plastic films. ASTMD-1004 details a method for determining tear resistance at low rates of loading. A test in ASTM D-1922 measures the force required to propagate a precut slit across the sheet specimen. ASTM D-1038 gives a method for determining tear-propagation resistance that is recommended for acceptance testing only. Tear resistance of rubber is force required to tear a specimen 1 inch in thickness under conditions outlined in ASTM D-624. For tear resistance of textiles, see ASTM D-1424.

Temperature-resistance. Relationship between the temperature of a material and its electrical resistance, expressed by a multivalue table, by a graph, by a calculated temperature coefficient, or by calculated values for constants in a standard mathematical equation. Provides information needed for design and use of resistance heating elements and precision resistors in electrical and electronic circuits.

Tensile strength. The ultimate strength of a material subjected to tensile loading. The maximum stress developed in a material in a tension test.

Thermal absorptivity. Fraction of the heat impinging on a body that is absorbed.

Thermal conductivity. Rate of heat flow in a homogeneous material, under steady conditions, through unit area, per unit temperature gradient in direction perpendicular to the area. Thermal conductivity is usually expressed in English units as Btu/ft^2/hr/°F for a thickness of 1 inch. (ASTM C-177)

Thermal diffusivity. Rate at which temperature diffuses through a material. Ratio of thermal conductivity to product of density and specific heat, commonly expressed in ft^2/hr.

Thermal emissivity. Ratio of heat emitted by a body to heat emitted by a black body as same temperature.

Thermal expansion. When heated or cooled, materials undergo a reversible change in dimensions that depends on the original size of the body and the temperature range studied. In addition to the change in dimensions, a change in shape may occur; i.e., the expansion or contraction may be anisotropic.

Thermal expansion, cubical coefficient of. Rate at which material increases in volume when heated. Unit increase in volume of material per unit rise in temperature over specified temperature range. Mean coefficient in mean slope of this curve over specified range of temperature.

Thermal expansion, linear coefficient of. Extent to which a material elongates when heated. Unit increase in length per unit rise in temperature over specified temperature range. Slope of the temperature to dilation curve at a specified temperature. Mean coefficient is mean slope between two specified temperatures. (ASTM D-1037)

Thermal insulating efficiency. Ratio of heat saved by an insulation to the heat that would be lost without insulation.

Thermal reflectivity. Fraction of heat impinging on a body that is reflected.

Transmittance, diffuse. Ratio of light scattered by material to light incident on it. Ratio depends on spectral distribution of incident energy. The analogous spectral ratio is diffuse luminous transmittance.

Transmittance, diffuse luminous. Ratio of light scattered by a material to light incident on it.

Transmittance, luminous. Ratio of light transmitted by material to light incident on it. Calculated from spectral transmittance and spectral luminosity, it is a function of spectral distribution of incident light energy. In test for haze, it is called *total luminous transmittance* to distinguish it from its diffuse component, *diffuse luminous transmittance.*

Transmittance, thermal. Rate at which heat is transmitted through a material by combined conduction, convection, and radiation. Overall coefficient of heat transfer. Term used particularly for textile fabrics and batting where heat transfer between opposite surfaces is not confined to conduction. For solid materials, thermal conductivity is a measure of thermal transmittance. (ASTM D-1518)

Transmittance, total. Ratio of light transmitted by material to light incident on it. The ratio depends on spectral distribution of incident light energy.

Ultrasonic. In ultrasonic nondestructive testing, pulses of sound energy are created by electrically exciting a crystal, called a *transducer*, which has piezoelectric properties, causing the crystal to vibrate. Sound energy is transmitted into the material being tested through a couplant, which can be oil, water, or glycerin. Sound that returns to the crystal as reflected energy is converted to electrical signals by the transducer and can be monitored on an oscilloscope. The time and amplitude relation of the reflected signals can be interpreted to show location of internal reflecting surfaces in three dimensions.

Viscosity. Resistance to flow exhibited within a material. Can be expressed in terms of relationship between applied shearing stress and resulting rage of strain in shear. Viscosity usually means "Newtonian viscosity," in which case the ratio of shearing stress to rate of shearing strain is constant. In non-Newtonian behavior (which occurs with plastics), the ratio varies with the shearing stress. Such ratios are often called the *apparent viscosities* at the corresponding shearing stresses.

Viscosity, absolute. In a fluid, the tangential force on a unit area of either of two parallel planes at unit distance apart when the space between planes is filled with the fluid in question and one of the planes moves with unit differential velocity in its own plane.

Viscosity coefficient. Shearing stress necessary to induce a unit velocity flow gradient in a material. The viscosity coefficient of a material is obtained from ratio of shearing stress to shearing rate.

Voltage ratio. Alternate term for surface-breakdown rate.

Volume, specific. Reciprocal of density.

Water-vapor permeability. Water vapor transmission through a homogeneous body under unit vapor pressure difference between two surfaces, per unit thickness.

Water-vapor transmission. Steady-state time rate of water-vapor flow through unit area of humidity at each surface.

Xeroradiography. A process utilizing a layer of photoconductive material on an aluminum sheet upon which is placed an electrical charge. After X-ray exposure, the electrical potential remaining on the plate in the form of a latent electrical pattern is developed by contact with a cloud of finely dispersed powder.

X-ray. Electromagnetic radiation, of wavelength less than about 500 Angstrom units, emitted as the result of deceleration of fast-moving electrons (*bremsstrahlung*, continuous spectrum) or decay of atomic electrons from excited orbital states (characteristic radiation). Specifically, the radiation produced when an electron beam of sufficient energy impinges upon a target of suitable material.

Yield point. The first stress in a material, usually less than the maximum attainable stress, at which an increase in strain occurs without an increase in stress. Only certain metals exhibit a yield point. If there is a decrease in stress after yielding, a distinction may be made between upper and lower yield points.

TERMS RELATED TO PRINTED CIRCUITS

Adhesion. *See* Bond strength.

Annular ring. A circular strip of conductive material completely surrounding a hole.

Artwork. An accurately scaled configuration used to produce a master pattern.

Base. The insulating support for printed wiring or printed circuit boards.

Base material. The insulating comprising the base.

Base plate. The preferred term is *base.*

Basic grid. *See* grid.

Blister, interlaminar. A void produced under the surface of a laminated material (a form of delamination).

Blister, foil. A localized separation between the base material and the conductive material.

Board thickness (total). The overall thickness of the base material and conductive material.

Bond strength. A measure (in pounds per square inch) of the force required to separate a layer of conductive material from the adjoining base by perpendicular tension applied to the surface of the conductive material.

Boss. The preferred term is *terminal area.*

Bow. The preferred term is *warp.*

Bridging, electrical. The formation of a conductive path between conductors.

Bulge. *See* Blister, interlaminar; Blister, foil.

Checking. A surface condition evidenced by hairline cracks.

Chemically deposited printed circuit. A printed circuit formed on a base by chemical reaction in the absence of an applied electrical field.

Chemically deposited printed wiring. A printed wiring formed on a base by chemical reaction in the absence of an applied electrical field.

Chemically reduced printed circuit. A printed circuit formed by the chemical reduction of a metallic compound.

Chemically reduced printed wiring. A printed wiring formed by chemical reduction of a metallic compound.

Component. A separable part of a printed circuit board assembly, which performs a circuit function (e.g., capacitor, resistor, transformer, transistor, etc.).

Component side. The side of a printed board upon which the components are mounted.

Conductive foil. A form of conductive material that covers one or both faces of the base material.

Conductive pattern. A pattern formed on a base from any electrically conductive material.

Conductor. A single conductive line forming an electrical connection between terminal areas.

Conductor layer. The layer of a printed board containing conductors.

Conductor layer No. 1. The conductor layer of a multilayered printed wiring board upon which the components are mounted.

Conductor pattern. A pattern of conductors in a conductor layer.

Conductor side. The side of a printed board containing conductors.

Conformal coating. An insulating protective coating applied on the completed board assembly, which conforms to the configuration of the object coated.

Connection. The means by which electrical contact is made to the conductive pattern.

Control drawing. The preferred term is *master drawing.*

Cover layer. The out layer or layers of a multilayered printed wiring board.

Crazing. A condition existing in the base laminate in the form of connected white spots or crosses on or below the surface of the base laminate, reflecting the separation of fibers in the glass cloth and connecting weave intersection.

Cross-hatching. The breaking up of large conductive areas by use of a pattern of voids on the conductive material.

Current-carrying capacity. The maximum current that can be carried continuously, under specified conditions by an individual conductor, without causing objectionable degradation of the electrical or mechanical properties of the printed board.

Definition. The fidelity of reproduction of pattern edge relative to the original master pattern.

Delamination. The separation of printed board material within the laminations.

Dents. Depressions in the copper foil that do not significantly decrease the foil thickness.

Dewetting. A condition wherein solder appears in droplets or balls which have withdrawn from a previously wet area.

DIP. Dual in-line package.

Dip-soldering. A process whereby printed boards are brought in contact with the surface of a static pool of molten solder for the purpose of soldering the entire exposed conductive pattern in a single operation.

Double-sided board. A printed board having conductive material on both sides.

Drag. A distortion of the conductor or base edge caused by improper cutting.

Dross. Oxides and other contaminants that form on the surface of molten solder.

Dual in-line package. An integrated circuit configuration.

Edge contact. *See* Printed contact.

Element. A constituent of a microcircuit or integrated circuit that contributes directly to its operation. (A discrete part incorporated into a microcircuit becomes an element of that microcircuit.)

Embossing. A process consisting of indenting a metal foil into an insulating base and removing the unwanted raised portions mechanically.

Etchant. A solution used to remove the unwanted portion of conductive material from a printed board by chemical reaction.

Etched printed circuit. A printed circuit produced by etching.

Etched printed wiring. A printed wiring produced by etching.

Etching. A process wherein a printed pattern is formed by chemical (or chemical and electrolytic) removal of the unwanted portion of conductive material bonded to a base.

Eyelet, flat flange. A piece of tubular metal having one and formed in a ring or lip with a flat cross section. The opposite end may be formed in a conical flare or rolled during installation.

Eyelet, funnel flange. A piece or tubular metal having one end formed in a conical flare of approximately 90° included angle. The opposite end is formed similarly during installation.

Eyelet, rolled flange. A piece of tubular metal having one end formed in a ring or lip with a circular cross section of approximately 180°. The opposite end is formed similarly during installation.

Fabrication holes. The preferred term is *tooling holes.*

Fabrication pattern. A duplicate made from the master pattern and used in fabrication of the actual board.

Feed-through. An electrical connection between conductive patterns on two-sided or multilayer printed wiring boards that has been created solely to completed or simplify a circuit.

Fiber exposure. A condition in which broken glass cloth fibers are exposed in machined or abraded areas.

Film integrated circuit. An integrated circuit consisting of elements which are films formed in situ on an insulating substrate.

Finger. The preferred term is *printed contact.*

Flow soldering. A process whereby printed boards are brought in contact with the surface of continuously flowing, circulated solder.

Flat-pak. *See* Integrated circuit.

Fluxing. The act of applying flux to the conductive paths and terminals prior to soldering.

Foil. The preferred term is *conductive foil.*

Grid. An orthogonal network consisting of equidistant lines providing the basis for an incremental location system.

Haloing. A condition existing in base laminates in the form of a light area around holes and/or other machined areas on or below the surface of the base laminate.

Haywire. A direct electrical connection between two points on a printed board added after the printed wiring is fabricated.

Hybrid microcircuit. A microcircuit consisting of elements which are a combination of the film and semiconductor types, or a combination of one or both of these types with discrete components.

Icicle. The preferred term is *solder projection.*

Inclusions. Foreign particles (usually lead) which may or may not be enclosed in the mass of copper foil.

Indexing holes. The preferred term is *tooling holes.*

Indexing notch. The preferred term is *tooling notch.*

Integrated circuit. A microcircuit consisting of interconnected elements inseparably associated and formed in situ on or within a single substrate to perform an electronic circuit function.

Interfacial connection. An electrical connection between conductive patterns on opposite sides of the base.

Jumper. See Haywire.

Land. The preferred term is *terminal area.*

Legend (marking). Lettering or symbols on the printed board indicating part number, component locations, etc.

Lot. All of the material manufactured at one time from one batch, forming a part of one contract or oder, and submitted for acceptance at the same time and place.

Lot number. The number assigned by the manufacturer to designate all of the material manufactured at one time from one batch.

Manufacturing holes. The preferred term is *tooling holes.*

Master drawing. A drawing showing the dimensional limits or grid location applicable to any or all parts of a printed circuit or printed wiring, including the base. *Note:* Litton G/CSD defines the printed circuit board fabrication drawing as the master drawing.

Master layout. A 4:1 scale translucent drawing containing a 0.025-in. grid. Dimensions, tolerances, all circuit details, and all holes are included.

Master pattern. A 1:1 scale pattern used to produce the printed wiring or printed circuit within the accuracy specified in the master drawing. *Note:* A master pattern is not artwork.

Measling. A condition existing in the base laminate in the form of discrete white spots or crosses below the surface of the base laminate, reflecting the separation of fibers in the glass cloth at weave intersection.

Metal-clad base material. An insulating base with metallic material bonded to one or both surfaces.

Microcircuit. A small circuit having a high equivalent circuit element density, considered as a single part composed of interconnected elements on or within a single substrate to perform an electronic circuit function. This excludes printed wiring boards, circuit card assemblies, and modules composed exclusively of discrete electronic parts.

Microcircuit module. An assembly of microcircuits (or microcircuits and discrete parts) designed to perform one or more electronic circuit functions, and constructed such that it is considered indivisible for the purposes of specification testing, commerce, and maintenance.

Monolithic integrated circuit. An integrated circuit consisting of elements formed in situ on or within a semiconductor substrate with at least one of the elements formed within the substrate.

Mounting hole. A hole for mechanical mounting of a printed board to the chassis or for mechanical attachment of components to the board.

Multilayer printed wiring. Alternate layers of conductors and insulating material bonded together, suitable for use as an assembly and component support and for electrically interconnecting electronic components.

Nonwetting. A condition whereby the surface has contacted molten solder but has had none of the solder adhere to it.

Overhang. The increase in conductor width caused by plating buildup or excessive etching.

Pad. The preferred terminology is *terminal area.*

Part side. *See* Component side.

Pinholes (printed wiring boards). Small holes, occurring as imperfections that completely penetrate the metal foil.

Pinholes (photographic film). Small clear areas in the black field of photographic images.

Pits. Small holes, occurring as imperfections, that do not completely penetrate the metal foil.

Plated printed wiring. Printed wiring that is plated with metal over the wiring.

Plated-through hole. An interfacial or interlayer connection formed by deposition of conductive material on the sides of a hole through the base.

Plating. A process consisting of the chemical or electrochemical deposition of metal on all or part of the conductive pattern, including the interlayer connections.

Plating-up. The process consisting of electrochemical deposition of the conductive pattern on the base material or through a hole, the surface of the base material or the walls of the hole having previously been made conductive.

Polarizing slot. The slot in the edge of a printed board used to assure proper insertion and location in the mating connector.

Preheating. An intermediate heating operation used to raise the temperature of the work above room temperature, thus reducing the thermal shock of an elevated temperature processing operation.

Pressed-powder printed circuit. A printed circuit formed from electrically conductive particles by application of pressure.

Pressed-powder printed wiring. A printed wiring formed from electrically conductive particles by application of pressure.

Pretinning. *See* Tinning.

Printed board. A general term including either printed circuit or printed wiring boards.

Printed circuit. A pattern comprising components, wiring, or a combination thereof, all formed in predetermined design in or attached to the surface or surfaces of a common base.

Printed circuit assembly. A printed circuit board on which separately manufactured component parts have been assembled.

Printed circuit board. A completely processed conductive pattern or patterns and preprinted component parts.

Printed component part. A component part in printed form, such as printed inductor, resistor, capacitor, transmission line, etc.

Printed contact. That portion of printed wiring used for the purpose of providing electrical connection by pressure contact.

Printed edge-board contacts. Patterns extending to the edge of a printed board intended for mating with external contacts.

Printed wiring. A conductor pattern used for the purpose of providing point-to-point electrical connections.

Printed wiring assembly. A printed wiring board on which separately manufactured component parts have been added.

Printed wiring board. A completely processed conductor pattern or patterns.

Printing. The act or art of reproducing a pattern on a surface.

Pull strength. The preferred term is *bond strength.*

Reference grid. *See* Grid.

Reference hole. The preferred term is *tooling hole.*

Reference notch. The preferred term is *tooling notch.*

Register. The relative position of one or more printed wiring patterns, or portions thereof, with respect to their desired location on the base material.

Register mark. A mark used to establish the registration.

Registration. The degree of register.

Resist. A material (ink, paint, metallic plating, etc.) used to protect the desired portions of the printed pattern from the action of the etchant, solder, or plating.

Schematic diagram. A drawing that employs graphic symbols to indicate the electrical connections and functions of a circuit, thus enabling the user to test, evaluate, and analyze the circuit.

Separable component part. A replaceable component part, the body of which is not bonded chemically to the base, excluding the effects of protective coatings, solder, and potting materials.

Single-sided board. A printed board with conductors on one side only.

Solderability. The property of a metal to be wetted or soldered.

Solder projection. An undesirable protrusion from a solidified solder joint or coating.

Substrate (of a microcircuit or integrated circuit). The supporting material upon or within which the elements of a microcircuit or integrated circuit are fabricated or attached.

Supported hole. A hole in a printed board containing plating, eyelets, standoff terminals, etc.

Tab. The preferred term is *printed contact.*

Terminal area. That portion of a printed circuit or wiring used for making electrical connections to the conductive pattern or component.

Terminal area clearance hole. A hole, opposite the component side of a multilayer printed board and perforated in each layer above the terminal area, which provides access to the terminal area after laminating.

Terminal hole. A hole within the terminal area.

Terminal pad. The preferred term is *terminal area.*

Through connection. See Interfacial connection.

Tinning. A solder coating of the conductive paths and terminals, primarily for minimization of in-process oxidation and enhancement of solderability.

Tooling hole. A hole placed in the base material to enable the base to be positioned accurately for processing.

Tooling notch. A notch placed in the edge of the base material to enable the base to be positioned accurately for processing.

Transferred printed circuit. A printed circuit in which the pattern is formed on a temporary base and subsequently transferred to the permanent base.

Transferred printed wiring. A printed wiring in which the pattern is formed on a temporary base and subsequently transferred to the permanent base.

Twist. Bending or curving distortion from the true or plane surface, in a direction parallel to a diagonal between two opposite corners of the sheet.

Undercut. The reduction of the cross section of a metal foil conductor caused by removal of metal by the etchant from under the edge of the resist or plating.

Unsupported hole. A hole within the terminal area containing no conductive material.

Unwetting. The preferred term is *nonwetting.*

Vapor-deposited printed circuit. A printed circuit formed by vapor deposition.

Vapor-deposited printed wiring. A printed wiring formed by vapor deposition.

Vapor deposition. A process consisting of condensation of a material from its gaseous state onto the insulating base.

Warp. The deviation from flatness of a board characterized by a roughly cylindrical or spherical curvature each that, if the board be rectangular, its corners or edges are in the same plane as the major surfaces of the board.

Wave soldering. See Flow soldering.

Weave exposure. A surface condition in which the unbroken woven glass cloth is not covered uniformly by resin.

Weave texture. A surface condition in which the unbroken fibers are completely covered with resin, yet exhibit the definite weave pattern of the glass cloth in the outer surface of the base.

Wetting. The free flow and spreading of solder on conductive paths and terminals to form an adherent bond.

Whisker. A slender metallic acicula that occurs after the printed board has been manufactured.

Wiring pattern. The preferred term is *conductor pattern.*

TERMS RELATING TO RFI/EMI

Absorption loss (attenuation). The property of a shield that measures the amount of energy reduction experienced by an impinging electromagnetic wave due to electrical dissipation in the shield caused by currents induced in it (by the impinging wave) flowing through the resistance of the shield material. (This energy is dissipated as heat.)

Antenna. A radiator designed specifically to radiate energy. An antenna is an *intentional* radiator.

Compression set. The permanent deformation of a material due to imperfect elasticity—i.e., its failure to return to its original shape after the compressing force is removed.

Deflection. The change in position of any part of a structure due to the application of stress. (In gasket applications, the deflection of the gasket surface due to the pressure per unit area applied to it is often of primary importance to the designer.)

Dipole antenna. An antenna in which the conductive portion is a pair of straight wires or rods of equal length, fiexed in line but separated by a small fraction of their length, and excited by a signal voltage connected between the short separation between them.

Elastomer. A polymeric material characterized by high elasticity—e.g., natural and synthetic rubbers.

Electric field strength. The magnitude, in volt/meter, of the force that would be exerted on a unit positive charge held at a specific distance from the specific point electric charge causing the electric field.

Electric flux density (displacement or induction). The product of the electric field strength and the permittivity of the medium.

EMC (electromagnetic compatibility). The ability of electrical equipment to operate in a given environment with other equipment, without causing intolerable EMI levels; or conversely, the ability of electrical equipment to operate normally in a given EMI environment.

EMI (*electromagnetic interference*). Undesired electrical energy that may after the normal functioning of a circuit that absorbs it.

EMP (*electromagnetic pulse*). Radiation or interference signals that take the form of pulses–i.e., very rapid changes in amplitude, from a baseline value to a peak value and back.

Far field (*radiation field*). The region surrounding a radiator in which the field strength (electric or magnetic) varies inversely as the first power of the distance from the radiator. The far field begins at the end of the near field and continues to infinity.

Fourier spectrum. The plot of energy amplitude vs. frequency for a signal. A sine wave produces a single point on the spectrum, while a nonsinusoidal periodic wave produces, a number of such points, at the fundamental frequency, and at each of the harmonic (integral multiples) of the fundamental that are present in the wave.

Galvanic corrosion. Alteration in the chemical properties of a metallic substance due to electric currents that flow in it. The currents are generally assumed to flow because the substance is in contact with a differential substance having higher or lower position on the atomic valence scale–i.e., a substance having a different galvanic potential.

Gasket. A structure designed to minimize or virtually eliminate the gap between mating surfaces.

Half-wave dipole antenna. A dipole antenna having a length, from end to end, that is one-half wavelength of the fundamental frequency of the exciting signal.

Intermodulation distortion. Interference effects caused by the generation, in nonlinear circuitry, of the sum and difference frequencies of two or more electrical signals having different frequencies.

Magnetic flux density (*induction*). The product of the magnetic field strength and the permeability of the medium.

Magnetic field strength. The magnitude, in ampere-turns/meter, of the force that would be exerted on a unit north magnetic pole held at a specific distance from the specific point magnetic charge causing the magnetic field.

Near field. The region surrounding a radiator in which the field strength (electric or magnetic) varies inversely as the cube or square of the distance from the radiator. Also known as the static and induction field regions surrounding the radiator. (In describing the fields surrounding an antenna, the near field may be assumed to have ended when the distance from the antenna is three to five wavelengths at the fundamental excitation frequency.)

Outgassing. The release from a material of occluded, absorbed chemically generated, or thermally generated gases.

Permeability. The property of a medium that measures the ratio of magnetic flux density to magnetic field strength, in Henry/meter.

Permittivity. The property of a medium that measures the ratio of electric flux density to electric field strength, in Farad/meter. The "dielectric constant" of a medium is directly proportional to its permittivity.

Plane of polarization (of an antenna, a field, or a radiator). The plane in which a conductor placed in the field will absorb maximum energy; the direction of the electric field in an electromagnetic wave.

Pressure seal. The ability of a joint between two members of a structure to prevent or reduce leakage through the joint due to a pressure differential between the medium on one side of the joint and the medium on the other side.

Radiator. A conductor, circuit, or conductive structure that radiates electromagnetic energy when the voltage applied to it (or the current flowing through it) changes.

Reflection loss (attenuation). The property of a shield that measures the amount of energy reduction experienced by an impinging electromagnetic wave due to reverse energy propagation from the shield back toward the radiator that generated the impinging wave (or away from the shielded space). Reflections are caused by the fields generated by currents induced in the shield by the impinging wave. Reflections may occur from both surfaces of a planar shield, and reflections from the inner surface may be re-reflected by the outer surface.

Repetition rate. The number of times per second that a periodic signal repeats itself. (The repetition rate is the inverse of the period.) In the case of sinusoidal signals, the repetition rate is called the *frequency*, in hertz.

Slot length. The longer of the two dimensions of a gap (assumed rectangular) in a shielding structure (usually, a gap in the closure of a seam joining two shield surfaces).

Weather seal. The ability of a joint between two members of an enclosing structure to prevent or reduce transfer, from the outside to the inside of the enclosure, of moisture (including humidity), particles, or dissolved contaminants.

INDEX